C0-ATY-435

WITHDRAWN

WITHDRAWN

FUEL CELLS

Modern Processes for the Electrochemical Production of Energy

FUEL CELLS

Modern Processes for the Electrochemical Production of Energy

Wolf Vielstich

Institut für Physikalische Chemie der Universität Bonn

Translated by

D. J. G. Ives

Birkbeck College, University of London

WILEY—INTERSCIENCE

a division of John Wiley & Sons Ltd.

LONDON   NEW YORK   SYDNEY   TORONTO

D. Paul University
CAMPUS LIBRARY

First published under Wolf Vielstich *Brennstoff-elemente Moderne Verfahren zur elektrochemi-schen Energiegewinnung* by Verlag Chemie, GmbH, Weinheim/Bergstr.

© Verlag Chemie, GmbH. 1965

English translation. Copyright © 1970 by John Wiley & Sons Ltd. All Rights Reserved. No part of this publication may be reproduced, stored in a retrieval system, or transmitted, in any form or by any means, electronic, mechanical photo-copying, recording or otherwise, without the prior written permission of the Copyright owner.

Library of Congress catalog card No. 76-114088

ISBN 0   471   90695 6

Set on Monophoto Filmsetter and Printed by
J. W. Arrowsmith Ltd., Bristol, England

321.359
V 661
329504

Preface to the German Edition

In recent years, electrochemical processes have assumed importance in relation to the direct conversion of heat, nuclear energy and, especially, of chemical energy into electrical energy. Galvanic cells which directly harness the chemical energy of fuels are of particular interest—these are called fuel cells. Success has been achieved in bringing various types of fuel cell to technical maturity, only because of the following contributory factors:

1. Extensive research in the fields of the kinetics and catalysis of electrochemical reactions.

2. Initial concentration on the development of the hydrogen–oxygen cell, attempts to effect direct electrochemical reaction of unreactive fuels, such as hydrocarbons or carbon, being relegated to the background.

3. Attractive possibilities of the application of fuel cells in space travel, regardless of expense.

The technical developments have, in turn, had a stimulating effect on fundamental research in a wider field, particularly on electrode reactions and catalytic processes.

An attempt has been made in this monograph to present a comprehensive but concise account of research and development in the field of the direct generation of electrical energy by electrochemical processes, to the stage achieved in 1964. The results of numerous hitherto unpublished investigations are included.

The first two chapters serve as an introduction. Ideas, terminology and basic physicochemical principles are summarized for the benefit of the non-specialized reader.

The most important fundamental research has been on the kinetics of electrode processes. These investigations are therefore treated separately, mainly from a technological aspect, in chapter III, which reports recent original work on numerous electrode reactions and may be welcome to a wider field of electrochemists than those directly interested in fuel cells.

The physicochemical and technological problems involved in the construction and operation of complete cells are described and discussed in chapters IV and V.

In addition to the known physical methods for the conversion of heat and nuclear energy into electrical energy, there are electrochemical methods of potential importance. Those which appear to be of greatest promise are reported in chapters VI and VII.

So as not to exclude fundamentally inseparable problems, electrochemical methods of storing electrical energy are treated in chapter VIII, and the separation of the isotopes of hydrogen accompanying the electrolysis of aqueous solutions is described in chapter IX.

In the closing chapter, the whole field is briefly reviewed, and possibilities of future application of the new sources of energy are discussed.

The stimulus for this book came from my association with the Ruhrchemie AG, Oberhausen-Holten ten years ago, and I thank Dr. H. Tramm, Dr. H. Spengler, and particularly Professor E. Justi of the Technischen Hochschule Braunschweig. A large proportion of the scientific investigation described was carried out in the Institut für Physikalische Chemie der Universität Bonn. I offer my sincere thanks to Professor W. Groth for his support and interest in the work. I give cordial thanks to Professor M. von Stackelberg for much advice and stimulating discussion.

The extensive experimental researches were made possible only by the perseverance and interest of my collaborators, cand.phys. H. Barth, cand.chem. M. Gutjahr, cand.phys. K. Hamann, Dr. D. Jahn, Dipl.-chem. E. Joachim, Dipl.-chem. A. Kutschker, Dr. H. Lauer, Dr. H. Schuchardt, Dipl.phys. D. Struck, U. Vogel and Dr. D. Wolf.

For support, I have to thank the Arbeitsgemeinschaft für Industrielle Forschung, the Bundesverkehrsministerium, the Deutschen Forschungsgemeinschaft, the Fraunhofer Gesellschaft zur Förderung der Angewandten Forschung e.V. and the Verband der Chemischen Industrie.

For individual sections of this book, I record thanks to my colleagues and co-workers; to Dr. G. Grüneberg for the chapters on heat and nuclear energy conversion, to Dr. A. Küssner for his section on the hydrogen membrane electrode, to Helmut Schmidt for dealing with high-temperature cells with solid electrolytes, and to Dr. R. P. Tischer for writing on fuel cells for medium and high temperatures. Dr. Tischer's investigations were carried out with the support of the Electric Storage Battery Company. I thank Dr. R. A. Schaefer of this Company for friendly agreement to publication of the results of this work.

The description of the present state of fuel-cell technology owes much to the release of yet unpublished material or to valuable discussion contributed by

Dr. H. Binder, Prof. H. Bode, Dr. G. H. J. Broers, Prof. R. Ch. Burshtein, Dr. J. Giner, Dr. W. T. Grubb, Dr. G. Grüneberg, Dr. R. G. Haldeman, Dr. R. Jasinski, Dr. K. Kordesch, Dr. A. Küssner, K. Leist, Dr. I. Lindholm, Dr. W. Mitchell, Dr. H. Mylius, Dr. L. W. Niedrach, Dr. H. G. Plust, Dr. P. Rüetschi, Dr. G. Sandstede, Dr. I. Trachtenberg, Dr. J. G. Tschinkel, Dr. J. L. Weininger and Dr. F. G. Will.

Not the last of my thanks are offered to numerous Companies for kind permission to reproduce original photographs (see Sources of Figures).

Finally, I record my thanks to the Publishers for their understanding and co-operation in the preparation of this book.

Bonn, April 1965. WOLF VIELSTICH

Prefatory Note by Academician A. N. Frumkin

Much has already been written, in original exposition and review, on the problem of the direct conversion of chemical into electrical energy. When a new book on this subject is presented I seek its justification in the treatment by the author of his material from several aspects, and in his special qualifications for doing so. In this book a distinguished specialist has written on the theoretical foundations and further of electrochemical kinetics. The author has wide experience in relation to all the chemical-technical problems in his field. He has so wide a knowledge of the practical applications of fuel cells that the overall result is a monograph of rare completeness.

The West European, American and Russian literature has been taken into account—if not entirely without omissions because of the rapid progress of research—in a manner which gives an objective picture of the present situation.

The book should be of value, not only to electrochemists but to all groups of research workers interested in energy conversion.

Moscow, Spring, 1965.   A. N. FRUMKIN

Preface to the English Edition

Nearly four years have elapsed since the completion of the original German edition of this book. In this interval, the installation of hydrogen–oxygen cells in space vehicles Gemini V and VII has demonstrated the technical advancement and reliability of these new sources of electrical power. Fuel cells are being used experimentally in other kinds of application. Swiss Television (PTT) powered the main television station (at altitude 7500 ft) serving the upper Valais district by means of a methanol–formate battery, during the winter of 1965/66 before the main electric supply transmission line had been completed. There is active interest in the possibilities of the new power sources for various military and civil purposes, but no impending breakthrough can yet be reported. Although it is now technically practicable to drive vehicles by fuel-cell systems or hybrid systems incorporating secondary batteries as well, production costs do not yet permit commercial exploitation. In other fields, however—such as the provision of power for trucks, signal and radio stations or any kind of electronic equipment—commercial development is within our grasp. It seems to me that the problem of special significance in future work is the development of electrode catalysts, effective in acid solutions and free from noble metals. Investigations in progress on tungsten carbide as anode catalyst and on phthalocyanin (in analogy with haemoglobin in living organisms) as an organic oxygen-reduction catalyst, are perhaps the most hopeful steps in this direction.

In recent years intensive effort has been directed to research and development in the field of the direct conversion of chemical energy into electrical energy, especially in the United States, but also increasingly in Europe. In accordance with the wishes of the publishers, I have attempted to present the stage of development attained in 1968, particularly in the technical aspects of the subject. This attempt is noticeable in the addition of more than 100 pages and

70 figures; Chapter V contains a new section on magnesium–air cells, to take into account the increasing importance of metal–air cells. In chapter IX a more detailed description has been given to cells with two oxygen electrodes, used for preparing pure oxygen from air. The contents of chapter X have been brought up to date and considerably extended, with a supplementary section on the subject of the electric vehicle.

The English version of the book in its present form would not have been possible without the unselfish and untiring efforts of the translator, my esteemed colleague Dr. D. J. G. Ives. He has not only translated the German text, including the numerous addenda, in an exact and objective manner, but has also pointed out certain errors and obscure formulations; part of the new version of the thermodynamic section of chapter II is from his pen. I am also indebted to Group Captain J. Gray Young for contributing to the section on biochemical cells. Dr. H. Schmidt has helped me with information and advice in preparing the addenda. The section of chapter IV dealing with medium- and high-temperature cells has been extended by Dr. R. P. Tischer. Finally, I acknowledge the valuable support from many colleagues who have eased the task of bringing this English edition as far as possible up to date, by making original photographs and as yet unpublished results available to me.

Bonn, August, 1968.   WOLF VIELSTICH

Contents

List of Symbols

A   ampère

atm   one atmosphere absolute pressure = 760 mm Hg = 760 Torr (N.B. There is no English equivalent to the German 'atü'—*Atmosphäre Überdruck*, except gauge pressure; thus, 1 atü + 2 atm absolute pressure)

c concentration, usually in mole l^{-1}

E electromotive force on voltage of a complete cell

$E°$ thermodynamic electromotive force of a complete cell, not necessarily the standard electromotive force, i.e. that appertaining to a reversible cell with all reaction participants in their standard states. If the latter is meant, it is clear from the context

E_{term} terminal voltage of a complete cell under working or other conditions

E_a activation energy

emf   electromotive force

F faraday

G Gibbs free energy

H enthalpy

hr   hour

i current

j current density

j_0 exchange current density

j_{00} standard exchange current density

M   molar, mole l^{-1}

N   normal, g equiv l^{-1}

n number of electrons involved in an electrode process, number of faradays transferred in a cell reaction

p	gas pressure (absolute)
R	gas constant, per mole
S	entropy
SCE	saturated calomel electrode
T	absolute temperature, °K
t^+, t^-	transference or transport numbers
V	volt
W	watt
α	transfer coefficient $(0 < \alpha < 1)$
ϕ	electrode potential
ϕ_0	equilibrium electrode potential
ϕ_{00}	standard electrode potential
η	$\phi - \phi_0$ overpotential, deviation from equilibrium potential
η_0	rest overvoltage, deviation of cell emf at $i = j = 0$ from the thermo-dynamic emf, $E°$
η_Ω	resistance overvoltage
η_D	charge-transfer overpotential
η_d	diffusion overpotential
η_r	reaction overpotential
η_{kin}	kinetic overpotential $(\eta_{kin} = \eta_D + \eta_d + \eta_r)$
η_r	is also used for reaction efficiency
η_{id}	ideal efficiency $(= \Delta G/\Delta H)$
η_{eff}	practical, or effective, efficiency
η_E	voltage efficiency
Ω	ohm

CHAPTER I

Introduction

The concept of the fuel cell is defined. The chief characteristics of this new kind of source of electricity are listed, and the principles of operation are illustrated in terms of the hydrogen–oxygen cell. A short review is given of historical developments which led up to the modern proliferation of research in this field in the fifties. A conspectus of types of fuel cell, including some examples of those operated in association with other chemical processes, concludes this introduction to methods for the direct conversion of chemical into electrical energy.

1.1 PRINCIPLES AND MODES OF ACTION

A fuel cell is a galvanic cell in which the chemical energy of a fuel is converted directly into electrical energy by means of electrochemical processes. Fuel and oxidizing agent are continuously and separately supplied to the two electrodes of the cell, at which they undergo reaction.

Fuel cells are thus primary cells that will supply electric current for as long as they are provided with the active material, or reagents, that they require. In contrast, traditional primary cells derive their electrical energy from active materials stored in solid electrodes, and they are therefore of limited capacity. This limitation applies equally to secondary cells, or accumulators, which, after discharge, require regeneration of the active electrode material at the expense of an external source of current.

Possible fuels for cells of this kind include, in addition to fossil fuels such as coal or naturally occurring hydrocarbons, such substances as may readily be derived from them, e.g. alcohols, aldehydes, carbon monoxide or hydrogen.

The new energy sources have the advantages, amongst others, of relatively low volume and weight, absence of moving parts, silence, and capacity to withstand overload. Efficiency lies between 30 and 75% and increases with decreasing load, in contrast with the performance of heat engines or internal combustion engines. The chief disadvantage, aside from the existence of many unsolved technological problems, is that the cheapest fuels, such as coal or natural gas, are particularly difficult to use electrochemically. Nevertheless, the new cells have such advantageous properties that they lend themselves to many applications, increasing in number as the technology advances. Some of these already exploited or in course of exploration are: to the provision of electrical power in space vehicles, submarine propulsion and traction of a variety of surface vehicles, to the provision of emergency standby electrical power sources (for hospitals, etc.), the maintenance of remote and seldom attended signal light and relay installations and power storage, to use in association with technical electrolytic processes to economize energy and to the utilization of low-grade 'fuels' of biological origin.

The direct conversion of the energy of the fuel into electrical energy in fuel cells avoids the heavy losses inseparable from the indirect conversion via heat and mechanical energy. It is the Gibbs free energy change, ΔG, accompanying the oxidation of the fuel at the absolute temperature T, related to the corresponding enthalpy and entropy changes, ΔH and ΔS, by the expression

$$\Delta G = \Delta H - T\Delta S$$

which determines the maximum possible electromotive force (emf) of the cell in terms of

$$E = -\Delta G/nF$$

where F is the value of the faraday, and n is the number of faradays transferred in the course of the reaction to which ΔG relates.

Fuel and oxidant are fed to appropriate electrodes, where, provided these electrodes possess the necessary electrochemical activity, they establish electrode potentials, the fuel and oxygen electrodes becoming respectively the negative and the positive terminals of the cell. If the terminals are connected together via an external circuit constituting an electrical load, electrons freed at the fuel electrode flow through the circuit to the other electrode of opposite polarity.

The hydrogen–oxygen cell can be used to illustrate fuel cell principles. Two platinum foil electrodes are immersed in a well-conducting electrolyte (e.g. a solution of sulphuric acid or potassium hydroxide), as in Figure I.1. One electrode (−ve) is supplied with hydrogen, bubbled around it through the solution, the other electrode (+ve) is similarly supplied with oxygen. An electrical potential difference of 0.9 to 1.2 volts can be measured between the electrodes, and if they are connected together with an external, resistive circuit, then for every molecule of hydrogen consumed, two electrons pass from the negative to the positive electrode, where they react with adsorbed oxygen. The distinction from the normal combustion of hydrogen and oxygen, as in the oxy-hydrogen flame, lies in the fact that the electrochemical reaction ('cold combustion') occurs at

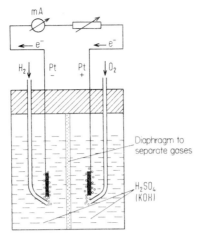

Figure I.1. Principles of the hydrogen–oxygen fuel cell.

two separated reaction sites. The operation of the cell with an acid electrolyte can be summarized as follows:

Negative electrode: $2H_{ad} \rightarrow 2H^+ + 2e^-$

Each hydrogen molecule brought to the electrode surface is dissociated into two atoms in virtue of the catalytic properties of the surface (see Figure I.2). These enter the solution as hydrogen ions, leaving behind two electrons, which pass through the external circuit to the positive electrode.

Positive electrode: $\frac{1}{2}O_2 + 2H^+ + 2e^- \rightarrow H_2O.$

The oxygen supplied to this electrode reacts with hydrogen ions from the electrolyte and the electrons to give water. The reaction scheme represented in Figure I.2 is that for the complete

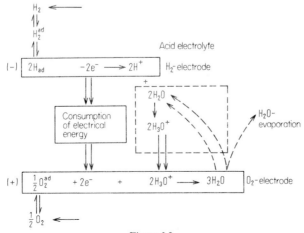

Figure I.2.

electrochemical reduction of oxygen. In practice, the operation of the oxygen electrode is complicated by the formation of hydrogen peroxide and metal oxide. Hydrogen ions consumed at the oxygen electrode are, of course, replaced by those generated at the hydrogen electrode, so that the overall reaction is

$$H_2 + \tfrac{1}{2}O_2 \rightarrow H_2O$$

This is the reverse of water electrolysis, in which hydrogen and oxygen are obtained by application of an external emf. As for every cell for the conversion of chemical into electrical energy, the maximum possible emf of the hydrogen–oxygen cell is that determined by the Gibbs free energy of the cell reaction. In the lower range of temperatures ($< 150°C$), however, this theoretical value of the emf is not attainable because oxygen ionization is accompanied by side reactions involving formation of hydrogen peroxide.

In general, the emf of fuel cells is of the order of 1 volt. Several cells are therefore normally connected in series to provide a useful terminal voltage. Such a combination of cells is referred to as a 'module'; a suitable number of modules either in series or in parallel, depending on the mode of application, constitutes the fuel cell battery. The *electrical output* from an electrochemical reaction depends on the charge transfer per reaction step. The reaction of 1 mole of hydrogen with half a mole of oxygen provides 2 faradays, or 53.6 A hr. For reactions with several charge-transfer stages, it often depends on the working temperature, the nature of the electrolyte, and whether the reaction goes to completion or leaves behind intermediate products, so that a lower electrical output is obtained than is theoretically to be expected. Other losses may occur by non-electrochemical decomposition of the reactant, or by chemical reaction between fuel and oxidant.

The *energy yield* is decisively determined by the rate of the cell reaction. This depends not only on the reversibility of the charge-transfer processes, but also on *mass transfer* to and from the site of reaction. The latter factor is complicated by the fact that the reaction zone is frequently situated in the interior of the ramified and fine-pore system of a catalyst electrode. During reaction, large concentration gradients can be established in the pores, particularly for two-phase systems (electrode/fuel dissolved in electrolyte), and if the reaction products are only sparingly soluble, precipitation and clogging of the pores may occur.

For the porous *diffusion electrodes*, used to bring reactive gases into electrochemical reaction (three-phase systems: electrode/electrolyte/gas), transport to the site of reaction (a three-phase junction) is usually fast enough even for heavy loading of the electrode ($> 100\,mA\ cm^{-2}$), but appropriate attention must be given to adequate rate of removal of any inert gases or gaseous reaction products.

Except when extreme working conditions (high temperatures and pressures) are used, the choice of catalytically active electrode materials is a decisive factor in the attainment of high current densities. The composition of the electrolyte can be such as to facilitate or hinder an electrode reaction, and the construction of the electrode must be such as to provide the greatest possible true surface area in the reaction zone, and simultaneously to secure as complete reaction of the active gas as possible.

I.2 HISTORICAL

The experiment by Grove[1] in 1839 can be regarded as the first electrochemical reaction of a fuel with oxygen in a galvanic cell. He electrolysed sulphuric acid with two platinum electrodes and found that the gases obtained, hydrogen and oxygen, were electrochemically active and established an open-circuit rest-voltage between the electrodes of about 1 volt. The current density that this 'electrolytic gas cell' could provide was, however, so small that it was of no practical use at all.

The first major step towards the development of a practical working fuel cell had its origin in an address by Ostwald[2] to the inaugural meeting of the Bunsen–Gesellschaft in 1894. He demanded the replacement of the heat engine by a fuel cell in which carbon would react with oxygen to produce carbon dioxide. Only by avoiding the Carnot efficiency limitation in this way could the poor efficiency of obtaining the energy of electricity from that of carbon (at that time about 10%) be sensibly improved, and better use be made of raw material such as coal. In following years, Nernst, Haber[3] and especially Baur and his school[3a], developed this idea.

Figure I.3 illustrates a cell developed by Baur and Preis[4] for the reaction of carbon with the oxygen in air to form carbon dioxide, operating at 1100°C. The high temperature was necessary because of the low reactivity of carbon.

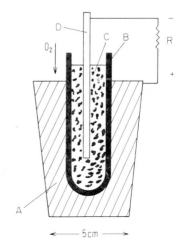

Figure I.3. Carbon–oxygen cell with solid electrolyte, after Baur and Preis (1938)[4]. A) Magnetite; B) Solid electrolyte (clay, CeO_2, WO_3); C) powdered carbon; D) carbon rod. Rest-voltage 0.7 volt; terminal voltage at 150 mA through resistance R, about 0.4 volt.

Only moderate current densities were attainable, and the cell was of short working life.

Many investigations were carried out of systems with molten salt and aqueous, as well as solid, electrolytes; the results of these earlier studies have been summarized in a review by Baur and Tobler[5]. As well as carbon, fuels such as hydrogen, carbon monoxide, acetylene, coal gas, petroleum and sugar were used. The general view was that the most desirable kind of system would be one which would operate at normal temperatures, using an alkaline electrolyte, a carbon–air electrode, and a metal electrode (for reaction of gaseous or liquid fuel) with a potential near to that of the hydrogen electrode. This is still true.

Good gas electrodes were already known before 1933. The so-called diffusion electrode for hydrogen was developed by Schmid[6] in 1923. It consisted of carbon with a finely porous surface layer of noble metal. The key idea was to supply the gas to one side of a plate electrode (or the inside of a tubular electrode) and allow it to come into contact with the electrolyte on the other side. Unless the gas was supplied to the diffusion electrode under pressure, it was essential to make the electrode hydrophobic to avoid flooding of the pore system with electrolyte. This problem had already been solved for the carbon diffusion electrode by 1930, in relation to the alkaline zinc–air cell[7]. The simplest method is to treat the carbon electrode with paraffin. Another approach to the prevention of flooding of porous metal electrodes had been studied as early as 1889. It involved the use of an 'electrolyte carrier' in the following way. Mond and Langer[7a] built a hydrogen–oxygen fuel battery with platinum sieves covered with platinum black. The sulphuric acid electrolyte was localized in a porous diaphragm (e.g. gypsum or asbestos) sandwiched between each pair of electrodes. The gaseous reactants were circulated over the electrode rear surfaces (cf. Figures I.4b and IV.26). Under stationary conditions, 0.73 volt at $3.5 \, \text{mA cm}^{-2}$ (2.5 amp per cell) was obtained. The anodic oxidation of fuels dissolved in electrolytes had also been previously studied (Taitelbaum, 1910[8]; Müller, 1928[9]). None of these various attempts, however, led to an economically satisfactory fuel cell.

Fuel cell research acquired a new impetus after the second World War. It became apparent that the direct use of carbon or coal presented very great technical difficulties, and attention was diverted to the group of 'indirect fuel cells' which would make use of substances obtainable from coal, such as hydrogen, carbon monoxide, or alcohols. For these, the electrochemical processes concerned proceed with useful velocities (current densities $> 10 \, \text{mA cm}^{-2}$ at room temperature).

The use of reactive fuels, together with recent advances in knowledge of electrode kinetics and heterogeneous catalysis, have led in the last ten years to some technically interesting types of cell. Most of these have in common that they use hydrogen, either directly or indirectly, i.e., the potential-determining reaction at the fuel electrode consists of the ionization of atomic hydrogen.

I.3   THE VARIOUS TYPES OF FUEL CELL

A survey of the large numbers of fuel cells that have now been described requires their division into groups, and is attended with difficulty because divisions tend to cut across each other, according to various properties and characteristics. A classification according to states of aggregation of the reactants appears to be the most rational method, since it distinguishes most clearly between methods of construction and operation. A further subdivision is made according to the kind of electrolyte used. Annexed to the three main groups of cells for gaseous, liquid and solid fuels are certain special types and combinations of cells.

1.3.1   Cells for Gaseous Fuels

Within this main group an additional subdivision is made in terms of the kind of oxidizing agent used: gaseous (e.g. $H_2 + \frac{1}{2}O_2 \rightarrow H_2O$), or liquid (e.g. $H_2 + H_2O_2 \rightarrow 2H_2O$).

a) Gaseous oxidizing agent

The principles of a cell for gaseous fuels have been illustrated earlier in this chapter in terms of the hydrogen–oxygen cell. Whereas chlorine is practically the only gaseous oxidant which comes into consideration in place of oxygen or air, many gases or gas mixtures can replace hydrogen as fuel. Thus, studies of cells with CO, CH_4, C_2H_4, C_2H_6, C_3H_8, CH_3OH (gaseous) and with mixtures of these gases with each other, as well as with hydrogen, have become well known.

The three most important types of cell for use with a gaseous fuel are represented formally in Figures I.4 and I.6 (Figure I.6, p. 9). In the arrangement

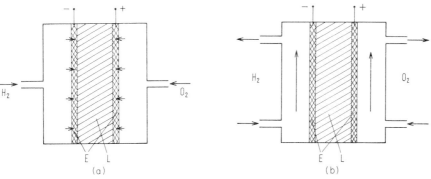

Figure I.4. Constructional diagrams of cells using gaseous fuels (in this case H_2 as fuel and O_2 as oxidant). (a) Gases under excess pressure; (b) Gases under normal pressure. E, Catalyst-electrodes; L, Electrolyte.

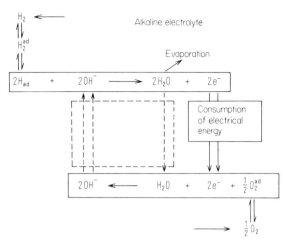

Figure I.5.

shown in Figure I.4a, the gases are forced under pressure into the pores of the catalyst electrode, and undergo ionization at the three-phase junction gas/electrode/electrolyte. According to the pH of the electrolyte, water as a reaction product dilutes the electrolyte either on the hydrogen side (alkaline electrolyte, cf. Figure I.5), or on the oxygen side (acid electrolyte, cf. Figure I.2, p. 3). The continuous removal of this water (perhaps by controlled evaporation) is one of the essential technical problems.

Cells of the type of Figure I.4a work with *aqueous* electrolytes between room temperature and 250°C. The porous electrodes need not be rendered hydrophobic because the excess gas pressure in each case prevents penetration of electrolyte through to the 'gas side'. Special features of electrode construction are necessary, however, to obviate gas bubbling through the electrode.

If the electrodes are rendered hydrophobic, for example by treatment with paraffin, the use of excess pressure becomes unnecessary; the gases can be supplied at normal pressure, or can be circulated over the electrode surfaces (Figure I.4b). If, instead of the normal aqueous electrolyte, some kind of solid matrix ('sponge') permeated by electrolyte is used to contain and localize the liquid phase, a construction such as that of Figure I.4b can be used without hydrophobic treatment of the electrodes. Between room temperature and about 100°C, ion-exchange membranes are available as 'electrolyte carriers'. According to the kind of fixed ions they contain, they can be charged either with OH^- or H^+ ions. The membranes are themselves electrolytes, and, in addition to mobile and exchangeable gegenions (counter ions), take up considerable quantities of water. Electrolyte-soaked asbestos membranes can be used at higher temperatures. For still higher temperatures, molten salts can be used, such as a mixture of alkali carbonates held in a matrix of magnesium oxide (500 to 700°C). Finally, there are gas cells with solid electrolytes; for these, the thinnest possible partition (of, for example, cerium oxide) is arranged between closely adjacent catalyst electrodes. Temperatures of around 1000°C are used.

b) Liquid oxidizing agent

Solutions of hydrogen peroxide, chlorine or bromine come into this category. Oxygen can react 'in liquid form' in so far as the electrolytes in circulating systems are continuously saturated with oxygen or air (Figure I.6, below). Since fuel electrodes are, by their nature, usually poisoned by oxidizing agents (then showing mixed potentials), it is usual for the cathode and anode regions to be separated by a diaphragm. Since the solubility of oxygen in water, or in acid or alkaline solutions, is low, hydrogen peroxide solutions are preferable for attaining higher current densities.

I.3.2   Cells for Liquid Fuels

Example: $CH_3OH + 3H_2O_2 \rightarrow CO_2 + 5H_2O$.

As with a liquid oxidant (H_2O_2), it is advantageous for liquid fuel to be dissolved in the electrolyte; however, when the metal of the oxygen electrode can also catalyse the fuel reaction, a diaphragm must be used to separate the electrode compartments from each other. The oxidizing agent can be supplied either in the form of a vapour, or dissolved in the electrolyte; appropriate forms of cell construction are shown in Figures I.6 and I.7. To illustrate the case of gaseous oxidants, in Figure I.6 hydrogen has to be replaced by, for example, oxygen, and H_2O_2 by the selected fuel. The reactants, pure or dissolved, can be

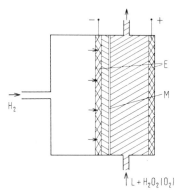

Figure I.6. The oxidizing component (H_2O_2 or O_2) is dissolved in the electrolyte; M, membrane; L, electrolyte.

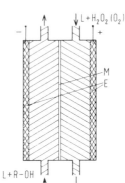

Figure I.7. Construction of a cell for liquid fuel (e.g. alcohol); oxidant dissolved in the electrolyte; M, Membrane; L, electrolyte.

supplied under pressure to the outer sides of porous catalyst electrodes, analogously to Figure I.4a; these less usual procedures are discussed in more detail in chapter IV.

I.3.3   Cells for Solid Fuels

Example: $C + O_2 \rightarrow CO_2$

Such a cell is indicated in Figure I.8. The solid fuel (carbon or metal) acts also as the electrode; an example, dealing with the reaction of carbon with atmospheric oxygen at 1100°C, has already been sketched in Figure I.3.

For the electrochemical oxidation of metals, such as zinc or aluminium, aqueous electrolytes at room temperature are effective. The zinc–air cell has been known as a primary cell for more than 30 years, but really falls outside the group of fuel cells. Indeed, cells using solid fuels hardly meet the definition of a fuel cell as previously formulated, since the necessity of periodic replacement of

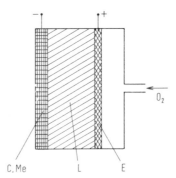

Figure I.8. Cell for solid fuels (carbon or metal). L, electrolyte; E, catalyst electrode for oxygen.

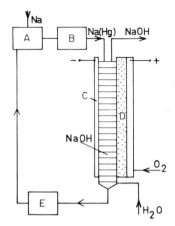

Figure I.9. Amalgam–oxygen cell for the electrochemical use of alkali metal (sodium), after Yeager[10]. A) Amalgamator, B) Heat exchanger; C) Metal anode with amalgam film; D) Porous O_2-diffusion electrode; E) Pump.

the electrodes means they must operate discontinuously. The *amalgam–oxygen cell* (Figure I.9), however, occupies a special position. A thin film of amalgam flowing down a vertical metal plate meets a counter-current of water passing between the amalgam film and a porous oxygen-diffusion electrode. The mercury releases the fuel (e.g. sodium), and the resulting caustic soda solution is discharged.

I.3.4   Cells Involving Continuous Recovery of Reactants (Regenerative Fuel Cells)

Regenerative fuel cells are those in which reactants are continuously regenerated from the products of the cell reaction. The cells, as such, are not novel in type and are included in the preceding classification; they consist of the combination of a cell with a source of energy capable of bringing about dissociation of a reaction product. Energy sources to be considered are (a) heat (cf. chapter VI), (b) light, (c) radiation, (d) chemical energy, (e) electrical energy (cf. I.3.5).

The reactants of a hydrogen–oxygen cell can be recovered and used again in the cell by the action of radiation (neutrons, γ-rays, electrons) on the product of the cell reaction, water (Figure I.10).

$$H_2 + \tfrac{1}{2}O_2 \rightleftarrows^\gamma H_2O$$

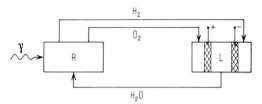

Figure I.10. Recovery of reactants by means of radiation (nuclear battery). R, regenerator; γ, radiation.

Reactants can also be regenerated by chemical means; this is the basis of the oldest type of regenerative cells, known as *redox cells*. In these, as indicated in Figure I.11, both substances, X and Y, concerned in the electrochemical process have their own regenerators, R(X) and R(Y). The oxidized component, X_{ox}, is reduced by a suitable fuel, B, in reactor R(X):

$$B + X_{ox} \rightarrow B_{ox} + X$$

The electrochemically active substance, X, reacts at the negative catalyst electrode, E.

Correspondingly, regeneration of the oxidant, Y (for example, by oxygen or air), occurs in reactor R(Y), and may be formulated:

$$O + Y_{red} \rightarrow O_{red} + Y$$

Clearly, the considerable increase in complication arising from the introduction of the intermediary substances X and Y is economically practicable only if the reaction rate is significantly increased, as compared with that of the direct reaction of B and O. It is self-evident that a cell may contain only one electrode which operates regeneratively.

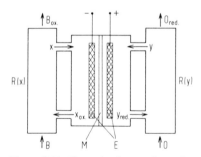

Figure I.11. General scheme of a redox cell. B, Primary fuel; O, primary oxidant; X and Y, electrochemically active substances; R(X) and R(Y) regenerators.

I.3.5   Cells in Combination with other Chemical Processes

In many cases the reactants required by a cell can be recovered from the product of the cell reaction by electrolysis; thus, the electrolysis of water provides the raw materials for the hydrogen–oxygen cell. However, a combination of electrolytic and galvanic cells, in the sense of a regenerative fuel cell along the lines of Figure I.11, is impracticable because the emf of the regenerative circuit would exceed the emf of the fuel cell itself.

There is, on the other hand, technical interest in the combination of the hydrogen–oxygen cell with the 'reverse' electrolytic cell in connexion with *electrochemical storage of energy* (Figure I.12). The gases from electrolysis of water can be stored under pressure and used to operate 'stand-by batteries' composed of hydrogen–oxygen cells.

In a number of electrolytic processes, the terminal voltage required can be reduced if the usual hydrogen-evolution electrode is replaced by an oxygen-dissolution electrode. This replacement is equivalent to connexion of a hydrogen–oxygen cell in series with the electrolysis cell. The lesser voltage requirement must, of course, be paid for by the loss of hydrogen as a product.

A series connexion of this kind is applicable to 'chlorine–alkali electrolysis' by the amalgam process. In the technique hitherto used, the amalgam is treated in a 'denuding cell' with water and graphite or iron catalyst, caustic soda and hydrogen being formed (Figure I.13a). If, however, an oxygen electrode is included with the amalgam in this secondary cell, an arrangement equivalent to that of Figure I.9 is established, with the oxygen electrode as the positive terminal (Figure I.13b). Hydroxyl ions are then formed at the oxygen electrode by the electrochemical reaction:

$$O_2 + 2H_2O + 4e^- \rightarrow 4OH^-$$

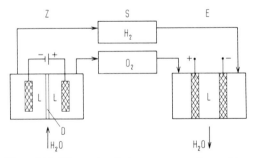

Figure I.12. Electrochemical energy storage by combination of water electrolysis and hydrogen–oxygen fuel cell. Z, electrolysis cell; S, storage for hydrogen and oxygen; E, hydrogen–oxygen cell; L, electrolytes; D, diaphragm.

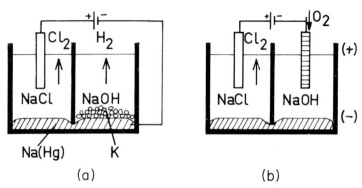

Figure I.13. Chlorine–alkali electrolysis by the amalgam process. (a) catalytic decomposition of the amalgam in the denuding cell; (b) amalgam–oxygen cell as a secondary cell for purposes of reducing the electrolysis voltage. K, catalyst (graphite on iron).

According to current density, the voltage required for the chlorine–alkali electrolysis is reduced by 1.0 to 1.5 volt; this corresponds with a saving of energy of 20–30%.

Finally, fuel cells can be used as *chemical reactors*, as, for example, in the oxidation or reduction of organic compounds, with a 'bonus' production of electric current. Thus, the anodic oxidation of isopropyl alcohol to acetone can be carried out in an alkaline electrolyte, an oxygen or air electrode being used as cathode:

$$CH_3.CHOH.CH_3 + 2OH^- \rightarrow CH_3.CO.CH_3 + 2H_2O + 2e^-$$

If the reduction potential of an organic compound is positive with respect to the potential of the hydrogen electrode, it can form the basis of a half cell which will operate as a cathode in combination with a hydrogen anode; thus the H_2–C_2H_4 cell[12] (cf. Figure III.48) provides an emf, in this sense, of several tenths of a volt. The ethylene is cathodically reduced principally to ethane. A similar reduction of cyclopropane to propane, with simultaneous production of current should also be practicable.

REFERENCES

1. W. R. Grove, *Phil. Mag.*, **14**, 127 (1839).
2. W. Ostwald, *Z. Elektrochem.*, **1**, 122 (1894).
3. F. Haber, *Z. Anorg. Allgem. Chem.*, **51**, 245 (1906).
3a. See *Elektrizität direkt aus Kohle*, Ber. Bunsenges, physik. Chemie, **4**, 129 and 165 (1897).
4. E. Baur and H. Preis, *Z. Elektrochem.*, **43**, 727 (1937).
5. E. Baur and J. Tobler, *Z. Elektrochem.*, **36**, 169 (1933).

6. A. Schmid, *Die Diffusionselektrode*, Enke, Stuttgart, 1923; *Helv. Chim. Acta*, **7**, 549 (1924).
7. G. W. Heise and E. A. Schumacher, *Trans. Elektrochem. Soc.*, **62**, 383 (1932); **92**, 173 (1947).
7a. L. Mond and C. Langer, *Proc. Roy. Soc.*, **46**, 296 (1889).
8. I. Taitelbaum, *Z. Elektrochem.*, **16**, 295 (1910).
9. E. Müller and S. Takegami, *Z. Elektrochem.*, **34**, 704 (1928); E. Müller, *Z. Elektrochem.*, **29**, 268 (1923).
10. E. Yeager, 'The Sodium Amalgam–Oxygen Continuous Feed Cell', in W. Mitchell, *Fuel Cells*, Academic Press, New York and London, 1963.
11. E. Justi and A. Winsel, *Fuel Cells-Kalte Verbrennung*, Steiner, Wiesbaden, 1962.
12. S. H. Langer and S. Yurchak, *J. Electrochem. Soc.*, **116**, 1228 (1969).

CHAPTER II

General Background

An introduction to the physicochemical basis of galvanic cells, specifically directed to the case of fuel cells, is given. The customary definitions and concepts are illustrated in relation to this new field of investigation.

A section on the thermodynamic calculation of emf is followed by a review of the various kinds of overvoltage. The factors that determine the energy efficiencies of fuel cells are explained.

The 'triangular wave cyclic potentiostatic scan method'*, specially suited to the qualitative study of electrode processes, is described in some detail in a section on electrochemical methods of investigation.

An introduction is given to the catalytic problems associated with fuel cells, together with an account of the relation between material transport and phase-boundary reaction at two-phase and three-phase electrodes.

II.1. THERMODYNAMICS

The world demand for electrical energy is largely satisfied by the chemical energy of fuels. The processes hitherto used involve the indirect conversion of the chemical energy into electricity by way of high-temperature heat energy and mechanical energy (Figure II.1). It is well known that considerable loss occurs in the conversion of high-temperature heat energy into mechanical work.

According to the second law of thermodynamics, the maximum work, w_{max}, obtainable from an ideally reversible heat engine, working by the natural flow

---

* Modern electrochemistry is regrettably notable for proliferation of esoteric neologisms; alternative but less self-descriptive names are 'triangular wave cyclic voltammetry' or the 'triangular wave potentiodynamic method'.

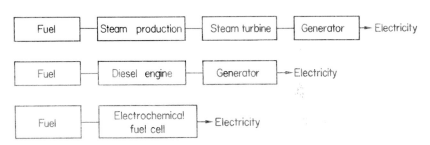

Figure II.1. Possible routes in the conversion of the chemical energy of fuels into electrical energy.

of heat from a higher temperature T_1 to a lower temperature T_2, depends on the temperatures between which heat is transferred:

$$w_{max} = \frac{T_1 - T_2}{T_1} \cdot q_1 = \eta_c q_1 \tag{II.1}$$

where q_1 is the heat absorbed at the higher temperature, and η_c is the so-called Carnot factor, $\eta_c = (T_1 - T_2)/T_1$. If the heat supplied to the engine at the higher temperature comes from a chemical reaction (for example, the combustion of a fuel), q_1 can be identified with the heat liberated by the reaction proceeding at constant pressure $(-\Delta H$, where ΔH is the enthalpy of reaction). It is clear that in principle only for $T_2 \to 0°K$ can an efficiency of 100% be approached. In practice, the optimum efficiency of power stations (steam turbine) is 45%; less for diesel (30%) and petrol (20%) driven generators.

The situation is quite different for the direct conversion of chemical into electrical energy by means of electrochemical fuel cells*. In principle, an efficiency of 100% or more (in relation to $-\Delta H$) is obtainable.

The maximum energy obtainable as work from a chemical reaction is determined by the Gibbs free energy of reaction:

$$\Delta G = \Delta H - T\Delta S \tag{II.2}$$

where ΔH and ΔS are respectively the enthalpy and entropy of reaction at the absolute temperature T. For the reaction to be spontaneous, and thus to be capable of providing a positive amount of work for external utilization, ΔG must be negative.

If the chemical reaction can be made the basis for the operation of a reversible galvanic cell, the maximum electrical work obtainable from the cell is related

---

* In the special case of 'regenerative fuel cells,' in which heat is used to recover the reactants from the reaction product (cf. chapter VI), the cell reaction is simply part of a cycle in which heat from a higher-temperature heat reservoir is converted into work. The efficiency of the combined unit is essentially limited by the Carnot factor[1].

to the Gibbs free energy of the cell reaction by

$$\Delta G = -nFE \tag{II.3}$$

where E is the electromotive force of the cell, F is the faraday, 96493 coul $(\text{g equiv})^{-1}$ (1 coul = 1 A sec) and n is the number of faradays transferred in the course of the electrochemical reaction to which ΔG relates. The negative sign indicates that a positive amount of electrical energy is being obtained in consequence of the Gibbs free energy loss accompanying the spontaneous cell reaction. If all reactants and products are in standard states (possibly not realizable in practice) the emf of the cell becomes the standard emf, $E°$, and is similarly related to the standard Gibbs free energy of the reaction:

$$\Delta G° = -nFE° \tag{II.4}$$

Since standard Gibbs free energies of formation of substances are commonly recorded in cal mole^{-1}, whereas $E°$ is measured in absolute volts, care must be exercised in relation to units. Since 1 joule = 1 V coul = 1 VA sec = 4.1840 cal,

$$E° = \frac{-\Delta G°}{23062n} \, V, \text{ with } -\Delta G° \text{ in cal} \tag{II.5}$$

Since, further, $(\partial \Delta G°/\partial T)_P = -\Delta S°*$, it follows that the temperature coefficient of the emf is given by

$$\left(\frac{\partial E°}{\partial T}\right)_P = \frac{\Delta S°}{23062n} \, V \, °K^{-1}, \quad \text{with } \Delta S° \text{ in cal } °K^{-1} \tag{II.6}$$

It can be seen that the emf of a cell increases with rise of temperature if the products of the cell reaction have greater entropy than that of the reactants. This is generally true for standard or non-standard emf or states of reactants and products, and in further development the superscript zero (indicating 'standard') will be dropped.

In terms of the useful energy obtainable from the combustion of a fuel, (in the most general sense), this signifies that, according to the sign of ΔS, more or less energy can be got by the electrochemical route than corresponds with the maximum calorific value $(-\Delta H)$ of the fuel.

If the entropy change, ΔS, of a reaction is positive, then the galvanic cell in which this reaction proceeds isothermally and reversibly has at its disposal not only the chemical energy, ΔH, but also (in analogy to a heat pump) a quantity of heat, $T\Delta S$, absorbed from the surroundings for conversion into electrical energy. The 'rule of molar balance' in relation to gas generation or consumption, is decisive in determining the sign of entropy change (cf. Tables II.1 and II.2).

---

* Since the first law may be expressed as $dU = TdS - PdV$, where U is the internal energy, and since $G = U + PV - TS$, so that $dG = dU + PdV + VdP - TdS - SdT$, it follows that $dG = -SdT + VdP$. Hence $(\partial G/\partial T)_P = -S$ and $(\partial G/\partial P)_T = V$. Since these relations apply to initial and final states, $(\partial \Delta G/\partial T)_P = -\Delta S$ and $(\partial \Delta G/\partial P)_T = \Delta V$.

Table II.1. Standard heats of formation and standard entropies of some substances at 25°C and 1 atm.

	ΔH° (kcal mole^{-1})	S° (cal$^\circ K^{-1}$ mole^{-1})
O_2	0	49.03
H_2	0	31.23
H_2O(liq)	-68.14	16.75
H_2O(gas)	-57.84	45.14
C	0	1.36
CO	-26.4	47.3
CO_2	-94.05	51.06

In the electrolytic gas reaction $H_2 + \frac{1}{2}O_2 = H_2O_{vap}$, the number of moles decreases from 3/2 to 1. Consequently, $\Delta S < 0$, that is, the emf of the H_2–O_2 cell, decreases with rising temperature. An example of $\Delta S > 0$ is provided by the partial combustion of carbon to CO; $C + \frac{1}{2}O_2 = CO^*$. The theoretical emf of the cells associated with these reactions are represented as functions of temperature in Figure II.2.

If (after Broers[2]) the 'ideal efficiency' of a reversible galvanic cell is related to the enthalpy of the cell reaction by

$$\eta_{id} = \Delta G/\Delta H = 1 - T(\Delta S/\Delta H) \tag{II.7}$$

then η_{id} amounts to 100% if the electrochemical reaction involves no change in the number of gas moles, i.e. when ΔS is practically zero. This is the case, for example, for the reactions $C + O_2 = CO_2$ (cf. the last in Table II.2 on p. 20, and Figure II.2).

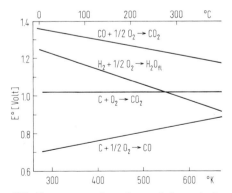

Figure II.2. Temperature dependence of the emf of some reactions.

* The entropy term is also positive for the lead accumulator: the Joulean heat arising from the internal resistance of the cell, however, swamps the cooling effect.

Table II.2. Thermodynamic data for some reactions at 25°C and 1 atm.

	$\Delta H°$ (kcal mole^{-1})	$\Delta S°$ (cal °K^{-1} mole^{-1})	$\Delta G°$ (kcal mole^{-1})	n	$E°$ (volt)	$\partial E°/\partial T$ (mV deg^{-1})	η_{id} (%)
$H_2 + \frac{1}{2}O_2 \rightarrow H_2O_{(\text{liq})}$	−68.14	−39.0	−56.69	2	1.23	−0.85	83
$H_2 + \frac{1}{2}O_2 \rightarrow H_2O_{(\text{gas})}$	−57.84	−10.6	−54.64	2	1.19	−0.23	94
$C + \frac{1}{2}O_2 \rightarrow CO$	−26.4	+21.4	−32.81	2	0.71	+0.47	124
$C + O_2 \rightarrow CO_2$	−94.05	+ 0.7	−94.26	4	1.02	+0.01	100
$CO + \frac{1}{2}O_2 \rightarrow CO_2$	−67.62	−20.7	−61.45	2	1.33	−0.45	91

The electrolytic gas reaction has, at 25°C, in consequence of the entropy term (equation II.7), an ideal efficiency of only 83% (reaction product liquid H_2O) or 94% (reaction product gaseous H_2O). On the other hand, the electrochemical conversion of C to CO has, thermodynamically, a possible efficiency, $\eta_{id} = 124\%$ or, at about 100°C, even of 200%.

A hydrogen–oxygen cell should, from the thermodynamic point of view, be operated at the lowest possible temperature. On the other hand, however, reaction velocity, and hence the current density available at a given terminal voltage, increases exponentially with temperature. The influence of kinetics on the effective efficiency of cells is discussed in section II.3.

The dependence of emf on pressure, as well as on temperature, requires consideration. In general, $(\partial \Delta G/\partial P)_T = \Delta V$, and only the large volume changes associated with the consumption of gaseous reactants and the formation of gaseous products need be considered. If, to a first approximation, ideal gas behaviour can be assumed, for each gas $V = RT/P$ per mole and for a reaction $\Delta V = \sum_i v_i RT/P$ where v_i represents number of moles of gaseous reaction participant, counted positive for products, negative for reactants. It follows that $\Delta G = \Delta G° + \sum_i v_i RT \ln p_i/p_i^0$ and hence

$$E = E° - \frac{RT}{nF} \sum_i v_i \ln p_i/p_i^0 \tag{II.8}$$

where E is the emf of the cell in which the reaction proceeds with gaseous participants at non-standard pressures, p_i, and $E°$ is the corresponding emf with all gases at the standard pressures p_i^0 (normally, 1 atm).

Introducing numerical values for R (8.3147 joule °K^{-1} mole^{-1}) and F (96493 coul (g equiv)$^{-1}$) and converting to common logarithms,

$$E = E° - \frac{1.9841 \cdot 10^{-4}T}{n} \sum_i \log p_i^{v_i} \tag{II.9}$$

For the reaction $H_2 + \frac{1}{2}O_2 \rightarrow H_2O$, it follows, for example, that

$$E(p_{H_2}, p_{O_2}) = E°(p_i = 1 \text{ atm}) + \frac{1.983 \cdot 10^{-4}T}{2} \log (p_{H_2} \cdot (p_{O_2})^{\frac{1}{2}}/p_{H_2O})$$

or, with p_{H_2O} constant at a given temperature

$$E(p_{H_2}, p_{O_2}) = E°(p_{H_2} = p_{O_2} = 1 \text{ atm}) + \frac{1.983 \cdot 10^{-4}T}{2} \log (p_{H_2} \cdot (p_{O_2})^{\frac{1}{2}}) \tag{II.10}$$

At 25°C, the factor to the left of the logarithmic term amounts to about 0.03 volt. If the pressures of H_2 and O_2 are each increased from 1 to 10 atm, then the theoretical emf becomes

$$E = 1.230 + 0.03 + 0.015 = 1.275 \text{ volt}$$

De Paul University
CAMPUS LIBRARY

II.2 ELECTRODE POTENTIALS AND CURRENT–VOLTAGE CURVES

The experimental *rest-voltage*, E, coincides with the emf, $E°$, calculated from the ΔG value only in exceptional cases. The reversible potential is not established at most half cells. The reasons may be the existence of a significant kinetic hindrance to the electrode process, or else that the process does not take place in the manner assumed in the thermodynamic calculation of $E°$. Thus, for many hydrogen-containing compounds, reaction proceeds by a preliminary cleavage of hydrogen. Under practical conditions the oxygen electrode establishes the theoretically expected potential only above 100°C. At lower temperatures, the bond of the O_2 molecule is not broken, and, on reduction, H_2O_2 appears as an intermediate product (cf. chapter III).

The thermodynamic emf values for a series of electrochemical reactions are given in Table II.3 (p. 28) and are compared with the experimental rest voltages. The hydrogen–chlorine cell shows good agreement with theory at lower temperatures. The H_2–H^+ and Cl_2–Cl^- reactions are both reversible at platinized platinum (the latter half cell is, of course, only applicable at temperatures for which the corrosion of platinum is negligible).

The rest-voltage of a cell depends, in general, on the electrode material and the kind of electrolyte (cf. section II.5).

If current is taken from a cell, potential drops are established at the electrodes and in the electrolyte. The various voltages are formally plotted as functions of current density in Figure II.3. The rest-voltage, E, at $i = 0$ is less than the thermodynamic emf $E°$, by a quantity η_0, called the 'rest-overvoltage'*.

The potential drop in the electrolyte increases proportionally to the current density:

$$i \cdot R_E = \eta_\Omega$$

(resistance overvoltage†).

The potential drop at the electrodes is due to kinetic hindrance (kinetic overpotential, η_{kin}). It is, however, necessary to distinguish[3],

1. the overpotential arising from charge-transfer (charge-transfer overpotential, η_D),
2. overpotential consequent on concentration change at the electrode by
2a. rate-limiting mass transfer (diffusion overpotential, η_d) or by
2b. hindrance to preliminary reaction (reaction overpotential, η_r).

When the contributions to overpotential are put together,

$$\eta = \eta_0 + \eta_\Omega + \eta_{kin}$$

---

* No name has hitherto been given in the literature to the deviation of the rest-voltage from the thermodynamically calculated emf. Based on the definition of 'rest-voltage', the author proposes the designation 'rest-overvoltage'.

† To minimize this voltage loss, the distance between the electrodes must be kept as small as possible, and an electrolyte with good conductance must be chosen.

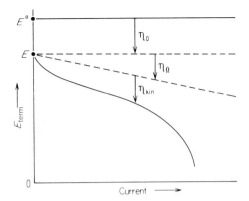

Figure II.3. Current–voltage curves for a cell. E^0, theoretical emf; E, rest-voltage; E_{term}, terminal voltage, η_0, rest-overvoltage; η_{kin}, kinetic overvoltage; η_Ω, resistance overvoltage.

with

$$\eta_{\text{kin}} = \eta_D + \eta_d + \eta_r$$

Since the kinetic overpotential, η_{kin}, does not increase linearly with rising current, the 'inner resistance' of a cell is a function of the load imposed upon it. If the demand of current from a cell is heavily increased, the terminal voltage falls rapidly towards zero. This collapse of voltage occurs, for example, if the electrochemical reaction attains a rate equal to that of mass transfer to or from the electrodes. The diffusion overpotential, η_d, then increases to a very high value (cf. Figure II.4).

Usually, the kinetic overpotentials, η_{kin}, at anode (negative terminal) and cathode (positive terminal) are very different, and the same applies to the rest-overpotentials, η_0, but it is not possible to gain this information from the plot of terminal voltage, E_{term}, as a function of current, as in Figure II.3. It has to be obtained by determining current–potential curves for individual electrodes; the method of measurement, involving the use of a reference electrode, is described in section II.4. A division of the current–potential curve for a cell into corresponding curves for anode and cathode is shown in Figure II.4. Electrode potentials are expressed relative to the thermodynamic potential of the anode, and both cathodic and anodic current densities are reckoned as positive. The potential drop due to the electrolyte resistance, η_Ω, is not represented, since the use of the reference electrode enables the electrode potentials to be directly determined*.

---

* Experimentally, the potential drop, η_Ω, cannot be completely eliminated without special methods because of the finite distance between the electrode and the Luggin capillary. This is, however, not at present significant.

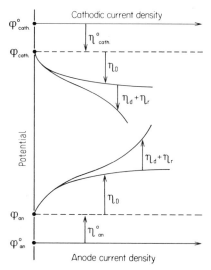

Figure II.4. Individual current–potential curves for anode and cathode potentials plotted against thermodynamic anode potential. η_{an}^0 and η_{cath}^0, contributions of anode and cathode to rest-overvoltage; η_D, charge-transfer overpotential; η_d, diffusion overpotential; η_r, reaction overpotential.

At low current densities, charge-transfer overpotential, η_D, predominates. With increasing loading of the electrode, the provision of reactant becomes the decisive factor (diffusion overpotential, η_d, and reaction overpotential, η_r). If the concentration of the reactant at an electrode falls to zero, the potential 'collapses'.

The form of the current–potential curve for a single electrode is given by the relation[3].

$$j = j_0 \left\{ c_1/c_1^0 \exp\left[\alpha \frac{nF}{RT}(\phi - \phi_0) \right] - c_2/c_2^0 \exp\left[-(1-\alpha)\frac{nF}{RT}(\phi - \phi_0) \right] \right\} \qquad (II.11)$$

where

c_1, c_2 = concentrations of electrochemically active substances at the electrode surface
c_1^0, c_2^0, = initial concentrations
ϕ = electrode potential
ϕ_0 = equilibrium potential
$\phi - \phi_0 = \eta$
= deviation from equilibrium potential, or general 'overpotential'
j_0 = exchange current density
α = transfer coefficient $(0 < \alpha < 1)$

If, during current flow, $c_1 = c_1^0$ and $c_2 = c_2^0$, the deviation from the equilibrium potential is attributable solely to the charge-transfer step, i.e., $\eta = \eta_D$, the transfer overpotential. If, under these conditions, the current density, j, is plotted against potential, the curve obtained is the resultant of two exponential curves ($j = j^+ - j^-$); its steepness is determined by j_0 and its shape by the magnitude of α. The transfer coefficient, α, is a measure of the degree of symmetry of the curves for the anodic and cathodic partial reactions. For $\alpha = 1 - \alpha = \frac{1}{2}$, the current–potential curves are completely symmetrical, always provided that the concentrations at the electrode surface remain unchanged (Figure II.5).

For $\phi = \phi_0$, or $\eta = 0$, one has $j = 0$, but $j^+ = j^- = j_0$, that is, there is a dynamic equilibrium. The greater the exchange current density, j_0, the faster and more precisely is the appropriate equilibrium potential, ϕ_0, established at an electrode.

For $|\eta_D| \gg RT/nF$*, the back-reaction becomes negligible, so that

$$j \sim j^+ = j_0 \exp\left\{\alpha \frac{nF}{RT}\eta_D\right\}$$

or

$$j \sim j^- = j_0 \exp\left\{-(1 - \alpha)\frac{nF}{RT}\eta_D\right\} \tag{II.12}$$

or

$$\eta_D = \frac{RT}{\alpha nF}\ln\frac{1}{j_0} + \frac{RT}{\alpha nF}\ln j = a + b \log j \tag{II.13}$$

This form of equation (II.11) is known as the 'Tafel equation'.

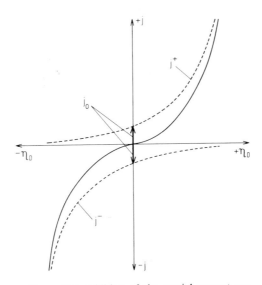

Figure II.5. Addition of the partial current–potential curves, $j^+(\eta_D)$ and $j^-(\eta_D)$, to form the charge-transfer over-voltage curve, $j(\eta_D)$, with $\alpha = \frac{1}{2}$. η_D, transfer over-voltage; j_0, exchange current density.

* $RT/nF \sim (25/n)\,\text{mV}$ at 20°C.

If $|\eta_D| \ll RT/nF$, a linear relation is obtained between current density, j, and overpotential, η_D.

$$j \sim j_0 \frac{nF}{RT}\eta_D \tag{II.11a}$$

The exchange current density, j_0, determines the slope of the current–potential curve in the vicinity of the equilibrium potential.

The temperature dependence of electrochemical reaction velocity, or of exchange current density, conforms to the usual Arrhenius relation:

$$j_0 \propto \exp(E_a/RT) \qquad (E_a = \text{activation energy}).$$

Charge transfer can be facilitated not only by favourable adjustment of potential and temperature, but by use of active catalytic materials, although, in general, the greater the activity of an electrode, the greater the hazard of its becoming poisoned.

Finally, consideration is required of the *concentration dependence* of the exchange current density, j_0, and of the equilibrium potential, ϕ_0. It can be deduced from equation II.11 that[4]

$$\ln j_0 = \ln j_{00} + \alpha \ln c_2^0 + (1 - \alpha) \ln c_1^0 \tag{II.14}$$

where j_{00} = standard exchange current density.
For the function $\phi_0(c_i)$, analogously to equation II.8, it follows that

$$\phi_0 = \phi_{00} + \frac{RT}{nF} \ln \frac{c_1^0}{c_2^0} \tag{II.15}$$

where ϕ_{00} = standard potential.
For $n = 1$ and 25°C, ϕ_0 changes by 59 mV for a ten-fold change of c_1^0/c_2^0.

II.3 THE EFFECTIVE EFFICIENCY

The maximum possible efficiency of a cell, relative to the heat of reaction, ΔH, of a fuel, has been identified in section II.1 with the *ideal efficiency*, $\eta_{id} = \Delta G/\Delta H$. At room temperature it is usually about 90%, decreasing with rising temperature (if $\Delta S < 0$). The practical efficiency, η_{eff}, is, however, only the same as η_{id} provided that (a) the rest-voltage of the cell corresponds with the theoretical emf and (b) the terminal voltage is practically independent of load. At the current densities which are of technical interest, overpotentials are to be expected at the electrodes, quite apart from the voltage drop due to the resistance of the electrolyte (cf. Figure II.3 and 4). The deviation from ideal behaviour under load can be expressed in terms of a *voltage efficiency*, η_E:

$$\eta_E = \frac{E_{term}(j)}{E^\circ} \tag{II.16}$$

η_E is a function of the voltage drop in the cell, as well as of any deficiency in maintenance of electrode potential due, for example, to the occurrence of a side reaction. By its nature, η_E is strongly dependent on the electrode materials used, on the conductance of the electrolyte, and on the geometry of the cell.

Finally, the *reaction efficiency*, η_r, is to be considered. As with accumulators, the electrical output of fuel cells is never 100%. With gas electrodes, some fuel gas or oxygen must be used for expulsion of unwanted residual gases (purging). Loss of some liquid reactants (e.g. hydrazine or hydrogen peroxide) by spontaneous decomposition can hardly be avoided. Some fuels, such as methanol or glycol, cannot always be brought into complete reaction because they are oxidized in several steps, the last of which may proceed less rapidly than the others. The fuel may then fail to be oxidized to the thermodynamically expected end-product. In general the effective efficiency may be regarded as composed of three contributions:

$$\eta_{eff} = \eta_{id} \cdot \eta_E \cdot \eta_r = \frac{\Delta G}{\Delta H} \cdot \frac{E_{term}(j)}{E^\circ} \cdot \eta_r \qquad (II.17)$$

Example: A hydrogen–oxygen cell operates at 25°C, and, with a load of 50 mA cm^{-2}, provides a terminal voltage of 0.85 volt. 95% of the hydrogen supplied to the porous diffusion electrode is consumed electrochemically. Then

$$\eta_{eff} = 0.83^* \cdot \frac{0.85}{1.23} \cdot 0.95 = 54.5\%$$

(relative to $\Delta H = -68.1$ kcal mole^{-1})

II.4   ELECTROCHEMICAL METHODS OF INVESTIGATION

The many methods that have been evolved for the study of electrode reactions cannot be individually described within the context of this book. The object of this section is to indicate the methods which are suitable for studying the problems arising in relation to fuel cells. Only the triangular wave cyclic potentiostatic scan method will be considered in greater detail, since it has proved especially valuable for the investigation of electrode materials and electrode reactions[6].

In section II.2 it has been explained that, to characterize the electrical properties of a cell, it is not sufficient to measure the rest-voltage and to take a current–voltage curve (characteristic curve, Figure II.3). It is, on the contrary, essential to determine the *steady state current–potential curves of the individual electrodes*, by use of a third, so-called reference electrode (Figure II.4). This allows the potential difference between anode (or cathode) and the reference electrode to be measured, without any demand for current (valve voltmeter, potentiometer), under varying conditions of load on the cell. The reference electrode, by appropriate use of a Luggin capillary, 'senses' the potential in the immediate

---

* See Table II.2.

Table II.3. Theoretical and experimental emf values for 25°C.

Fuel	Reaction	ΔG° (kcal mole^{-1})	E° (Volt)	Anode material	Cathode material	Electrolyte	E_{exp} (V)
Hydrogen	$H_2 + \frac{1}{2}O_2 \rightarrow H_2O_l$	-56.69	1.23	Pt	Pt	H_2SO_4	1.15
	$H_2 + Cl_2 \rightarrow 2HCl$	-62.70	1.37	Pt	Pt	HCl	1.37
Propane	$C_3H_8 + 5O_2 \rightarrow 3CO_2 + 4H_2O$	-503.2	1.085	Pt	Pt	H_2SO_4	0.65
Methane	$CH_4 + 2O_2 \rightarrow CO_2 + 2H_2O$	-195.5	1.06	Pt	Pt	H_2SO_4	0.58
Carbon monoxide	$CO + \frac{1}{2}O_2 \rightarrow CO_2$	-61.45	1.33	Ra–Cu	Ag	KOH	1.22
Ammonia	$NH_3 + \frac{3}{4}O_2 \rightarrow \frac{3}{2}H_2O + \frac{1}{2}N_2$	-80.8	1.17	Pt	Pt	KOH	0.62
Methanol	$CH_3OH + \frac{3}{2}O_2 \rightarrow CO_2 + 2H_2O$	-166.8	1.21	Pt	C	KOH	0.98
Formaldehyde	$CH_2O + O_2 \rightarrow CO_2 + H_2O$	-124.7	1.35	Pt	C	KOH	1.15
Formic acid	$HCOOH + \frac{1}{2}O_2 \rightarrow CO_2 + H_2O$	-68.2	1.48	Pt	Pt	H_2SO_4	1.14
Hydrazine	$N_2H_4 + O_2 \rightarrow N_2 + 2H_2O$	-143.9	1.56	Ra–Ni	C	KOH	1.28
Zinc	$Zn + \frac{1}{2}O_2 \rightarrow ZnO$	-76.05	1.65	Zn	C	NaOH	1.45
Sodium	$Na + \frac{1}{2}H_2O + \frac{1}{4}O_2 \rightarrow NaOH$	-71.84	3.12	Na(Hg)	Ag/C	KOH	2.10
Carbon	$C + O_2 \rightarrow CO_2$	-94.26	1.02	C	CuO/Cu$_2$O	Na$_2$CO$_3$	1.03[a]

[a] Measured[5] at 750°C.

vicinity of the electrode under test, including in the measurement the least possible contribution from the voltage drop in the electrolyte (η_Ω)[7].

The electrode best adapted to reference purposes is the saturated calomel electrode (SCE), with an intervening KCl cell–electrolyte junction. It is, however, often desirable to use a reversible hydrogen electrode, operating in a solution identical with the cell electrolyte. If, in this case, the usual bubbling of hydrogen from a cylinder is to be avoided, it is satisfactory to mount two platinum electrodes in the reference electrode compartment. These, supplied from a suitable dc source, are used to electrolyse the solution at a current density of about 1 mA cm^{-2}, so that the lower electrode evolves hydrogen. If this electrode is platinized, it comes with sufficient accuracy to the reversible hydrogen potential appropriate to the solution (provided that the pH of the solution remains unchanged)*.

The individual current–potential curves for single electrodes do not have to be determined by the use of a normal, complete cell. It is adequate to use, in conjunction with the electrode under study, a 'counter electrode' (gegenelectrode) consisting of a foil at which hydrogen or oxygen (or chlorine) is evolved. The electrode compartment concerned is suitably separated from the main experimental cell by means of a sintered-glass disc or diaphragm.

The contribution of resistance overpotential† (η_Ω) and charge-transfer overpotential (η_D) to the overall current–potential characteristic curve can be determined by a current pulsing method[8], by which the load on the cell is periodically interrupted. With a suitable choice of interruption frequency $(> 30 \, \text{sec}^{-1})$, the difference between the terminal potentials with and without load corresponds to the sum of resistance and charge-transfer overpotentials. The interruption time must be short enough to ensure that concentration conditions at the test electrode remain practically unaltered.

Many methods have been developed to eliminate the contribution of mass transfer to inner resistance. In *steady state* methods, transport is greatly improved by forced convection. For reactants dissolved in the electrolyte, bubbling with inert gas is effective. Over and above these, rotating or vibrating electrodes are advantageous; use of the spinning disc electrode[9] allows convective diffusion to be controlled in a well-defined manner.

The three most important *non-steady state* methods[3] use galvanostatic or potentiostatic pulses, or a sine wave potential of small amplitude (a few mV). The pulse methods allow potential–time or current–time curves to be extrapolated to zero time. Immediately after connexion is made, mass transfer is always faster than the electrochemical reaction because of the very steep concentration gradient.

---

* J. Giner, *J. Electrochem. Soc.*, **111**, 376 (1964).
† With porous electrodes, it is not only the potential drop in the electrolyte between the Luggin capillary and the electrode surface which must be considered, but also that in the pores filled with electrolyte.

The *completeness of a reaction*, as, for example, the oxidation of methanol via formaldehyde and formic acid to carbon dioxide, can most simply be investigated by means of galvanostatic potential–time curves. The current density must be fixed at a value low enough to avoid premature decrease to zero of the fuel concentration at the surface of the electrode. Use of a known quantity of fuel leads to information on the course of reaction in terms of the quantity of electricity used (cf. chapter III).

The triangular wave cyclic potentiostatic scan method

The use of a triangular potential–time signal for the investigation of electro-chemical processes is not new. The name 'triangular wave' (Dreieckspannung) implies a potential which increases linearly with time and, after attaining a maximum value, decreases linearly with time, and so on with regular periodicity. For application in the present field, the period should be variable between 10^{-3} seconds and 10 minutes. A similar kind of time-dependent voltage is used in oscillographic polarography, but is applied to the terminals of the cell. In many investigations it is desirable to vary the potential of the electrode under study to and fro between that of hydrogen deposition on the one hand, and oxygen evolution on the other. Bogdanovski and Shlygin[10] have studied the processes of anodic oxidation of alcohols and aldehydes in this way.

If a strictly linear change of the potential of the experimental electrode with time is to be secured, the use of a *potentiostatic circuit* and a suitable reference electrode in association with the electrolysis cell is essential. Satisfactory current–potential diagrams can be obtained only with an electronic potentio-stat, which for 'potential velocities' of only 30 to 100 mV sec^{-1} must have a resolution time of 10^{-1} to 10^{-2} sec if the sharp current peaks which may occur are to be satisfactorily resolved. A potentiostatic triangular wave method of this kind was first used by Knorr and Will[11] for the study of oxygen and hydrogen charging of noble-metal electrodes.

Apparatus and electrolysis cell

Figure II.6 gives a schematic representation of the potentiostatic circuit associated with the elec-trolysis cell[12]. The upper input terminal of the potentiostat is connected via the triangular potential wave generator and a variable source of dc voltage to earth. The lower input terminal goes to the reference electrode and thence to the electrolyte via the Luggin capillary, opposite to the experi-mental electrode, which is also connected to earth.

The potentiostatic circuit operates by controlling the current through the cell so that the voltage at the input of the potentiostat is always zero, i.e. the potential difference between reference and experimental electrodes, $\Delta\phi_{eff}$ must accurately follow the imposed triangular potential $\Delta\phi_{appl}$. The two potential differences must 'compensate' to zero. With special potentiostats, currents of up to 1000 mA can be controlled, with response (resolution) time of 10^{-5} sec.

An oscillograph, or an XY recorder, is suitable for registration of the current–potential diagrams. The triangular potential is taken directly to the input of the X (horizontal) plates of the oscillograph. The current through the cell is measured in terms of the potential difference across a resistance, in the output of the potentiostat, and this is fed to the Y (vertical) amplifier of the oscillograph. This leads to the display of a steady state current–potential diagram.

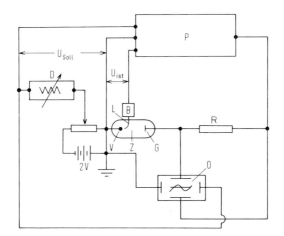

Figure II.6. Experimental device for the triangular wave cyclic potentiostatic scan method. Z, electrolysis cell; V, experimental electrode; G, counter electrode; B, reference electrode; D, triangular potential wave generator; P, potentiostat; O, oscillograph; R, resistance; L, Luggin capillary.

In contrast to the method of taking stationary current–potential curves, the potentiostatic triangular wave method involves the traverse of each potential, ϕ, within a predetermined potential range, $\Delta\phi$, at a constant potential–time scan rate $d\phi/dt(10^{-2}$ to 100 V/sec^{-1}). Accordingly, non-steady state electrode processes, such as the charging of a surface with hydrogen or oxygen, are registered in the current–potential diagram.

A cell for this type of measurement is shown in Figure II.7. Experimental electrode and counter electrode are separated by a sintered-glass partition. Another, coarser, glass frit is used to displace

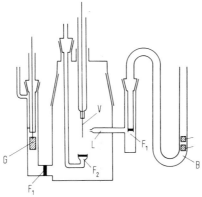

Figure II.7. Electrolysis cell. V, experimental electrode; G, counter electrode; L, Luggin capillary; B, reference electrode; F_1, sintered-glass partitions; F_2, sintered-gas bubbler.

air by means of argon or nitrogen. The potential of the test electrode is sensed by the Luggin capillary leading to the reference electrode. For the latter, a hydrogen electrode in the same solution is used (see above); the two platinum foils serving this purpose are connected to a dc source, and hydrogen is evolved at the lower platinized electrode at about 1 mA cm^{-2}.

Example: Figure II.8 shows two diagrams (with $d\phi/dt = 63.2$ and 632 mV sec^{-1}) obtained with smooth platinum in 1N H$_2$SO$_4$. The abscissae represent the potential of the test electrode relative to the reference electrode (reversible hydrogen). The axis of ordinates represents current in terms of the 'differential capacity', C^d. This is defined by

$$i/(d\phi/dt) = i \, dt/d\phi = dQ/d\phi = C^d$$

The capacity, C^d gives the charge, dQ, transferred for the potential change, $d\phi$, and is thus a pseudocapacity (in contrast to the double-layer capacity, see below). A comparison of the two diagrams shows that, within the velocity range studied, the C^d values are independent of $d\phi/dt$. With increasing potential scan rate, however, this remains the case only as long as the electrode reactions proceed sufficiently rapidly relative to the rate of potential change. For very much slower scan rates (< 30 mV sec^{-1}), then, according to the degree of purity of the electrolyte, a decrease of C^d is observed owing to poisoning of the platinum electrode.

Similarly, the *activity of the electrode* declines if the potential is allowed to reach only the beginning of the oxygen charging region (800 to 1000 mV). If, on the other hand, the potential interval is chosen to extend from the start of hydrogen evolution to that of oxygen evolution (Figure II.8), the otherwise very sensitive noble-metal electrodes provide highly reproducible diagrams. This is also

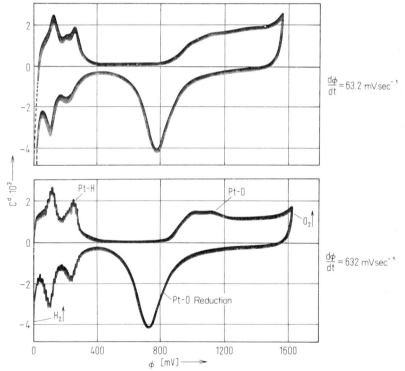

Figure II.8. Current–potential diagram for smooth platinum in 1N H$_2$SO$_4$. Abscissae: electrode potential relative to the hydrogen electrode in the same solution. Ordinates: differential capacity, $C^d = i/(d\phi/dt)$.

the case if reactive substances are added to the electrolyte. Obviously, the electrode surface is regenerated by the formation and destruction of an oxygen layer.

Along the anodic (upper) section of the curve, the hydrogen covering the surface is first of all oxidized, and at about +350 mV the platinum is free of hydrogen. In the 'double-layer region' which follows, the only current which flows is that required for charging the double layer, and this is usually negligible in comparison with reaction currents. At about +800 mV, the formation of a chemisorbed oxygen layer* begins. At about +1600 mV oxygen evolution sets in.

Along the cathodic return (lower) section of the curve, the oxygen covering the electrode is reduced with an overpotential of more than 100 mV, a greater part of it in the double-layer region. At 350 mV, the surface begins to cover itself with atomic hydrogen again, and finally, hydrogen evolution begins at the reversible hydrogen potential.

The removal and formation of the hydrogen layer shows two well-marked peaks. These correspond to at least two kinds of firmly adsorbed hydrogen. The potentials at which the peaks occur, and their shapes, are little influenced by the potential scan rate. If the potential region included in the measurements is extended somewhat in the cathodic direction (hydrogen evolution), then a third well-developed hydrogen peak appears on the anodic curve. This corresponds to the anodic oxidation of molecular hydrogen.

II.5 FUEL CELLS AND CATALYSIS

The electrodes of a fuel cell have the principal function of making electrochemical reactions possible at interfaces between phases by donating or accepting electrons. The electrode material is not itself concerned in the chemical reaction†. In this sense, every electrode at which any appreciable reaction occurs is a catalyst electrode. In practice, however, the name 'catalyst electrode' is used only for those electrodes which can promote a desired reaction on a technically interesting scale. For low-temperature fuel cells (<100°C), adequate catalytic activity of electrode materials is an essential consideration.

The catalytic activity of an electrode depends on a number of properties of the electrode material.

The three most important functions are:
1. Chemisorption of the reactants at the electrode surface.
2. Facilitation of the interface reaction by the dissociation of adsorbed molecules into atoms or by the cleavage of reactive groups.
3. Lowering of the activation energy of charge transfer.

In addition, the affinity of the electrode for the components of the electrolyte has a considerable influence. Thus, the adsorption of the reaction product, acetaldehyde, has a hindering effect on ethanol oxidation at platinum (cf. III.3), and the adsorption of anions passivates hydrogen oxidation at high overpotentials (cf. III.1.3).

Since a large heat of adsorption of reactants facilitates interface reactions, it is the metals which strongly adsorb hydrogen, such as palladium, platinum

---

* The different kinds of oxygen layers are discussed in section III.2.1.

† An exception is, for example, the classical carbon–oxygen cell, in which the fuel, carbon, simultaneously acts as anode.

and finely-divided nickel (skeleton-nickel)*, which are good catalysts both for hydrogen deposition and hydrogen oxidation. A fundamental relation between metal structure and catalytic properties is mainly dependent on the relation between the energy levels of the metal electrons and the 'exchange levels' of the adsorbate in the chemisorbed layer.

As in heterogeneous catalysis, besides the composition of the catalyst, its *state of subdivision* and its surface structure are important. The greatest possible disorder and distortion of crystalline structure is desirable in the surface.

More recently, the influence of nuclear radiation on the catalytic behaviour of electrodes has been studied. For this purpose, radioactive elements have been incorporated in the electrode material, to secure the maximum irradiation at the site of reaction. This kind of procedure is different from that involved in the conversion of nuclear energy (chapter VII), since the extent of nuclear reaction concerned is small in comparison with the electrochemical current density.

There are three conceivable effects of irradiation:

1. Electrochemically active radicals, such as H, OH and HO_2 may be formed.
2. The surface may become markedly roughened, with, particularly, an increase in the number of high-energy sites.
3. The irradiation may catalyse electrode reactions.

A catalytic effect of irradiation has not yet been unambiguously demonstrated (cf. chapter III).

In addition to specific catalytic properties, a catalyst electrode must possess good *electronic conductivity*. This is important in relation, for example, to oxide catalysts. Thus Bacon[13] improved his nickel oxide–oxygen electrode by addition of lithium; the incorporation of lithium atoms into the semi-conducting nickel oxide lattice led to the formation of Ni^{3+}, in consequence of the electronic influence of Li^+ on nickel ions, and greatly improved the electrical properties of the material.

Components of the electrolyte can also exert a catalytic effect. Base-catalysed dehydrogenation provides an important example—the presence of OH^- ions may facilitate, or even be essential to, the reaction which precedes the electro-chemical step.

In addition to its dependence on the above variables, the transfer reaction is influenced by *temperature* and by *potential*. Increase of potential always involves, besides increase of current density, a loss in terminal emf. The decrease in emf with rise of temperature (usually $\partial E°/\partial T < 0$; cf. Table II.2) is, on the other hand, more than compensated for by the increase in reaction rate.

At temperatures above 200–300°C, catalysis begins to assume a comparatively minor role, but this considerable advantage is offset by greater difficulties associated with working conditions. The properties of materials, such as

---

* Highly porous, active nickel, cf. chapter III.1.

resistance to corrosion, and types of cell construction, have to meet more stringent requirements.

The results of catalyst research are discussed in some detail in chapters III and IV.

II.6 THE RELATION BETWEEN MASS TRANSFER AND REACTION AT TWO-PHASE AND THREE-PHASE ELECTRODES

An introduction to general principles involves a discussion of how mass transfer and reaction at two-phase and three-phase systems are related, and of how this leads to the establishment of principles for the construction of two-phase and three-phase electrodes.

The treatment of two-phase systems is relatively simple, under the following three headings:

1. Electrode/reactive component (solid, liquid or gas) dissolved in the electrolyte.
2. Gas dissolved in (adsorbed at) electrode material/electrolyte.
3. Electrode material as fuel/electrolyte.

On the other hand, the treatment of three-phase systems is both technically and mathematically extremely difficult[21,31,45-47]. Similarly, three different arrangements must be distinguished:

1. The so-called 'gas diffusion electrode', in which gas under excess pressure is forced through a porous electrode.
2. The hydrophobic, porous diffusion electrode supplied with reactive gas at normal pressure.
3. The electrode in which the reactive gas or gas mixture under normal pressure is in contact with a porous catalyst layer associated with a 'spongy' electrolyte carrier, or matrix, or with a solid electrolyte.

II.6.1 Two-Phase Electrodes

The most straightforward procedure of dissolving the reactive substance (gas, liquid or solid) in the electrolyte* and immersing a suitable, catalytically active electrode in the solution has already been frequently applied[14-16]. The solution of the reactive substance guarantees an ideal degree of fine dispersion, and, provided the solubility is adequate, mass transfer to the interface is greatly facilitated; with sufficiently high charge-transfer velocities, large current densities (> 300 mA cm^{-2}) can be attained[17,18].

---

* Insoluble liquids can be emulsified, and solids used as suspensions.

Since, in general, gases are but slightly soluble in liquids, promotion of mass transfer by stirring or streaming of the electrolyte is essential for good current-passing capacity of the electrode. The same applies to suspensions of solid substances.

For the case of smooth, flat electrodes in stationary or moving electrolytes, the usual mathematical expressions[19,20] are applicable for calculation of the transport of matter to the electrode.

For any cell geometry, the change of concentration with time for each volume element is given by

$$dc/dt = D \operatorname{div} \operatorname{grad} c - (\vec{v}, \operatorname{grad} c) = 0 \tag{II.18}$$

For $v = 0$ (without convection), and linear diffusion to a planar electrode, it follows from this that the limiting diffusion current ($c = 0$ at the electrode surface) is

$$j_{\lim} = nFD(\partial c/\partial x)_{x=0} = nFDc^0/\delta_N \tag{II.19}$$

where c^0 = initial concentration and δ_N is the thickness of the Nernst diffusion layer. The thickness of this layer in a resting electrolyte is time-dependent (Figure II.9):

$$\delta_N = \sqrt{\pi Dt} \tag{II.20}$$

For convective diffusion, a stationary concentration gradient is attained, and the thickness of the diffusion layer is then calculated in terms of the Prandtl boundary layer, δ_{Pr}[20]:

$$\delta_N = (D/v)^{\frac{1}{3}}\delta_{Pr} \approx \tfrac{1}{10}\delta_{Pr} \tag{II.21}$$

where v is kinematic viscosity.

Example: For laminar flow along a flat plate of length l

$$\delta_N = 3l^{\frac{1}{2}} . D^{\frac{1}{3}} . v^{\frac{1}{6}} . v^{-\frac{1}{2}}$$

where v is velocity of flow outside of the boundary layer.

For the rotating disc of infinite extent, this becomes[27]:

$$\delta_N = aD^{\frac{1}{3}} . v^{\frac{1}{6}} . \omega^{-\frac{1}{2}}$$

where $a = 1.805[0.8934 + 0.316(D/v)^{0.36}]$ and ω = angular velocity.

It is frequently necessary to reduce the overpotential of an electrode by roughening the surface (as by platinization) or depositing on it a porous surface layer ('Sparelektrode')[21]. Whereas for smooth electrodes at least some degree of homogeneity of current density can be assumed, the behaviour of a very

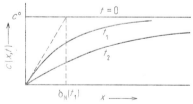

Figure II.9. Concentration contours for an electrode in stationary electrolyte at various times, t, after switching on a constant current; x = distance from the electrode surface.

rough or quite porous surface is very complicated. According to the calculations of Saidenmann[22,22a–c], mass transfer to a porous electrode gives, in consequence of the spatial distribution of current densities and concentrations in the pores, a linear current–potential plot, such as would be observed experimentally if the discharge overpotential were not too large. The larger part of the current is passed by the interphase reaction in the larger pores closer to the surface.

A second kind of two-phase electrode is represented by the system *electrolyte/ gas dissolved in (adsorbed at) metal (electrode)*.

The most simply constructed and, at the same time, technically most interesting electrode is a membrane of palladium–silver alloy used as a hydrogen electrode[23] (cf. section III.1.7). The membrane, free from pin-holes or pores, is supplied on one side with gaseous hydrogen; the other side, in contact with electrolyte, is polarized at a potential positive with respect to the equilibrium hydrogen electrode. Hydrogen diffusing through the foil is anodically oxidized. Pure palladium is unsuitable for this purpose because it becomes mechanically deformed when charged with hydrogen. The discontinuous lattice expansion due to the formation of the β-hydride phase in pure palladium makes it brittle and porous. Palladium alloys with not less than 25% of silver do not form a β-phase[23a]; they remain mechanically stable and are therefore very suitable for the construction of pore-free hydrogen-diffusion electrodes. In the absence of hindrance to the transfer of hydrogen between phases, the anodic current density that can be attained is a function only of the thickness of the palladium membrane. At 20°C, with a hydrogen pressure of 1 atm and a foil thickness of 0.1 mm, the limiting current density amounts to 120 mA cm^{-2}.

Use of a smooth metal membrane involves strong hindrance to the transfer of hydrogen from the gaseous to the metallic phase. Contact between the metal surface and a granular, hydrogen-absorbing substance (cf. section III.1.7) can, however, so facilitate the *interphase transfer* that the limiting diffusion current appropriate to the foil thickness can be attained. On the electrolyte side, a roughening of the surface, as by platinization, similarly assists hydrogen transfer. Hydrogen chemisorbed on *Raney-nickel* (or skeleton-nickel, cf. section III.1) or on platinum can also be utilized in two-phase reactions.

It was discovered by Müller and Schwabe[40] that an inactive metal wire dipping into a solution of formic acid containing a suspension of finely-divided platinum, palladium or other noble metal would record a potential. With sufficient movement of the wire or the electrolyte, this potential would assume a value characteristic of the solution and the suspended catalyst, being established by contact between the wire and the catalyst particles. This principle has since been brought to attention again by Schwabe[40] and has been applied by Sokolsky and co-workers[41]. Podvyazkin and Shlygin[42] have used a system of this kind to determine, by anodic and cathodic charging curves, the hydrogen content of Raney-nickel and of platinum- and palladium-black (cf. section III.1.5).

Recently Gerischer[24] and Boutry, Bloch and Balaceanu[44] have proposed the use of such combinations of *contact electrode and catalyst suspension* for the construction of fuel cells*. Gerischer[24], by using Raney nickel and Ag/Al_2O_3 suspensions in 1M NaOH solution at room temperature, succeeded in attaining current densities of more than 100 mA/cm^{-2} for the reaction of hydrogen or oxygen, respectively. Sheet or gauze Au, Pt, Pd, Ag, Cu, stainless steel and Ni served as contact electrode. Repeated contact between the catalyst particles and the compact metal electrode involves, by transfer of electrons from the particles to the electrode, oxidation of part of the hydrogen adsorbed by the catalyst. A hydrogen stream passed through the electrolyte serves to renew the 'charge' of the catalyst particles with hydrogen. When small quantities of catalyst are used, the current–potential curves are dependent on the number and size of the catalyst particles, as well as on the rates of streaming of hydrogen or oxygen. For higher concentrations of catalyst and sufficient rates of gas supply, the discharge of the catalyst particles becomes the slowest step (cf. Figure III.8). Boutry, Bloch and Balaceanu[44] studied hydrogen oxidation in sulphuric acid by using a suspension of platinized active carbon (5% Pt). According to Gerischer[24], this kind of process is applicable to the oxidation of dissolved organic fuels such as methanol or glycol.

Another original device[25,†] uses a *rotating copper cylinder*, with an axis approximately in the plane of the electrolyte–air surface. The part of the cylinder not immersed in the electrolyte continuously adsorbs oxygen, which is electrochemically reduced as the metal surface comes round into the electrolyte. As the copper surface enters the gas phase, it is covered with a thin film of electrolyte, which offers some resistance to oxygen transport to the surface; higher current densities are attained by removal of the liquid film with a rubber scraper. Measurements indicate that current density is little dependent on rate of rotation (>10 rpm), but it is sensitive to oxygen partial pressure. For $p_{O_2} = 0.844$ atm, 26 mA/cm^{-2} was attained at 32°C. From the dependence of current on oxygen partial pressure, it is concluded, in agreement with Evans[26], that the transport of oxygen through the oxide film formed at the metal surface is the rate-determining step.

For completeness it is necessary to mention a third group of two-phase electrodes; that involving *self-consuming electrodes*. In practice, this is concerned with carbon or metals as fuels. Diffusion hindrance does not occur at a dissolving electrode surface—although this is not the case for the amalgam electrode, for which in practice, a thin amalgam film flowing down a vertical surface is used (cf. section V.1). Electrode processes can, however, be affected by passivation phenomena (for example with gallium as a 'fuel electrode').

---

* This principle has also been studied by Sokolsky (A. Frumkin, personal communication).

† Recently Bonnemay and co-workers have constructed a laboratory hydrogen–oxygen cell with rotating electrodes on a common axis (M. Bonnemay, G. Bronoël, E. Levart and A. A. Pilla, *Revue Energie Primaire*, Brussels, 1965, III, p. 67; *Proc. J. Intern. d'Étude des Piles à Combustible.*)

II.6.2   Three-phase Electrodes

By 'three-phase electrode' is to be understood a porous body (cylinder, disc or plate) which simultaneously acts as catalyst for the electrochemical reaction (cf. section II.5) and electronic conductor, in the pores of which a gas phase (fuel gas or oxygen) and a solution phase (electrolyte) are both present. This does not signify, however, that in every case a three-phase boundary, gas/liquid/solid, is formed in the pores of the electrode; frequently the electrolyte—if only in the form of a very thin film—wets the whole of the pore walls in the gas-filled part of the electrode.

II.6.2.1   The Gas Diffusion Electrode

The name 'gas-diffusion electrode' was given by Schmid[28] to a system in which reactive gas is supplied under pressure to a porous electrode partition which separates gas and electrolyte phases from each other. If gas pressure and average pore diameter are suitably adjusted in relation to each other, the electrolyte fills only part of the system of pore channels.

Figure II.10 indicates the relationship between an *ideal* pore and an aqueous electrolyte. The gas pressure, p, is balanced on the electrolyte side by the hydrostatic pressure, p_h, and the capillary pressure, p_k. The capillary pressure varies with capillary radius in accordance with

$$p_k = \frac{2\sigma}{r} \cos \alpha \qquad\qquad (II.22)$$

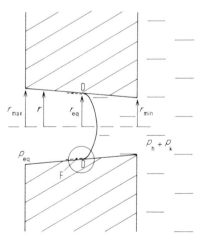

Figure II.10. Section through an ideal pore. F, electrolyte film; D, 'Three-phase boundary'; p_{eq}, equilibrium gas pressure; p_h, hydrostatic pressure; p_k, capillary pressure; r, pore radius.

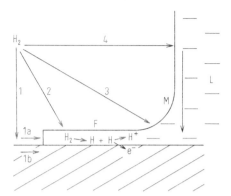

Figure II.11. Simplified scheme of the reaction
paths at the wall of a pore, in terms of the ionization of H_2. F, film; M, meniscus; L, electrolyte.

and is also dependent on surface tension, σ, and the contact angle. The location
of the meniscus is determined by the radius, $r = r_{eq}$, appropriate to the equilibrium pressure, $p = p_{eq}$.

In earlier discussions, the site of reaction was identified with the so-called
'three-phase junction', D (in Figure II.10, the circumference at the pore wall
described by the meniscus of the electrolyte). For aqueous electrolytes, however,
the pore walls on the gas side of the meniscus are covered with a thin liquid
film and there is, in general, no genuine three-phase junction. The electrochemical reaction must therefore occur in a reaction zone of finite extension.

Possible reaction routes are indicated in Figure II.11* in terms of the example
of H_2 ionization.

1. H_2 from the gas phase is chemisorbed on bare pore wall and is dissociated into atoms; (1a), surface
   diffusion of atoms to the film, ionization, diffusion of the hydrated H^+ ions towards the meniscus;
   (1b), solution of hydrogen in the metal, diffusion in the metal, ionization at the interface elec-
   trode/film.
2. Diffusion of H_2 molecules through the film, dissociation and ionization at the pore wall.
3. Diffusion of H_2 molecules through the outer part of the meniscus.
4. Diffusion of H_2 molecules into the electrolyte-filled pore.

The contributions of the various reaction routes to the total current have
recently been much studied experimentally and theoretically; it is certain that
they are dependent on electrode material, electrolyte, and in particular on the
working temperature.

At higher temperatures, the reaction paths 1a and 1b are favoured[29,30]. At room temperature,
and with electrode materials in which the reactive gas has but a low solubility, it has been shown
experimentally that paths 2 and 3 account for more than 90 % of the total current[31], the contribution
of path 1 being practically negligible. It follows from this that, at room temperatures or slightly
higher, it is the diffusion of gas molecules through a thin liquid layer that is rate determining.

* Figure II.11 shows, enlarged and formalized, that section of Figure II.10 enclosed in a circle.

For less active metals, hindrance to the transfer reaction in the case of overpotentials less than 0.4 volt also plays a part. For metals such as skeleton-nickel and palladium, which readily dissolve hydrogen, the above partition between reaction paths has not yet been demonstrated.

The potential drop in the film is decisive in determining the extension of the effective reaction zone at the outer limit of the meniscus. It is produced by current flow in the electrolyte along the electrode surface, and is a function of film thickness, the specific resistance of the electrolyte, and the diffusion coefficient of the reacting gas. It has the consequence that, even for a constant film thickness, current density is a function of distance along the pore.

In the electrochemical reaction of hydrogen at the phase boundary, a gradient of H^+ ion concentration is established in the direction of the meniscus (i.e. parallel to the electrode surface). Consequently, the potential at the outer end of the film is positive relative to that close to the meniscus.

The larger the deviation from the equilibrium potential, the more is the effective reaction zone extended, which means that, despite the diffusion-limited reaction, current increases with increase of potential. Will[31] has computed this relationship for the case of a wire partially dipped into an electrolyte*. With the relations

$$di = -2\pi r j \, dx$$

$$d\phi/dx = ip/2\pi r\delta \tag{II.23}$$

where i is total current; $j = f(\phi, x)$ is current density; ρ, specific resistance of electrolyte; $\delta(x)$, thickness of film, and with the simplifying assumptions that

1. ρ = constant

2. ϕ only $f(x)$

3. $\eta > 0.1$ volt

the current–potential relation

$$i = 4\pi r (FDC^\circ/\rho)^{\frac{1}{2}}\sqrt{\eta} \tag{II.24}$$

is obtained.

The general current–potential relation (without limitation 3, that $\eta > 0.1$ volt) reads

$$i = 4\pi r (FDC^\circ/\rho)^{\frac{1}{2}}\{\eta - RT(1 - \exp^{(-2F\eta/RT)})/2F\}^F. \tag{II.24a}$$

This equation agrees well with experimental results for the $H_2/Pt/H_2SO_4$ system. Weber, Meissner and Sama[25] carried out simultaneous independent qualitative studies with O_2† in alkaline solution, and came to similar conclusions. However, their treatment did not envisage the existence of a liquid film outward from the meniscus.

Experiments by Bonnemay[33], with capillaries of varying diameter, confirmed the view expressed by Will. Bonnemay found that the current was not proportional to r^2 (i.e. to increasing cross-sectional area of the capillary), but to r; it was thus proportional to the area of the capillary wall covered with a film.

Burshtein[34] studied porous electrodes made from nickel of grain size 50 to 100 μ in 7N KOH, finding that current density increased linearly with $\sqrt{A \cdot \phi}$, where A is the surface area of the electrolyte-free macropores, and ϕ is the cross-section of the electrolyte-filled micropores.

As in the calculations of Will, very simplified models have been used in other theoretical treatments of the gas diffusion electrode‡. Borucka and Agar[35] have used a partially immersed plate as an electrical analogy for the meniscus-gas-electrode. Gorin[30] and Wagner[29] have given closer attention to gas transport via the metal phase (reaction path 1b, Figure II.11). Justi and Winsel[21]

---

* The situation in a partially wetted pore is similar to that of a partially immersed wire ('inverse pore'). Difficulties of measurements with a single pore[32] can thus be avoided.

† Recently Will has reported (ECS Pittsburgh 1963, Ext. Abstr. No. 181) an investigation of the system $Pt/O_2/8NH_2SO_4$. The results agreed with calculation over the whole potential range, except for deviation from $i \propto \sqrt{\eta}$ low current density, attributable to discharge hindrance.

‡ Cf. also I. G. Gurevich and R. de Levie, *CITCE Meeting*, Moscow, 1963, No. 2.14 and 2.15, as well as K. Micka, *Amer. Chem. Soc., Fuel Cell Symp.*, 1963, p. 125; *Z. Naturforschung* **19a**, 611 (1964); *Coll. Czech. Chem. Comm.*, **29**, 1998 (1964).

consider that surface diffusion of hydrogen atoms (reaction path 1a) is the rate-determining step for H_2 ionization at Raney-nickel. According to Will[31], however, their mathematical model is highly oversimplified*.

Electrochemical reaction at an electrode surface sets up a concentration gradient within a thin liquid layer—in the case of the hydrogen electrode in alkaline solution, a gradient of OH^- ions, or in acid solution, a gradient of H^+ ions. The provision of OH^- ions or of water to the electrolyte-filled section of the pores may, at high current densities, become partly potential-determining. If part of the gas passes unused through the electrode, the three-phase zone is set into pulsating movement, and mass transfer is facilitated. If the current density at a diffusion electrode is increased by means of an appropriate change in electrode potential, a *limiting current* density is attained, of magnitude determined by the slowest transport process. A further increase in potential leads to decomposition of the electrolyte (hydrogen or oxygen evolution; chlorine evolution). The limiting current density falls to a low value if, by decrease of working pressure, the pore system of the electrode becomes flooded with electrolyte.

The flooding of the pore system can be discouraged by the following precautions:

1. Use of a gas pressure adequate to maintain continuous free passage of gas through the pores. Even in the region of the limiting current density, only a small proportion of the gas ($< 10\%$) is brought into electrochemical reaction.
2. A possibility first exploited by Bacon[13] is to construct the electrode of two layers differing in porosity—on the gas side, for example, with a pore size of 30 to 60 μ and on the electrolyte side, 10 to 20 μ. Because of surface tension in fine capillaries, the electrode layer on the electrolyte side is completely filled with solution, and menisci are established in the region between the finer- and coarser-pored layers (Figure II.12). By this means, practically complete consumption of gas can be achieved.
3. A third effective device is hydrophobic treatment. Parts of the surface of the larger pores are covered with water-repelling molecules. Excess pressure of the working gas can be dispensed with.

II.6.2.2   The Hydrophobic Gas-diffusion Electrode

The hydrophobic gas electrode, which separates gas and electrolyte phases from each other and operates at atmospheric pressure, has been known for a long time in the form of the carbon air-diffusion electrode.

It is used in some commercial types of zinc–air cells ('Füllelemente', water activated cells)[36]. Just before current is required the cell is filled with alkaline electrolyte (4–6 N NaOH). Atmospheric oxygen diffusing through the porous carbon (positive electrode) is reduced to OH^- ions in contact with the electrolyte.

* Winsel has recently extended his theory. See A. Winsel, *Adv. Energy Conv.*, 3, 427, 677 (1963).

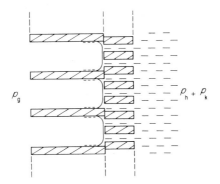

Figure II.12. Simplified diagram of a two-layer electrode. If p_k^E is the capillary pressure on the fine-pored side, and p_k^W that on the wide-pored side, then the fine-pored layer is impermeable to gas if $p_h + p_k^E > p_g > p_h + p_k^W$.

The technique of hydrophobic treatment has required considerable study. Since the agents used have no electronic conductivity, the hydrophobically treated electrodes must have the following apparently contradictory properties:

1. The surface must be effectively hydrophobic;
2. the electronic conductivity of the electrode must not be adversely affected;
3. the catalytic activity of the electrode must be preserved.

If organic substances are to be dissolved in the electrolyte, and if the electrode material consists wholly or partly of metal, the demands on the agent for producing hydrophobic properties become more exacting. Teflon has proved satisfactory for electrodes to be used in acid solutions.

For the treatment of electrodes to be used in alkaline or neutral electrolytes, a solution of, for example, paraffin, polystyrene or polyethylene in an organic solvent (benzene, decalin) can be used, in which the electrode is immersed for a certain time. Composition of the solution, immersion time and temperature must be chosen so that the resulting covering of the pore walls with electrolyte-repelling molecules meets the three conditions mentioned above.

A second method[37] involves incorporation of the hydrophobic material during the preparation of the electrode itself. Thus, by mixing active carbon with polyethylene powder and hot-pressing at a temperature above the softening point of the plastic, a satisfactorily hydrophobic and active electrode material can be obtained, with adequate electronic conductivity. With the correct choice of the plastic/carbon ratio, this method affords electrodes which will not become flooded with electrolyte even when the latter contains alcohol. The electrode is usually built up of two layers. The metal catalyst on the electrolyte side is incorporated in a metal gauze or perforated screen for current conduction and, because of its metal content, is somewhat hydrophilic; this is a desirable feature (cf. Bacon's electrode and Figure II.12).

Micropores of an electrode too small to admit the molecules of the impregnating material with which the electrode has been treated remain free from this material and become completely filled with electrolyte by the action of strong capillary forces (contact angle $<90°C$). This occurs, for example, with active carbon electrodes.

The walls of the macropores, on the other hand, are always more or less covered with the impregnating material (contact angle $> 90°$, e.g., for paraffin/water, about $110°$[37a]). If at some region in a pore the influence of such covering predominates, an inverted meniscus will be formed (in contrast to the meniscus formed in a hydrophilic pore), and this has the effect of stabilizing the location of the three-phase zone[37b]. Under constant conditions (gas pressure, temperature, etc.) the zone will then undergo displacement only if the chemical nature of the impregnating material changes. This material, taking the oxygen electrode as an example, in contact with electrolyte, oxygen and H_2O_2, may, over a long period, become oxidized. Should this lead to a decrease of the contact angle below $90°$, the meniscus will move towards the gas side. This tendency becomes more marked under conditions of strong cathodic loading of the electrode, since this favours the production of H_2O_2.

II.6.2.3.   Gas Electrodes for use with Solid or Semi-solid Electrolytes

Solid electrolytes have some special advantages for use in fuel cells. For instance, the removal of reaction products from the electrolyte is considerably simplified; in the case of the hydrogen–oxygen cell, the water produced can be got rid of in a simple way by drainage or evaporation. It is clearly an advantage to avoid liquid electrolytes in cells to be used for energy generation or storage under gravity-free conditions. The use of solid electrolytes in membrane form allows the interelectrode distance to be kept very small, with the advantage of compact cell construction. Such membranes also act as practically ideal barriers for separation of, for example, the working gases of the hydrogen–oxygen cell. Finally, the use of solid or semi-solid electrolytes eases the technical problems associated with the construction of three-phase electrodes; highly porous and yet 'leak-free' electrodes are no longer necessary because the electrolyte itself is 'fixed'.

Genuine solid electrolytes that have been used are solid salts (discs or cylinders of Na_2CO_3) or metal oxides that conduct by transport of oxide ions[38] (e.g. zirconium calcium oxide) (cf. section IV.2). By 'semi-solid electrolyte' is to be understood a system consisting of a liquid electrolyte held in a *porous matrix*. A matrix of magnesium oxide is suitable for high-temperature cells, since it will absorb molten alkali carbonates[2].

For ordinary temperatures, ion-exchange membranes are commonly used[39]. The three-dimensional network of the membrane contains fixed ions of one kind, and the more or less mobile gegenions, or counter ions, are exchangeable with other ions of similar charge. A content of about 50% of hydrational water confers gel-like properties on the membrane, and its conductance can be increased by uptake of excess electrolyte.

Recently, membranes of asbestos saturated with electrolyte have proved especially advantageous (cf. section IV.1.2).

In this case, the practical problems of electrode construction and of effecting contact between electrode and electrolyte, remain essentially the same. The simplest method is to press a metal gauze (if necessary, platinized or silvered) against the solid electrolyte. A second procedure is to use a granular or powdered catalyst layer, compressed against the electrolyte by means of a metal gauze or perforated sheet. The most effective method is to bring a thin, porous metal layer into contact with the solid electrolyte, or the matrix of the semi-solid electrolyte (cf. chapter IV).

Mass transfer of the reactive gas to the electrode/electrolyte interface is ensured by all these arrangements. A factor decisive for satisfactory performance of cells of this type, besides catalytic activity (at lower temperatures), and good contact between electrodes and electrolyte, is transport of charge through the electrolyte, i.e. its specific electrical resistance.

REFERENCES

1. H. A. Liebhafsky, *J. Electrochem Soc.*, **106**, 1068 (1959); A. J. de Bethune, *J. Electrochem. Soc.*, **107**, 937 (1960).
2. G. H. J. Broers, *High Temperature Galvanic Fuel Cells*, Diss. Univ. Amsterdam, 1958.
3. e.g. K. J. Vetter, *Elektrochemische Kinetik*, Springer Verlag, 1961.
4. H. Gerischer, *Z. Elektrochem.*, **54**, 362 (1950).
5. E. Justi and A. Winsel, *Fuel Cells—Kalte Verbrennung*, p. 10, Steiner, Wiesbaden, 1962.
6. W. Vielstich, *Z. Instrumentenk.*, **71**, 29 (1963).
7. R. Piontelli, *Z. Elektrochem.*, **59**, 778 (1955).
8. K. Kordesch and A. Marko, *J. Electrochem. Soc.*, **107**, 480 (1960).
9. V. G. Levich, *Disc. Faraday Soc.*, **1**, 37 (1947); D. Jahn and W. Vielstich, *J. Electrochem. Soc.*, **109**, 849 (1962).
10. C. A. Bogdanovski and A. I. Shlygin, *Z. Physik. Chem.* USSR, **33**, 1769 (1959); **34**, 57 (1960).
11. C. A. Knorr and F. G. Will, *Z. Elektrochem.*, **64**, 258 and 270 (1960).
12. W. Vielstich and H. Gerischer, *Z. Physik. Chem.*, New Ser. 4, 10 (1955).
13. F. T. Bacon, *Ind. Eng. Chem.*, **52**, 301 (1960); *Brit. pat.* 725661 (1. 1. 1954).
14. C. Kaiser, *D.P.* 114740 (1899).
15. I. Taitelbaum, *Z. Elektrochem.*, **16**, 295 (1910).
16. E. Müller and S. Takegami, *Z. Elektrochem.*, **34**, 704 (1928).
17. H. Krupp, H. Rabenhorst, G. Sandstede and G. Walter, *J. Electrochem Soc.*, **109**, 533 (1962).
18. M. E. Indig and R. N. Snyder, *J. Electrochem. Soc.*, **109**, 1104 (1962).
19. V. G. Levich, *Physikalisch-chemische Hydrodynamik*, Staatsverlag für Physikalisch-mathematische Literatur, Moscow, 1959.
20. W. Vielstich, *Z. Elektrochem.*, **57**, 464 (1953).
21. E. Justi and A. Winsel, *Fuel Cells—Kalte Verbrennung*, p. 245, Steiner, Wiesbaden, 1962.
22. Saidenmann, cf. V. S. Bagotzky, in *Arbeiten der* 4. *Konferenz uber Elektrochemie*, Moscow, 1956, p. 737; also cf. J. S. Newmann and Ch. W. Tobias, *J. Electrochem. Soc.*, 109, 1183 (1962).
22a. G. Gurevich and V. S. Bagotzky, *Electrochim. Acta*, **9**, 115 (1964).
22b. E. A. Grens and C. W. Tobias, *Electrochim. Acta*, **10**, 761 (1965).
22c. J. Euler, *Chem. Ing. Techn.*, **37**, 626 (1965).
23. A. Küssner, *Z. Elektrochem.*, **66**, 675 (1962); *Z. Physik. Chem.* New Ser. 36, 383 (1963).
23a. G. Rosenhall, *Ann. Physik* (5), **24**, 297 (1935); *U.S. Pat.* 2773561 (11. 12. 1956) Silver Palladium Film for Separation and Purification of Hydrogen.
24. H. Gerischer, *Ber. Bunsenges. Physik. Chem.*, **67**, 164 (1963); J. Held and H. Gerischer, *Ber. Bunsenges. Physik. Chem.*, **67**, 921 (1963); and *DAS* 1 162433 (11. 8. 1961).
25. H. C. Webber, H. P. Meissner and D. A. Sama, *J. Electrochem. Soc.*, **109**, 884 (1962).

26. U. R. Evans, Oxidation of Metals, *Rev. Pure Appl. Chem.* **5**, 1 (1955).
27. D. P. Gregory and A. C. Riddiford, *J. Chem. Soc. (London)*, 1956, p. 3756.
28. A. Schmid, *Die Diffusions-Gas-Elektrode*, Enke, Stuttgart, 1923; *Helv. Chim. Acta*, **7**, 169 (1933).
29. C. Wagner, *Polarization Characteristics of Gas Electrodes in Stagnant Electrolytes*, 1957.
30. E. Gorin and H. L. Recht in G. J. Young, *Fuel Cells—Brennstoffelement*, Krauskopf, Wiesbaden, 1962.
31. F. G. Will, *J. Electrochem. Soc.* **110**, 145 and 152 (1963).
32. R. Buvet and M. Guillon, *CITCE-Meeting Rome*, 1962, Abstract No. 3.21 and 3.22.
33. M. Bonnemay, G. Bronoël and E. Levart, *Electrochim. Acta (London)*, **9**, 727 (1964).
34. R. Chs. Burshtein, V. S. Markin, A. G. Pshenichnikov and V. A. Chismadgev and Y. G. Chirkov, *Electrochim. Acta (London)*, **9**, 773 (1964).
35. A. Borucka and J. N. Agar, *CITCE-Meeting Rome*, 1962, Abstract No. 3.20.
36. *Ullmanns Encyklopadie der Technischen Chemie*, Vol. 7, p. 769, Urban and Schwarzenberg, Munich 1956.
37. G. Grüneberg and H. Spengler, *Dechema-Monographie*, 1961, *Belg. Pat.* 592862 (1959).
37a. F. M. Fowkes and W. D. Harkins, *J. Amer. Chem. Soc.*, **62**, 3377 (1940).
37b. J. Jindra and J. Mrha, *Coll. Czech. Chem. Comm.*, in the press.
38. E. Baur and H. Preis, *Z. Elektrochem.*, **43**, 727 (1937); J. Weissbart and R. Ruka, *J. Electrochem. Soc.*, **109**, 723 (1962).
39. W. T. Grubb and L. W. Niedrach, *J. Electrochem. Soc.*, **107**, 131 (1960).
40. E. Müller and K. Schwabe, *Z. Elektrochem.*, **35**, 165 (1929); *Kolloid-*No. 52, 163(1930).
41. D. V. Sokolsky and W. A. Druz, *Ber. Akad. Wiss. USSR* **73**, 949 (1950); D. V. Sokolsky and G. D. Zakymbaeva, *Proc. Acad. Soc. USSR*, **124**, 153 (1959); D. V. Sokolsky, *CITCE-Meeting*, Moscow, 1963, No. 4, 46.
42. Y. and A. Podvyazkin and A. I. Shlygin, *J. Physik. Chem. USSR* **31**, 1305 (1957).
43. K. Schwabe, *Z. Elektrochem.*, **61**, 744 (1957).
44. P. Boutry, O. Bloch and J. Balaceanu, *Extrait des C.R. hebd. Seances Acad. Sci. t.* 254, p. 2583 (1962).
45. L. G. Austin, *Symposium on Recent Advances in Fuel Cells*, Amer. Chem. Soc., Chicago, Sept. 1961, p. 91.
46. H. B. Urbach, *Symposium on Recent Advances in Fuel Cells*, Amer. Chem. Soc., Chicago, Sept. 1961, p. 105.
47. M. Eisenberg and L. Fick, *Symposium on Recent Advances in Fuel Cells*, Amer. Chem. Soc., Chicago, Sept., 1961, p. 121.

CHAPTER III

Electrode Reactions

In this chapter a summary is given of the results of fundamental research on single electrodes, in relation to reaction mechanisms, catalytic properties of electrode materials, types of current–potential curves, and the influence of electrolyte, pH, temperature, etc. In the chapters which follow, the technical problems of the construction of electrodes and the operation of complete cells are treated, and the properties of various types of battery are described.

Hydrogen and methanol, with the intermediate oxidation products of the latter (formaldehyde, formic acid) are given special attention, not only because of their technical significance, but because knowledge won by studies of the electrochemical oxidation of these substances has contributed greatly to the understanding of the electrode processes involved in the reaction of many other fuels.

In the consideration of oxidizing agents, oxygen and hydrogen peroxide naturally call for discussion in detail.

Hitherto, little has been known about electrode reactions at somewhat elevated ($>100°C$) or high ($>400°C$) temperatures. These are accordingly discussed, in relation to the appropriate cells, in chapter IV.

III.1 HYDROGEN ELECTRODE

The most recent views on the mechanism of the hydrogen electrode reaction, including pH dependence and behaviour at high overpotentials, are presented. In further sections, the influence of the catalytic activity of the electrode material and the effects of oxygen admixture are discussed. Finally, Küssner describes the pore-free hydrogen electrode of palladium–silver alloy that he developed conjointly with Wicke.

III.1.1   Reaction Mechanism

In aqueous solution, the hydrogen reaction proceeds in the *acid* region according to

$$H_{ad} + H_2O \rightleftharpoons H_3O^+ + e^- \tag{III.1}$$

but in *alkaline* solution,

$$H_{ad} + OH^- \rightleftharpoons H_2O + e^- \tag{III.2}$$

In the *neutral* pH region, both reactions can occur simultaneously*.

According to present views, the overall anodic processes represented by equations III.1 and III.2 are considered to take place in the following partial reaction steps:

(a) Transport of molecular hydrogen to, and adsorption on, the electrode surface, either from the gas phase or from the electrolyte;

$$H_2 \text{ or } H_2 \text{ (dissolved)} \rightarrow H_{2,ad}$$

(b) Hydration and ionization of the adsorbed hydrogen, which it is considered may proceed by two alternative reaction paths;
   (i) dissociation of the molecules into atoms,

$$H_{2,ad} \rightarrow H_{ad} + H_{ad} \qquad \text{('Tafel reaction')}$$

followed by hydration and ionization at discrete sites on the electrode surface,

$$H_{ad} + H_2O \rightarrow H_3O^+ + e^- \qquad \text{('Volmer reaction')}$$

   (ii) hydration and partial ionization, proceeding in a quasi-single step ('electrochemical' or 'Heyrovsky–Volmer' or 'Horiuti–Volmer' mechanism),

$$H_{2,ad} + H_2O \rightarrow [H_{ad} \cdot H_3O^+] + e^- \rightarrow H_{ad} + H_3O^+ + e^-$$
$$H_{ad} + H_2O \rightarrow H_3O^+ + e^- \tag{III.3}$$

(c) Transport away of H_3O^+ ions.

The above division of the possible reaction steps corresponds with the reaction mechanisms denoted Tafel–Volmer (T–V) and Heyrovsky–Volmer (H–V).

---

* At high *cathodic* current densities (10–100 mA cm^{-2}) reaction III.2 predominates, since the provision of protons by dissociation is not nearly fast enough[1].

　　The introduction of the Heyrovsky reaction as an identifiable reaction step was regarded by some investigators as somewhat arbitrary and problematical. In relation to this, it is relevant that the calculation of current–potential curves for the alternative reaction sequences gives an unambiguous distinction between H–V and T–V only if the Tafel reaction is rate-determining. This can be seen from the graphical representation in Figure III.1, due to Vetter[2]. In this, the curves are calculated by means of the equations developed by Vetter[2] for the H–V reaction sequence:

$$j = 2j_{0,2}\, \exp\!\left(\frac{\alpha_2 \cdot F \cdot \eta}{RT}\right) \cdot \frac{1 - \exp\!\left(-\dfrac{2F}{RT} \cdot \eta\right)}{1 + \dfrac{j_{0,2}}{j_{0,1}} \exp\!\left\{-\dfrac{(1 + \alpha_1 - \alpha_2)F}{RT} \cdot \eta\right\}} \tag{III.4a}$$

or

$$j = -2j_{0,1}\, \exp\!\left(\frac{-(1 - \alpha_1)F \cdot \eta}{RT}\right) \cdot \frac{1 - \exp\!\left(\dfrac{2F}{RT} \cdot \eta\right)}{1 + \dfrac{j_{0,1}}{j_{0,2}} \exp\!\left\{\dfrac{(1 + \alpha_1 - \alpha_2)F}{RT} \cdot \eta\right\}} \tag{III.4b}$$

where $j_{0,1}, \alpha_1$ and $j_{0,2}, \alpha_2$ are the respective parameters for the Volmer and Heyrovsky reactions.

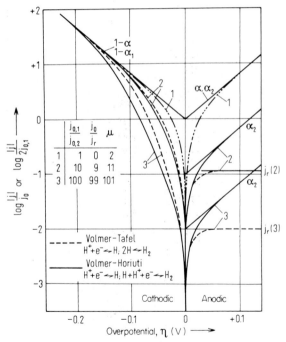

Figure III.1. Calculated curves for anodic and cathodic overpotentials in dependence on $\log|j|/j_0$ or $\log |j|/2j_{0,1}$ for the H_2, H^+ electrode (after Vetter[2]). Continuous curves: for the Volmer–Horiuti (Heyrovsky–Volmer) mechanism, $\alpha_1 = \alpha_2 = 0.5$. Discontinuous curves: for the Volmer–Tafel mechanism, $\alpha = 0.5$. $\mu = $ Stoichiometric number.

Also shown in Figure III.1 are the results obtained by Breiter and Clamroth[3] for the reaction sequence $T-V$, for the special case of $\alpha = 0.5$, by use of the relation

$$\eta = -\frac{RT}{2F}\ln\left(1 - \frac{j}{j_r}\right) + \frac{RT}{0.5F}\sinh\frac{j}{2j_0(1 - j/j_r)^{\frac{1}{2}}} \tag{III.5}$$

where j_r = recombination (or dissociation) limiting current density.

Only in the case of T–V (Tafel reaction rate-determining) is the comparison of theory and experiment simple, because of the existence of a limiting current in the anodic region. There can also be a limiting anodic current, however, in the case of hindered adsorption[4].

For smooth platinum electrodes, completely immersed in hydrogen-saturated electrolyte, the anodic limiting current is reached at an overpotential of between 100 and 300 mV (cf. section III.1.3). For electrodes of this kind that have been highly activated (for example by short periods of H_2- and O_2-evolution), anodic current is transport-controlled, except at very high rates of stirring or flow of electrolyte[4–6]. In relation to porous electrodes, especially gas-diffusion electrodes, diffusion processes in the pores must always be considered (cf. II.6.2). Because of potential drop in the pores, anodic current continues to increase with rising overpotential until the oxygen-charging region of electrode potential is reached (cf. equation II.24 and section III.1.3).

The *anodic dissolution*, $H_2 \rightarrow 2H^+ + 2e^-$, can only occur if the molecular hydrogen is chemisorbed by the electrode metal, since this is the essential condition for the dissociation of molecules into atoms. Platinum metals, iron, cobalt, nickel (cf. section III.1.5) are effective in this way. The *cathodic deposition* of hydrogen, however, is possible on any electrode, as, for instance, on carbon, lead or mercury. A kinetically-controlled cathodic limiting current is only conceivable as a function of strongly hindered desorption of the molecular hydrogen which is formed; this is not taken into account in equations III.4 and III.5. A transport-controlled *cathodic limiting current* can occur in the electrolysis of acidic electrolytes when the concentration of H_3O^+ ions falls to zero at the electrode surface. Current limitation in alkaline aqueous solutions is not a possibility (cf. equation III.2 and Figure III.2). With increasing current density, the increasing stirring effect of gas evolution is adequate for the transport away of the OH^- ions formed[7].

III.1.2 pH-dependence of the Hydrogen Reaction

The pH-dependence of the exchange current density of the hydrogen electrode has hitherto been studied only over restricted ranges of pH, mainly those relating to strongly acid and strongly alkaline solutions. It was of interest to measure the exchange current density at a catalytically active metal over the whole pH range, and to compare the results with the predictions of theory. As will be shown, the exchange current density must pass through a minimum with increasing pH, corresponding with the transition from the reaction in acid solution (equation III.1) to that in alkaline solution (equation III.2).

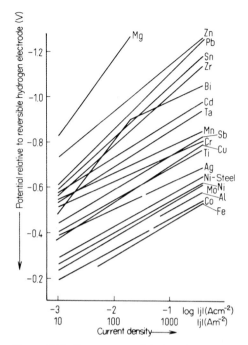

Figure III.2. Current–potential curves for cathodic hydrogen evolution at smooth electrodes in 6N NaOH solution at 25°C (after Zholudev and Stender[8]). For corresponding curves relating to porous electrodes of lesser overpotential, see chapter VIII.

According to equation II.14, the concentration-dependence of the hydrogen electrode reaction in the acid region is given by:

$$\log j_0 = \log j_{00}^{H^+} + \alpha \log a_{H^+} \qquad \text{(III.6)}$$

but in the alkaline region by

$$\log j_0 = \log j_{00}^{OH^-} + (1 - \alpha) \log a_{OH^-} \qquad \text{(III.7)}$$

where $j_{00}^{H^+}$ and $j_{00}^{OH^-}$ are the standard exchange current densities in acid and alkaline solutions, respectively. In acid solution, the exchange current density and hence the performance of the electrode must decrease with falling hydrogen ion concentration. For alkaline solution, the OH^- ion concentration enters the rate equation, and exchange current density increases with rising pH. It is therefore to be expected that j_0, at constant ionic strength, will pass through a minimum as reaction III.2 replaces reaction III.1.

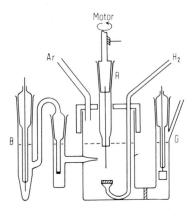

Figure III.3. Electrolysis cell with rotating electrode for the determination of H_2/H^+ exchange current densities (after Joachim, Lauer, Struck and Vielstich[9]). R, rotating experimental electrode (wire); G, counter electrode; B, reference electrode. Argon and hydrogen gas supplies.

For the experimental determination of the pH-dependence of exchange current density, by obtaining stationary current–potential curves in the vicinity of equilibrium (± 10 mV), Joachim, Struck, Vielstich and Lauer have used the cell illustrated in Figure III.3. To exclude diffusion overpotential (cf. II.2), a spinning electrode was used (rotational speeds up to 15 000 rpm.). A stream of argon over the electrolyte served to keep the system oxygen-free. The pH values of the solutions were established with H_2SO_4 or KOH, and the ionic strength was kept constant, in each case by appropriate additions of K_2SO_4.

Figure III.4 shows the variation of exchange current density as a function of pH for smooth* and platinized platinum and Raney-nickel. The exchange current densities were in each case calculated, according to equation II.11a, from the slope of the current–potential curve at the equilibrium potential.

Up to pH 11, the curve through the experimental points agrees quite well with expectations from equations III.6 and III.7. Logarithmic plotting gives, from the slopes of the asymptotes for acid and alkaline regions, a transfer coefficient of 0.5. It is noteworthy that, above pH 11, the experimental points for smooth and platinized platinum no longer follow the theoretical curve, but the points for Raney-nickel do so very well. This shows that the platinum electrode loses its catalytic activity for the hydrogen reaction in strongly-alkaline solutions. It is obvious that the presence of OH^- ions affects the platinum–hydrogen bonding to a considerable extent. The amount of adsorbed hydrogen is hardly altered by increase of pH. For rhodium also, a marked falling off in j_0 is found between pH 13 and 15[9].

* To obtain reproducible current–potential curves for smooth platinum, the electrode was activated before each measurement with anodic and cathodic current pulses (each 30 sec at 10 mA cm^{-2}).

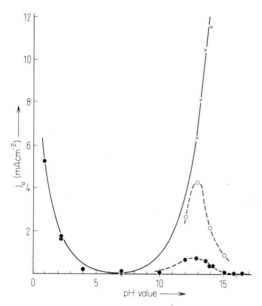

Figure III.4. Dependence of the hydrogen exchange current density on pH of electrolyte at 20°C (after Joachim, Lauer, Struck and Vielstich[9]). ●, Smooth platinum; ○, platinized platinum, ×, Raney-nickel.

Similar curves are obtained with KCl added as supporting electrolyte, but the absolute values of the exchange current densities are smaller. Similarly, deviating results are obtained with buffer solutions because of the influence of the foreign ions.

III.1.3   Anodic Current–potential Curves at High Overpotentials

Current–potential curves for the ionization of molecular hydrogen have been studied mainly by Russian and German authors. The experimental results and their significance have been reviewed by Frumkin[10].

    The general form of the anodic current–potential curve is represented in idealized form in Figure III.5. The curve can be divided into *five sections*:

(a) At low overpotentials, an exponential increase of current, corresponding with pure *charge-transfer overpotential* is to be expected. Experimentally, however, this type of increase is observed only with relatively inactive electrodes (Figure III.6, electrode S 20). With platinum, the current density at small overpotentials is very strongly dependent on stirring[10] (compare Figures III.10 and 11 with Figure IX.7.). With freshly activated, smooth platinum, current densities up to 10 mA cm^{-2} can be attained at 20°C and 20 mV overpotential[9].

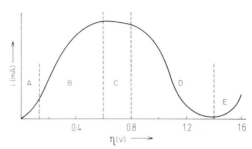

Figure III.5. Schematic current–potential curve for the ionization of molecular hydrogen. A, Region of predominant charge-transfer overpotential; B, additional diffusion overpotential; C, passivation due to anion adsorption; D, passivation due to oxygen adsorption; E, decomposition of the electrolyte, with gas evolution.

(b) At higher current densities, mass transfer assumes significance, and the current–potential curve becomes linear. In the case of gas-diffusion electrodes, for $\eta > 50$–100 mV, current density increases linearly with $\sqrt{\eta}$ (Figure III.6, electrode 76; cf. equation II.24). As explained in section II.6.2, this is a function of the potential drop parallel to the walls of the electrolyte-filled pores. The same $\sqrt{\eta}$ dependence as for a porous nickel electrode (Figure III.6) has been found by Will[11] for a partially-immersed platinum wire (inverse pore); cf. section II.6.2 and Figure III.7.

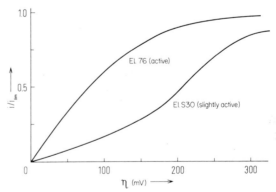

Figure III.6. Stationary anodic current–potential curves of nickel–DSK, see section IV.1.* Hydrogen electrodes (after Grüneberg[12]) in 6N KOH; electrode S 30, slightly active electrode, 20°C, $pH_2 = 2.5$ atm (total), limiting current density, $j_{lim} = 38$ mA cm^{-2}; electrode 76, strongly gassing, 68°C, $pH_2 = 3.5$ atm (total), $j_{lim} = 324$ mA cm^{-2}.

* DSK electrode = double skeleton catalyst electrode.

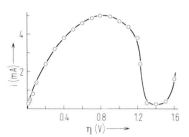

Figure III.7. Potentiostatically-determined current–potential curve for a
platinum wire partially dipping into
$1N$ H_2SO_4 at 25°C, with a hydrogen
stream over the electrolyte (after Will[11]).

(c) Depending on electrolyte composition, the current density sooner or later
reaches a maximum with increasing overpotential, and then decreases
again. Because of the *chemisorption of anions*, the binding energy of hydrogen
atoms falls away with increasing overpotential; the determination of charging curves[13] and measurements of adsorption potentials[14] confirm this
interpretation.

The passivating action of ions decreases in the sequence $I^- > Br^- >$
$Cl^- > SO_4^{2-}$ [10]. Dolin and Ershler[15] have shown by means of ac measurements that it is the adsorption $H_2 \rightarrow 2H_{ad}$, and not the ionization $H_{ad} \rightarrow$
$H^+ + e^-$, that is hindered by the chemisorption of anions.

(d) In the potential range $\eta = +600$ to $+800\,mV$, charging of the electrode
surface with oxygen begins (cf. Figure III.10 ff.), with displacement of
adsorbed anions. With increasing overpotential, the metal–oxygen bonding
becomes so strong that the electrode becomes passive in relation to hydrogen, and the anodic current falls to zero. In the system studied by Will[11],
the fall of current density begins at about $800\,mV$ (Figure III.7); with a
rotating platinum gauze electrode, it sets in at between 500 and $700\,mV$
(Figure IX.7.). In the range of negative slope, Tahlinger and Volmer[16] have
observed, under certain conditions, periodic fluctuations of the potential
of a platinum electrode. At constant current, the electrode oscillated
between 'active' and 'passive' states; our own investigations have confirmed
this observation.

(e) When the electrode potential is made still more positive, finally decomposition of the electrolyte sets in, with evolution of oxygen or chlorine, and
the anodic current density increases steeply.

The results of measurements show that the ionization of molecular
hydrogen at smooth sheet metal, or porous diffusion electrodes, attains a
limiting anodic current at an overpotential of around $300\,mV$ (cf. Figures
III.6 and IX.7). The magnitude of the limiting current is influenced by the

anions of the electrolyte[10]. At high current densities, transport processes begin to exert partial control. Overpotential can be diminished by increase of pressure of hydrogen.

If an electrode is brought into contact with a *catalyst suspension* (cf. section II.6), the hydrogen content of which is kept constant by the passage of a stream of hydrogen, then a linear current–potential curve can be observed, extending to an overpotential of about 1 volt (Figure III.8). According to Gerischer, the linear rise is attributable to the ohmic potential drop arising from the poor metallic contact between the catalyst particles and the electrode proper. The high transfer resistance, and the short contact time (50 to 200 μsec), have the further consequence that the catalyst granules are not held at the potential set by the potentiostat. This is why the steep falling away of current does not occur before 1500 mV (potential of the electrode proper), and oxygen evolution is deferred until 2V is reached.

At a potential of > 0.8 V, it is often the case that the contact electrode already carries a layer of chemisorbed oxygen. Yet anodic oxidation is still possible, since it can occur at the suspended catalyst particles, and the electrons freed in the process are easily able to tunnel through the oxygen layer.

III.1.4   Catalytic Activity of Electrode Materials

From numerous experimental investigations, mainly concerned with cathodic hydrogen overpotential, it is known that only a few metals are able to catalyse the hydrogen reaction adequately, and thus establish reversible hydrogen potentials (cf. Figure III.2). It is essential for this that hydrogen be adsorbed on the surface in the form of atoms.

Two phenomena are included under the general heading of *adsorption*: '*physical*' adsorption of molecules under the influence of van der Waals forces, and '*chemical*' adsorption. The latter is usually associated with dissociation of molecules, and formation of chemical bonds between the surface of the solid and the 'chemisorption layer'[18]. It is naturally essential, if this is to occur, that the heat of chemisorption should meet the energy requirements of the dissociation. The transition of an adsorbate molecule from the state of 'physical' adsorption to that of 'chemical' adsorption is usually associated with a certain activation energy. This is the reason why, on occasion, the expression 'hindered adsorption' is used. In the interests of unambiguous nomenclature, in the whole of the following text, the phenomenon of 'chemical' adsorption will be called

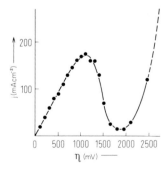

Figure III.8.   Potentiostatically-determined current–potential curve for a smooth platinum foil electrode in contact with suspended Raney-nickel, with hydrogen bubbling (after Vielstich[17]), 10g Raney-nickel/200 cm^3 1N KOH, 20°C.

'chemisorption', whereas 'physical' adsorption will be simply referred to as adsorption (in the narrower sense).

Adsorption and chemisorption are distinguishable by their heat effects. Heats of adsorption usually amount to $1-10$ kcal mole^{-1} of adsorbate. Heats of chemisorption, on the other hand, are in the range of $10-100$ kcal mole^{-1}. Heats of adsorption thus attain the magnitude of heats of evaporation of the adsorbate, but adsorption occurs to a notable extent only in the vicinity of the appropriate boiling point. Adsorption equilibria are established reversibly, and heats of adsorption do not change radically with the extent to which the surface is covered. Chemisorption equilibria, on the other hand, are generally established only with difficulty, because of the activation energy of the transition from the adsorbed state. As a rule, heat of chemisorption decreases markedly with increasing fraction of the surface covered. This is partly because of inhomogeneities of the surface (different crystal faces differ considerably in their adsorbing powers), and partly because of inter-action between the chemisorbed atoms.

Thus for iron, cobalt and nickel, the heats of chemisorption of hydrogen lie between 30 and 35 kcal mole^{-1}. The heat of chemisorption on nickel decreases from 32 kcal mole^{-1} for the free surface to 12 kcal mole^{-1} for the covered surface.

Only metals capable of chemisorbing hydrogen are suitable for establish-ment of the reversible hydrogen potential. This facility is confined to metals with incompletely filled d-orbitals in their electronic structures, and is therefore special to the 8th Group metals of the periodic table[18a].

At room temperature, the platinum metals, especially in finely-divided form (platinum black, palladium black), will establish the reversible hydrogen potential over the whole pH range; skeleton-nickel (defined below) will do so in alkaline electrolytes*. For operation at elevated temperatures, ($>150°C$), the activities of porous nickel or iron are adequate.

The chemisorption capacities of finely-divided metals for hydrogen can be considerable. Podvyazkin and Shlygin[19] determined the *hydrogen capacities* of layers of powdered metals held between platinum gauzes (cf. section II.6.1); for this purpose they used cathodic and anodic charging curves (Figure III.9). The hydrogen contents corresponded with the following quantities of electricity:

platinum black	11 A hr kg^{-1}
palladium black	165 A hr kg^{-1}
Raney-nickel (Ni:Al $= 1:1$)	450 A hr kg^{-1}

Intercomparison of these figures requires essential differences in physical conditions to be taken into account. For platinum, the value calculated from the charging curve corresponds approximately with a monatomic coverage of the metal with hydrogen atoms. Uptake of hydrogen into the lattice is impossible in this case. Palladium, on the other hand, can accommodate hydrogen, not only in a chemisorbed layer, but also in its crystal lattice. The equilibrium concentra-tion at room temperature and 760 mm hydrogen pressure amounts to about 0.7 H/Pd (cf. section III.1.7).

---

* Recently it has been observed that also in acid electrolyte non-noble metals can catalyse the oxidation of H_2. With tungsten carbide current densities of $50-100$ mA cm^{-2} were obtained. (F. A. Pohl and H. Böhm, Fuel Cell Meeting, Brussels, 1969.)

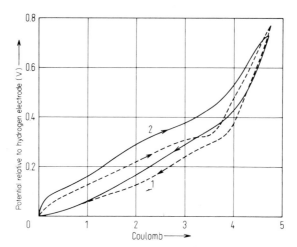

Figure III.9. Anodic and cathodic charging curves for adsorbed hydrogen on platinum black (after Podvyazkin and
Shlygin[19]). Electrode potential as function of H_{ad} coverage in
coulomb. 0.1 g of Pt-black in (1) 0.1 N H_2SO_4; (2) 0.1 N NaOH.
Catalyst layer held between fine platinum screens.

For Raney-nickel, the H/Ni ratio is about 1:1 (cf. also measurements of
Pilkuhn and Winsel[20]). According to Tompkins[21], the hydrogen in Raney-
nickel (as for platinum) is not dissolved, but adsorbed in monatomic form at the
large internal surface of the porous material.

Raney-nickel is highly porous nickel prepared from powdered or coarsely granular Raney-alloy
(Al : Ni, preferably in ratio 1 : 1) by dissolving out the aluminium in warm alkali. The freshly
activated catalyst has a large internal surface area, and is strongly pyrophoric. It can therefore be
kept dry only under oxygen-free conditions.

Raney-nickel is stable in contact with aqueous solution only at ph > 9. In neutral or slightly
acid solutions, it dissolves with formation of $Ni(OH)_2$ and hydrogen; Travers and Aubry[22] reported
in 1938 that Raney-nickel will establish the reversible hydrogen potential from pH 5 to strongly
alkaline solutions[23].

The hydrogen content of Raney-nickel determines not only its catalytic activity[24] but also its
conductivity. According to Sokolsky and Dzhardamalieva[25], the quantity of adsorbed hydrogen
increases with rising temperature and OH^- ion concentration.

As a catalyst for the hydrogen electrode in fuel cells, Raney-nickel can be used
either (exploiting the transfer principle, cf. section II.6.2) as a suspension in the
electrolyte, in conjunction with a corresponding contact electrode for current
flow, or, according to the suggestion of Justi and Winsel[26], in the form of a
massive gas-diffusion electrode (on the DSK principle, cf. section IV.1). In
the DSK electrode, mechanical stability is conferred on the Raney-nickel by a
supporting framework of nickel.

The Raney process, which can be applied to Al/Ni, Mg/Ni or Zn/Ni alloys,
is not the only method which can be used to obtain nickel in highly dispersed
and catalytically active form.

Thus, for example, Horner[27] has shown that, by addition of anhydrous $NiCl_2$ to molten sodium under naphthalene, highly active nickel can be obtained with better catalytic properties than freshly activated Raney-nickel. A particularly simple technical method of preparation consists of mixing nickel powder with Al, Zn or Mg powder, pressing, sintering, and finally dissolving out the inactive constituent with warm alkali. This process gives a mechanically stable electrode of catalytically active nickel (MSK electrode, cf. section IV.1).

The essential requirement for catalytic properties is not so much the use of Raney-metals as such, but rather the attainment of a disperse surface structure for the nickel. In the text which follows, the name skeleton-nickel, already customary in the Russian literature, will be used as a generic name for porous, pyrophoric nickel catalysts.

Nickel boride, Ni_2B, also provides a form of active skeleton-nickel[28,29]. Recently, Jasinski[30] studied the properties of these non-pyrophoric catalysts for hydrogen oxidation. Current densities of more than $100\,mA\,cm^{-2}$ were obtained at room temperature. Besides nickel, cobalt, or mixtures of the two metals, are suitable (cf. section IV.111).

Jung and Kröger[31] have shown that sintered bodies of nickel, cobalt, or mixtures of these two metals can, on treatment with alkaline borohydride solution, 'build in' hydrogen reversibly, with change of lattice structures (cf. section IV.111).

For the practical application of highly active catalyst electrodes to, for example, the oxidation of hydrogen at ordinary temperatures, it is essential to consider the influence of foreign gas and electrolyte composition on the activity of the electrode. The susceptibility to *poisoning* is very variable from one system to another, and so is *passivation* by specifically adsorbed anions (cf. section III.1.3). Thus, for example, hydrogen oxidation at platinum in strongly alkaline solution is considerably hindered by adsorption of OH^- ion. At skeleton-nickel, by contrast, no effect of the kind can be detected (Figure III.4). For the choice of suitable electrode materials, theoretical considerations are not enough.

For a qualitative study of *half-cell properties*, the triangular wave potential scan method described in section II.4 has been chosen. Figures III.10 and III.11 show the current–potential diagrams obtained in this way for the most important reaction-catalysing metals[17].

Figure III.10 shows, in the left-hand column, the formation and destruction of H and O layers on smooth electrodes of Pt, Pd, Ir, Rh and Au, in N_2-saturated $1N\ H_2SO_4$, observed with a scan rate of $1V\,sec^{-1}$. The right-hand column shows, for the same metals, similar diagrams (recorded at $50\,mV\,sec^{-1}$) under conditions of hydrogen bubbling around the electrodes.

The general form of these surface layer diagrams (left-hand column) including the current peaks in the hydrogen region, has already been discussed in section II.4. In contrast to Figure II.8, the diagram for platinum in Figure III.10 shows a third anodic peak. This is attributed to the oxidation of molecular hydrogen, formed cathodically on crossing the hydrogen potential; this potential region was not included in Figure II.8.

The form of the curves in the hydrogen region is very variable from metal to metal. The amounts of adsorbed hydrogen were assessed by Will and Knorr[32] from similar diagrams ($8N\ H_2SO_4$ as

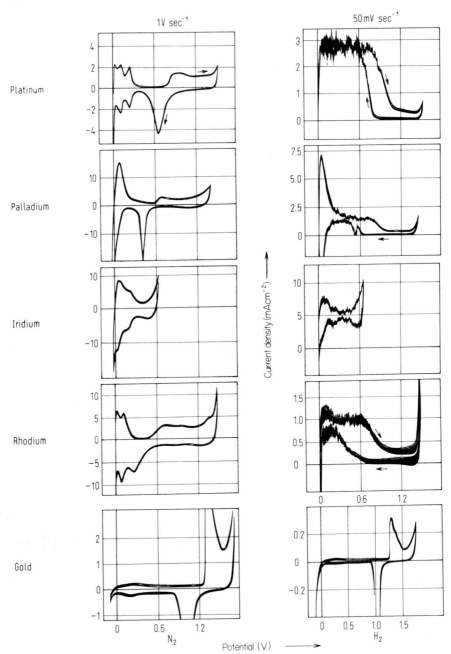

Figure III.10. Potentiostatic current–potential curves for various smooth electrode metals in 1_N H_2SO_4 at 20°C (after Vielstich[17]). Left column: N_2-saturated electrolyte. Right column: H_2-streaming of the electrodes. Scan rate, 1 volt sec^{-1} or 50 mV sec^{-1}.

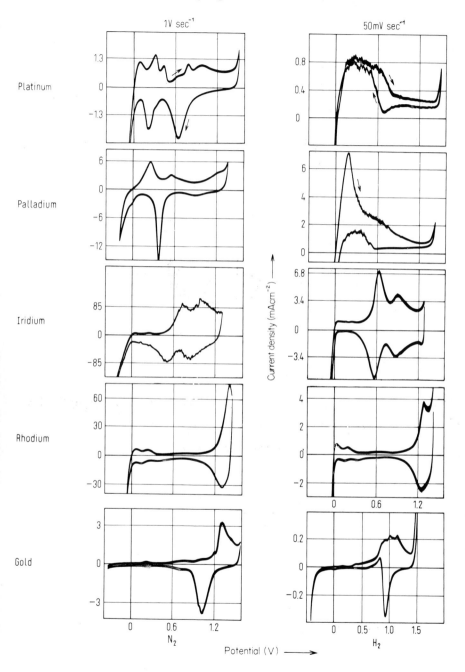

Figure III.11. Potentiostatic current–potential diagrams for 1N KOH; conditions of measurement as in Figure III.10 (after Vielstich[17]).

electrolyte). The relative hydrogen coverages at $\eta = 0$ were Pt:Ir:Rh:Au = 100:45:62:3. The hydrogen adsorbed by platinum corresponds with an approximate monatomic layer[32].

Hydrogen bubbling (right-hand column of Figure III.10) leads to curves characteristic of rate-limitation by diffusion. At Pd electrodes extensive hydrogen dissolution (see above) dominates the current-potential diagram; the anodic current falls away very steeply towards the oxygen-charging region. For gold, the anodic current is initially extremely small, and traces of platinum as impurity give a simulation of the platinum curve on a reduced scale (cf. Figure III.15).

On the 'cathodic return' curve, hydrogen oxidation first begins when at least part of the oxygen layer has been destroyed. For iridium, reactivation is not attained until the reversible hydrogen potential is reached, because of the very stable oxygen layer. *Platinum appears to be the most suitable metal for use in acid solutions.*

Figure III.11 shows the corresponding diagrams obtained with 1N KOH as electrolyte. The behaviour of gold is similar to that in acid solution. The hydrogen charging of Ir and Rh is considerably less than in H_2SO_4, whereas for Pt and Pd it is of the same order of magnitude.

With hydrogen bubbling, the anodic oxidation at Rh is surprisingly slow, but Ir, despite its low hydrogen coverage, has good catalytic properties. The influence of oxygen charging, or of adsorbed anions, is different in the two electrolytes. Thus, transfer from H_2SO_4 to KOH as electrolyte narrows the potential range within which the platinum surface is active for H_2 oxidation.

A mixed palladium–platinum catalyst for the anodic oxidation of hydrogen and of methanol has been proposed; Grimes, Murray and Spengler[33a] found the optimum weight ratio to be Pd:Pt = 3:1.

III.1.5   The Influence of Oxygen Additions to Hydrogen

The electrochemical behaviour of hydrogen–oxygen mixtures has not hitherto been extensively investigated. In what follows, reference is made to the experimental results of Bennewitz and Neumann[34] (on the potentials established at platinum as a function of gas composition) and of Rozental and Veselovsky[35–37] (on the gas reactions at platinum and gold electrodes).

From the practical point of view, the interest of 'electrolytic gas electrodes' lies in their use for direct conversion of nuclear energy into electrical energy (cf. chapter VII).

Radiolysis of aqueous solutions provides 'electrolytic gas' (oxyhydrogen), a mixture of hydrogen and oxygen, the energy of which is available for electrochemical use by means of a hydrogen–oxygen cell. If the gas mixture is taken first to an oxygen electrode, consisting of a material which does not catalyse hydrogen oxidation (e.g. carbon or silver), the major part of the oxygen can be reduced at the selectively working electrode. The residual gas, however, still has a considerable content of oxygen, and must, in spite of this, be utilized at a hydrogen electrode.

Bennewitz and Neumann[34] studied the establishment of potentials at bright and platinized platinum in 1N H_2SO_4 saturated with hydrogen–oxygen mixtures of varying composition. For this they used a rotating platinum cylinder with a hydrogen electrode in the same solution for reference purposes.

Their experimental results are collected in Figure III.12. At the platinized electrode, the gas present in deficiency of the stoichiometric ratio was practically completely consumed. In the range $p_{H_2} > 0.66$ (p = mole fraction) the electrode

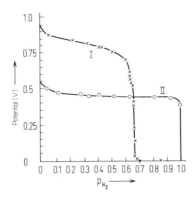

Figure III.12. Electrode potential of a rotating platinum cylinder in 1N H_2SO_4, saturated with various H_2–O_2 gas mixtures (after Bennewitz and Neumann[34]). I, Platinized platinum; II, bright platinum.

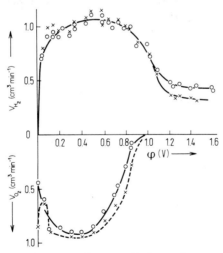

Figure III.13. Hydrogen oxidation and oxygen reduction (with only one gas at a time) at a rotating platinum gauze electrode in 1N H_2SO_4 at 20°C (after Rozental and Veselovsky[36]). ○, Gas consumption; ×, current in $cm^3 min^{-1}$, in dependence on electrode potential, ϕ. Partial pressures of hydrogen and oxygen correspond with a stoichiometric electrolytic gas mixture; total pressure = 1 atm; added inert gas, nitrogen.

therefore worked as a hydrogen electrode; in the range $p_{H_2} < 0.66$, as an oxygen electrode. At bright platinum, the velocity of the reaction $H_2 + \frac{1}{2}O_2 \rightarrow H_2O$ is small in comparison with diffusion rate. An appreciable consumption of

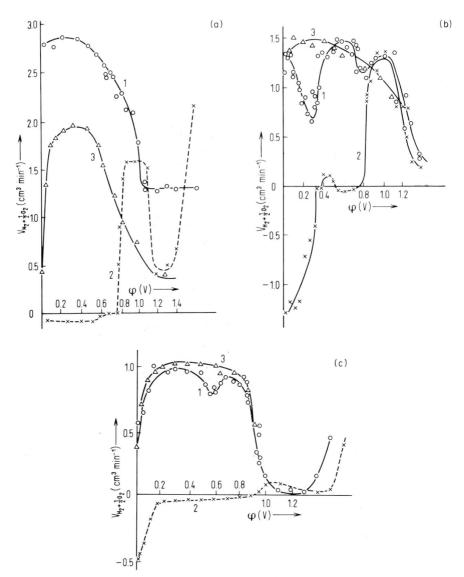

Figure III.14. Reaction of a stoichiometric electrolytic gas mixture at a rotating platinum gauze electrode in (a) 1N H_2SO_4, (b) 1N $HClO_4$, (c) 1N KOH (after Rozental and Veselovsky[36]). 1, Gas consumption; 2, current; 3, gas consumption calculated from the separate measurements (hydrogen oxidation and oxygen reduction of Figure III.13).

oxygen occurred only at $p_{H_2} > 0.95$; over almost the whole range a hydrogen–oxygen mixed potential was established.

Rozental and Veselovsky used a rotating gauze electrode (700 rpm) of platinum[35,36] or gold[35,37], to investigate the electrochemical reaction of a stoichiometric electrolytic gas mixture as a function of electrode potential. Current and gas consumptions were both measured and were also compared with data obtained under the same conditions for separate hydrogen oxidation and oxygen reduction.

The most important results obtained with *bright platinum*[36] in H_2SO_4, $HClO_4$ and KOH are collected in Figures III.13 and III.14. Gas and current consumptions in cm^3 min^{-1} for hydrogen oxidation or oxygen reduction in 1N H_2SO_4 are plotted in Figure III.13. In the potential region between 150 and 800 mV, the anodic current density amounts to about 22 μA cm^{-2} (corresponding with 10^{-2} cm^3 hr^{-1} cm^{-2}). Whereas the oxygen reduction current and gas consumption correspond well with each other, for hydrogen oxidation at $\phi > 1.1$ volt, gas consumption is about 10–20% greater than the current calculated in cm^3 min^{-1}. The authors do not comment on the significance of this effect.

The data obtained with electrolytic gas are displayed in Figure III.14. The experimental curves for the consumption of gas (1) are to be compared, first with the observed currents (2), and then with the gas consumption (3), calculated in accordance with Figure III.13 on the assumption that hydrogen oxidation and oxygen reduction have no effect on each other.

The dependence of current on potential (2) shows that for $\phi < 0.8$ volt oxygen reduction predominates, but for $\phi > 0.8$ volt, it is hydrogen oxidation which determines the direction of current flow. The calculated gas consumption (3) for the case of 1N H_2SO_4 as electrolyte is considerably lower than that observed (1), but for 1N $HClO_4$, it is, in certain potential regions, higher, whereas for 1N KOH the curves are almost coincident.

From this it can be concluded that, in KOH, hydrogen oxidation and oxygen reduction occur independently of each other, whereas in sulphuric acid, hydrogen oxidation is promoted, and, in perchloric acid as electrolyte, it is partially repressed.

The velocity of the electrolytic gas reaction at $i = 0(0.6 < \phi < 0.85$ volt) in 1N KOH, 1N $HClO_4$, 1N H_2SO_4 has the relative values 1:2:3.

At active, smooth gold*, the velocity of hydrogen oxidation in 1N H_2SO_4 approximately corresponds with that found for smooth platinum (Figure III.15). As in the case of platinum, the calculated curve for hydrogen consumption lies below that actually observed for $\phi > 0.9$ volt. A similar effect is observed for oxygen reduction at $\phi < 0.2$ volt.

---

* Probably there were traces of platinum on the gold surface (cf. section III.1.4). The catalytic properties can be very greatly altered simply by a short electrolysis in a cell with a counter electrode of platinum[17].

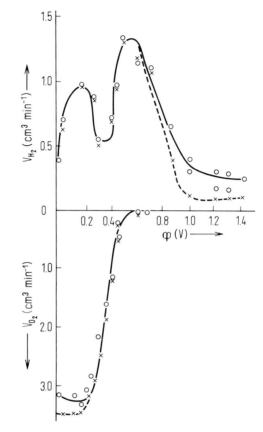

Figure III.15. Hydrogen oxidation and oxygen reduction at a rotating gold gauze electrode in $1N$ H_2SO_4 at 20°C (after Rozental and Veselovsky[37]). ○, Gas consumption; ×, current in cm^3 min^{-1} in dependence on electrode potential.

Supplied with electrolytic gas, the gold electrode (Figure III.16) allows hydrogen oxidation and oxygen reduction to proceed almost independently for $\phi < 0.4$ volt. For $\phi > 0.5$ volt, total gas reaction is greater than that calculated from the measurements conducted with each gas separately (cf. Figure III.15); hydrogen oxidation is specially accelerated.

A gold electrode heated to 850°C in a stream of oxygen becomes passive to hydrogen.

Rozental and Veselovsky's work[35–37] has recently been extended by Struck[37a]; similar experiments with smooth, rotating electrodes in potassium hydroxide have provided the following additional information.

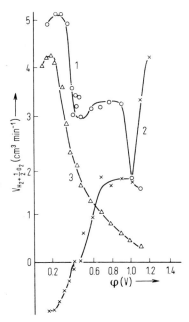

Figure III.16. Reaction of a stoichiometric electrolytic gas mixture at a
rotating gold gauze electrode in 1N
H_2SO_4 (after Rozental and Veselovsky[37]). 1, Gas consumption; 2, current;
3, gas consumption calculated from
measurements with individual gases as
in Figure III.15.

At electrodes of platinum, palladium or palladium–silver (80:20 atom %)
supplied with stoichiometric electrolytic gas ($2H_2 + O_2$), oxidation of hydrogen
and reduction of oxygen proceed independently within an error of $\pm 10\%$.
At potentials more positive than the oxygen reduction potential, oxygen admixture has a strong effect on the hydrogen oxidation current. For gas mixtures
containing hydrogen in excess of the stoichiometric ratio, anodic current
increases linearly up to 100% of hydrogen, and the total current is approximately equal to the sum of the currents due to the individual components.
No hydrogen reaction has been observed at gold electrodes in alkaline solution.

The results have shown that it is possible to construct a cell with catalytically
selective electrodes that will provide a current, even though both electrodes are
supplied with one and the same gas mixture. For this purpose Struck has used
electrodes consisting of granular carbon, compressed between nickel grids. On
the anodic side, active carbon containing 5 wt. % of Pd, or 1 % Pd–Pt (4:1) was
used; on the cathodic side, silvered active carbon served as catalyst. This pair

of electrodes, immersed in an electrode vessel containing potassium hydroxide solution and streamed from beneath with H_2–O_2 gas mixture (80:20 to 95:5), provided a terminal voltage of 500 to 600 mV at a continuous loading of 0.3 to 0.6 mA cm^{-2} over several hours.

III.1.6   Activation by Addition of a Radioactive Isotope to the Electrode Material

Braun[39] and Crossley[40] long ago suggested that the activity of gas-diffusion electrodes might be enhanced by the action of short-wavelength radiation, or of radiation from radioactive substances. Since then, many authors have studied the effects of rays on the activity of catalysts but their results are very discrepant.

For instance, Clark, McGrath and Johnson[41] found a slight increase in the activity of platinum catalysts for the oxidation of SO_2 in the presence of water vapour.

Harker[42] reported that the uptake of hydrogen from the gas phase by Cu and Pd was accelerated by irradiation. On the other hand, Emmett and Jones[43] found that the velocity of electrolytic gas combustion at a platinum catalyst was not affected; Schwab, Sizmann and Todo[44] even found that the catalytic activity of Cu for ethylene hydrogenation was diminished by irradiating the catalyst with α-particles.

Clarke and Gibson[45] recorded an unequivocally positive effect in relation to the Fischer–Tropsch synthesis; γ-irradiation (^{60}Co) of the catalyst improved the conversion by 40 per cent under otherwise similar conditions. Although the working temperature was 280°C, the activity did not decrease during the period of experiment (300 hr).

Salcedo and Lang[46] have recently investigated the effect on hydrogen oxidation of incorporation of ^{63}Ni and ^{14}C isotopes into electrodes. Porous electrodes of carbon or nickel were impregnated with the α- or β-emitters by dipping into appropriate solutions. The salts used in these solutions were not soluble in the alkaline electrolyte used for the electrochemical measurements. The activity of the electrodes so prepared amounted to 5 to 30 μc cm^{-2}.

The impregnated electrodes supplied with hydrogen showed, at 500 mV overpotential, a three- to four-fold greater current density than electrodes not furnished with radioactive sources, but the current-carrying capacity of the electrodes was in any case relatively poor (25 mA cm^{-2} at 60°C). Impregnation with a non-radioactive nickel salt would probably have led to a similar increase in activity.

In this connexion reference may be made to the investigation by Feates[33] of platinum in 0.8N H_2SO_4. At passive platinum*, the overpotential for hydrogen evolution was reduced by about 50 mV by γ-irradiation. If, however, the platinum was activated in a stream of hydrogen at 250°C, the influence of irradiation was no longer observed. It is obvious that attempts to improve the activity of already active hydrogen-diffusion electrodes by use of radioactive isotopes are unlikely to succeed[47]. The same applies to oxygen electrodes (cf. section III.6).

* Previously heated in air at 250°C.

III.1.7   The Hydrogen-Membrane Electrode*

III.1.7.1   *Chemisorption and Absorption of Hydrogen by Transition Metals*

Most transition metals are able to chemisorb hydrogen on their surfaces. In contrast to the physical adsorption of molecular hydrogen observable at low temperatures (20–40°C), this chemisorption involves *dissociation* into atoms[48,49]. The corresponding activation energy is so slight that chemisorption equilibrium is very rapidly established, even at room temperature. The chemisorbed atoms are mobile by surface diffusion independently of each other[50]. Entropy determinations indicate that, collectively, they behave like a two-dimensional gas[51].

This behaviour, unusual in comparison with other chemisorption layers, can best be understood in terms of a wave-mechanical concept of protons, with a certain tunnelling probability between corresponding ψ functions of neighbouring chemisorption centres (cf. ref. 52).

Chemisorbed hydrogen can participate in chemical and electrochemical reactions with relatively small activation energy.

Independently of *chemisorption at the surface*, numerous transition metals can *absorb hydrogen into their metal lattices*. According to the heat of absorption, distinction can be made between endothermally and exothermally absorbing metals.

The equilibrium hydrogen concentration in the case of *endothermally* absorbing metals is extremely small at room temperature (10^{-6} to 10^{-10} H/Me)[53], but increases with rising temperature:

$$L = L_0(p_{H_2})^{\frac{1}{2}} \exp(-\Delta H/RT) \qquad \text{(III.8)}$$

where L = atom ratio, H/Me; ΔH = heat (enthalpy) of solution, and L_0 is a constant. Examples of endothermally absorbing metals are Fe, Co, Ni, Cu, Pt.

Exothermally absorbing metals absorb hydrogen first into the homogeneous metal phase. At sufficiently low temperatures, and on attainment of a certain limiting concentration, a so-called metallic hydride may be formed, with one or more hydride phases identifiable by X-ray diffraction. They usually show approximately stoichiometric compositions, but in some cases considerable deviations may occur[54,55]. Corresponding to their exothermic formation, the metallic hydrides are vulnerable to thermal decomposition. As a rule, their decomposition pressures are unique functions of temperature in accordance with

$$p = p_0 \exp(-\Delta H/RT)$$

where p_0 is a constant. Examples from this group of metals are U, Ti, Th, Ce, V, Ta, Pd.

Metals and hydrides are distinguished from each other by their densities, and often by their types of lattice. Lattice distortions in the course of hydride

* By A. Küssner, Research Laboratory of the Siemens A.G., Erlangen, Germany.

formation give rise to very strong internal stresses in the metal structure. In the case of palladium, for example, this leads to the formation of fissures and rifts[56,57]. The pressures in the interior of the metal structure produced in this way have been estimated as 10^3 to 10^4 atm[58]. In the case of uranium the density difference is so large and the resulting internal stress arising from hydride formation is so great, that the hydride crumbles away from the compact metal in the form of a fine powder[58a]. The pressure hysteresis in the formation and decomposition of hydrides can be related to density differences and internal stresses of this kind[58,58b].

During exothermic uptake of hydrogen into a *homogeneous* metal phase, there is a *continuous* lattice expansion. In contrast to the phenomena associated with hydride formation, the internal stresses resulting from this continuous lattice expansion can normally be taken up by the elasticity of the metal. In the series of palladium–silver alloys, the range of existence of the β-hydride phase* is progressively restricted as the silver concentration is increased until, at about 23 atoms per cent of silver[56a], it has, at room temperature, completely vanished[59,60]. Only continuous hydrogen absorption into the homogeneous metallic phase is then observed. This alloy can be charged with hydrogen without loss of mechanical stability and without the development of rifts[56].

III.1.7.2   The Mechanism of the Catalysed Phase-boundary Reaction of Hydrogen

All of the metal–hydrogen systems concerned have in common a kinetic hindrance to the establishment of equilibrium at temperatures below about 200°C. This is unequivocally due to the *slow transfer* of hydrogen through the metal surface, and not to a slow rate of concentration equalization within the interior of the metal by solid state diffusion. A suitable choice of working temperature therefore allows measurements of hydrogen chemisorption at the surface of a hydride-forming metal to be made without disturbance by absorption of hydrogen into the interior of the metal[48].

It has long been known that a coating of palladium black on the surface of compact metallic palladium accelerates the uptake of hydrogen by the latter[62]. It is also known that alternating electrolytic oxidation and reduction, heating to redness in air[63], or prolonged silent electric discharge[64] exert a surface

---

* Palladium crystallizes in a cubic face-centred lattice, with a lattice constant of 3.883 Å. A small quantity of hydrogen (up to 0.9 atom % at 30°C) is exothermally soluble in this lattice; it remains homogeneous and is the so-called α-phase. The lattice constant is continuously increased thereby up to 4.894 Å. At this concentration, a so-called β-phase becomes stable; it also has a cubic face-centred lattice, but the lattice constant is 4.017 Å.

At 30°C the formation of this β-phase is completed at a composition of $PdH_{0.6}$[61]. In the two-phase region, between the named limiting concentrations, the equilibrium pressure of hydrogen remains constant. Further hydrogen can be dissolved in the fully-developed β-phase. The concentration of this additional dissolved hydrogen increases in proportion to the logarithm of the equilibrium pressure[61,67].

'loosening effect' that expedites the attainment of equilibrium with gaseous hydrogen. The cause of this is not so much the increase in the geometrical surface area, as at first supposed[65], but rather the generation of 'catalytically active' rifts[66].

It has been repeatedly observed that the reaction between gaseous hydrogen and compact palladium proceeds the faster the more *cracks* of molecular dimensions are present in the metal[54,68].

Gaseous hydrogen, by gas kinetic diffusion, can hardly penetrate into cracks which measure <ca.100 Å across, but chemisorbed atomic hydrogen can readily do so by surface diffusion. In this way, *chemisorbed* atoms in the interior of such a crack can pass smoothly and continuously into the state of atoms *absorbed* in the metal lattice. The direct transfer of hydrogen across the gas–metal interface, with its high activation energy, can thus be avoided. The crack can therefore formally be described as an 'active centre' for the 'catalysis' of a normally kinetically-hindered phase-boundary reaction[66].

In the process of surface diffusion into the interior of a crack, it is not only the potential energy of the hydrogen atoms that changes, but also their mobility. The activation energy of surface diffusion gradually approaches that of solid-phase diffusion as the crack narrows, and the distance between its walls approaches the normal separation of the lattice planes. At the same time, the activation energy for transfer across the pore walls decreases, and eventually also becomes equal to that of solid-state diffusion when the crack has shrunk in dimensions to the normal interlattice plane distance. The activation energy of the phase-boundary reaction at one crack wall is thus lowered by the close approach of the other crack wall. It can easily be seen that a similar effect may arise from the close approach of, instead of the other crack wall, any hydrogen-chemisorbing surface. This might be the surface of a hydrogen-chemisorbing, but not hydrogen-absorbing metal, i.e., a metal not permeable to hydrogen. For instance, a porous coating of copper or platinum may activate a palladium surface for the absorption of hydrogen from the gas phase, whereas a coherent film would, on the contrary, block it. Numerous hydrogen-chemisorbing substances (e.g., Fe, Co, Ni, Cu, platinum metals[66] and metallic hydrides[69]) in the form of a powder layer or porous coating in contact with compact Pd, Ta or Ti, can considerably accelerate the absorption of gaseous hydrogen by these metals, or, in general, initiate it.

This activation method is applicable to hydrogen-absorbing alloys as well, and is of technical interest in relation to the palladium–silver alloy (77:23) mentioned above.

III.1.7.3   The pore-free Hydrogen-membrane Electrode[66,70,71]

A membrane of a hydrogen-absorbing metal or alloy can be brought into thermodynamic equilibrium with gaseous hydrogen at room temperature with the help of a hydrogen-transfer catalyst on one of its faces. If, at the same time, the other face is in contact with a suitable electrolyte, it is active as a hydrogen electrode. This pore-free hydrogen electrode is not only applicable to the electrochemical use of hydrogen in fuel cells, but also to the preparation of hydrogen in its purest state. In contrast to porous electrodes, no three-phase boundary is involved. This carries a simplification in operation because there is no particular requirement in relation to working pressure. Thus a small tubular electrode of 3 mm diameter and 0.1 mm wall-thickness has been operated with hydrogen at pressures varying between 100 mm and 40 atm. Hydrogen-containing mixtures can also be used, without any fear of reaction between the foreign gas in the mixture and the electrolyte. For instance, a palladium–silver membrane anode has been operated over nine months at 250°C with town gas containing carbon monoxide. The only cleaning procedure for the gas was the absorption of unsaturated hydrocarbons by means of a charcoal column[71a]. Here, of course, a forced convection of the gas mixture is necessary to avoid the establishment of a diffusion barrier due to a hydrogen-depleted layer of the foreign gas.

Hydrogen-membrane electrodes can be made in any shape or size. Powdered layers or porous coatings of any hydrogen-chemisorbing substance are of use for activation of the gas side of a membrane electrode. Certain restrictions apply to electrodes for technical application, because of requirements of mechanical strength and resistance to possible corrosive attack by foreign gases or air. Suitably activated membrane electrodes are absolutely stable to CO_2, NH_3 and oxygen, and can be kept in air indefinitely[71].

III.1.7.4   *Examples of Application*

(a) The preparation of hydrogen[66,72,73].

Passage of cathodic current at a membrane electrode causes liberation of hydrogen on the gas side. If the membrane is impermeable to other gases, as in the case of palladium–silver alloy, the hydrogen is ultra-pure.

Figure III.17 shows the production of hydrogen in cm^3 min^{-1} as a function of current density at the gas side of a suitably activated, tubular membrane electrode. At low current densities, the electrolytically-deposited hydrogen diffuses practically quantitatively to the inside of the tube, where it is liberated. At higher current densities, an increasing proportion of hydrogen is evolved in bubbles from the outer surface of the tube. The membrane electrode can be used for the preparation of hydrogen at higher pressures, but under these conditions a back reaction becomes perceptible. Some hydrogen is evolved, without any current passing, at the outer surface of a tubular electrode supplied with hydrogen under pressure[66].

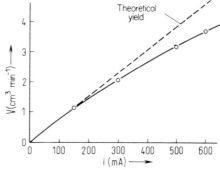

Figure III.17. Hydrogen production at a tubular membrane electrode of the type of Figure III.18.

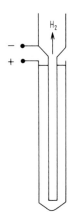

Figure III.18. Apparatus for the preparation of hydrogen. Pd–Ag (75:25) tube; 3 mm diameter; 0.1 mm wall thickness; depth of immersion, 90 mm; temperature 23°C; electrolyte, 5N NaOH.

An arrangement of this kind is suitable when small quantities of very pure hydrogen are required over prolonged periods, as for instance in gas chromatography. A suitable form of electrolysis cell will operate with a small quantity of electrolyte. The membrane electrode can, in consequence, be used with advantage for the preparation of the purest deuterium from heavy water on the laboratory scale. Well-dehydrated potassium carbonate has been used as an electrolyte salt in this case.

(b) Utilization of hydrogen for fuel cells.

The results of electrode-potential measurements (Luggin capillary) at a tubular membrane electrode carrying various cathodic or anodic current densities, are shown graphically in Figure III.19. There is an overpotential in each case, dependent on the state of the electrode surface. These overpotentials can be

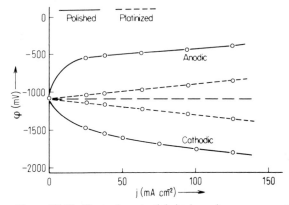

Figure III.19. Electrode potentials in dependence on current density. Membrane Pd–Ag (75:25), 0.1 mm thick. Saturated calomel reference electrode; temperature, 23°C; electrolyte 5N NaOH; hydrogen pressure, 760 mm.

reduced by, for example, a layer of platinization. Probably in this case the same mechanism of transfer of hydrogen atoms between the metal lattice and the chemisorption layer is operative as in the case of the catalysed interphase reaction with gaseous hydrogen.

Despite the use of different hydrogen-transfer catalysts and of various methods of activation, a certain *anodic* current density cannot be exceeded at a given membrane electrode under given experimental conditions. It can be shown that this value is determined by the diffusion resistance of the membrane itself. The limiting anodic current varies inversely as membrane thickness[71]. Higher current densities are therefore obtainable only by use of thinner membranes, or by operating at higher working temperatures. A membrane electrode 12 μ thick (Pd–Ag, 77:23; Johnson, Matthey & Co., Ltd.) will pass a current density of 1.2 A cm^{-2} at 30°C.

From the limiting anodic current density attained at a Pd–Ag (77:23) membrane electrode at 30°C, the average diffusion coefficient of hydrogen in the alloy is calculated to be 3.5×10^{-7} cm^2 sec^{-1}. The diffusion coefficient is, however, strongly dependent on the hydrogen concentration in the metal, in line with the non-ideal solubility[56b]. Figure III.20 shows Fick's law diffusion coefficient, measured with very low concentration gradients, plotted against hydrogen concentration. These values are in accordance with earlier measurements at very low[74] and very high[75] concentrations. The average value between equilibrium and zero concentrations is 3.35×10^{-7} cm^2 sec^{-1}, which agrees well with the value calculated from the limiting anodic current density. This means that the catalytic activation of the gas side of the membrane is apparently 'ideal', and that the kinetic hindrance of the phase-boundary reaction with gaseous hydrogen plays no part, compared with

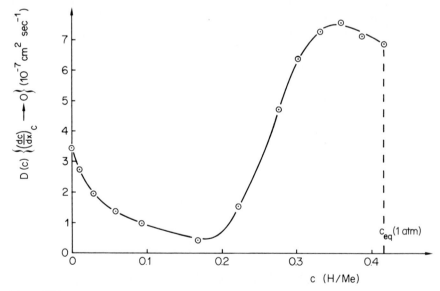

Figure III.20. Plot of Fick's diffusion coefficient of hydrogen in a Pd–Ag (77:23) membrane against the hydrogen concentration; 30°C (after Küssner[56b]).

the diffusion resistance of the membrane. The distance between catalytically active cracks on the gas side of the membrane must therefore be small compared with the membrane thickness.

According to measurements by Wicke and Bohmholdt[75], the diffusion coefficient of hydrogen in this alloy at higher concentrations ($0.34 \, H/M < C_H < 0.40 \, H/M$) is $6 \times 10^{-7} \, cm^2 \, sec^{-1}$ at 25°C. The diffusion coefficient is therefore not independent of hydrogen concentration.

Oswin and Chodosh[71] have studied the temperature-dependence of current–potential curves for hydrogen anodes of Pd–Ag (75:25) in KOH. Judging from the limiting current density of 400 mA cm^{-2} at 25°C, the membrane must have been about 0.025 mm thick; it was activated on both gas and electrolyte sides. The following overpotentials were measured at 300 mA cm^{-2}:

Temp. (°C)	25	50	80	100	150	200
η (mV)	+350	+170	+110	+ 75	+ 35	+ 15

The limiting current density at 200°C was quoted as 3.5 to 4 A cm^{-2}.

III.1.7.5   Material Problems

Hydrogen membrane electrodes have been made of pure palladium[63a,72,76,77,77a], but this material is hardly suitable. Because of the discontinuous lattice expansion that accompanies the formation of the β-phase it becomes brittle and porous[56,57], but, as already mentioned, the β-hydride phase is no longer stable at room temperature in a palladium–silver alloy containing more than 25 atom per cent of silver[59,60]. This alloy remains mechanically stable after repeated charging with hydrogen and is a suitable material for membrane electrodes.

For the diffusion coefficient, D, of hydrogen in homogeneous Pd–Ag–H alloys over the composition range Pd:Ag = 75:25 to 50:50, the following equation is approximately valid for vanishingly small hydrogen concentrations:

$$D = D_0 \exp(-5700/RT) \qquad \text{(III.9)}$$

where

$$D_0 = \exp(-1.61 - 14.5c_{Ag})^{74}$$

(c_{Ag} = weight concentration of Ag $\sim$ mole fraction)

In this concentration range the alloy with 25 atom % of silver has the largest hydrogen diffusion coefficient. Since the activation energy of the solid-state diffusion is about the same for all the alloys this statement is valid, independent of the temperature.

The equilibrium concentration of hydrogen at 1 atm pressure is also largest for the 25 atom % silver alloy. The larger the equilibrium concentration of hydrogen, the greater is the concentration gradient to 'drive' hydrogen diffusion.

For both of these reasons, it appears preferable to choose an alloy composition at which the stability of the hydride phase at room temperature just falls to zero. This occurs at an alloy composition of 23 atom % of silver[56a]. Unfortunately, membrane electrodes of this material may undergo *irreversible plastic* deformations in the course of long-continued anodic tests with frequent changes of load[77a]. Small, irregular outgrowths may appear on the surface, deranging the contact between the membrane surface and the activation layers. The membrane contracts, and may even crack. It has been shown that these irreversible deformations are caused by stresses set up within the homogeneous alloy.

When this palladium–silver alloy absorbs hydrogen, it expands continuously, approximately in proportion to the hydrogen content[59,60], until the enlargement reaches about 2.4% at an equilibrium pressure of hydrogen of 1 atm[77b]. If there is a hydrogen concentration gradient within the sample—as there must be for every membrane electrode under load—there is also a density gradient, with a consequent gradient of solid-state pressure that sets up an internal stress. Attempts must be made to ensure that these stresses in the homogeneous alloy are accommodated by elastic (i.e. reversible) deformations. This can be accomplished by hardening the alloy through addition of a third component[77a,77c]. Experiments on palladium–silver electrodes containing small concentrations of carbon or boron[77a] have shown that these additions affect the concentration-dependent diffusion coefficient of hydrogen in a characteristic way, but the average value for the whole concentration range of hydrogen is lowered. It seems that the undesirable plastic deformation can be obviated by a 50% increase in the hardness of the alloy; this involves no more than a 20% reduction in the limiting anodic current.

III.2   METHANOL, FORMALDEHYDE AND FORMIC ACID (FORMATE)

The electrochemical oxidation of methanol, formaldehyde and formic acid was studied in some detail nearly 40 years ago by E. Müller and collaborators[78–80]. Since then, it has been known that all three substances, dissolved in an acid electrolyte, can be oxidized at relatively high current densities (> 10 mA cm^{-2} at 20°C) at platinum electrodes in the potential region between the hydrogen and oxygen potentials. Pavela[80a] has provided further information on the oxidation of methanol, recognizing the significance to the anodic process both of dehydrogenation and of the oxygen layer on the electrode. He specified analytical procedures for the simultaneous estimation of formaldehyde, formate and carbonate. Pavela found that, for alkaline solutions, the limiting current is strongly dependent on OH^- ion activity, and he observed the occurrence of potential oscillations with current densities greater than 100 mA cm^{-2}, at platinized platinum electrodes in an acid solution (1N H_2SO_4, 0.25 m

CH_3OH, $25°C$). It is of interest that he referred to glycol, isopropyl alcohol and ethanol as easily oxidizable alcohols.

Intensive efforts to develop methanol–oxygen[81] and methanol–hydrogen peroxide[82] cells have given fresh stimulus to the study of the electrode reactions concerned[83–98].

The anodic oxidation of methanol is accompanied by the appearance of consecutive oxidation products as intermediates. The overall course of reaction, in acid or in alkaline solution, can accordingly be formulated as follows:

$$CH_3OH \rightarrow CH_2O + 2H^+ + 2e^-$$

$$CH_2O + H_2O \rightarrow HCOOH + 2H^+ + 2e^- \qquad \text{(III.9a)}$$

$$HCOOH \rightarrow CO_2 + 2H^+ + 2e^-$$

$$CH_3OH + 2OH^- \rightarrow CH_2O + 2H_2O + 2e^-$$

$$CH_2O + 3OH^- \rightarrow HCOO^- + 2H_2O + 2e^- \qquad \text{(III.9b)}$$

$$HCOO^- + 3OH^- \rightarrow CO_3^{2-} + 2H_2O + 2e^-$$

Two electrons are released to the electrode per reaction step for each molecule of fuel.

Methanol and its successive oxidation products have the following properties in common:

1. The thermodynamically calculated potentials are not established at any known electrode materials, either in acid or in alkaline electrolytes.

   At potentials less than about $+500$ mV with respect to the reversible hydrogen potential, reaction of the organic substance is initiated by a dehydrogenation. At higher anodic potentials, reaction proceeds at the expense of OH or O radicals or atoms formed on the electrode surface.

2. By suitable choice of temperature, pH values and electrode materials, dehydrogenation of the organic fuel will occur without flow of current with evolution of hydrogen. As a result the electrode potential is displaced towards the reversible hydrogen potential.

3. In acid electrolytes at high current densities, potential oscillations (up to 300 mV) are observed.

4. In chloride-containing acid solutions, oxidation no longer occurs at potentials below $+1.4$ volt.

In the oxidation of formaldehyde in alkaline solution, increase of anodic current density is accompanied by increasing evolution of hydrogen. This reaction, surprisingly, is also catalysed by silver electrodes.

III.2.1 Mechanisms of Reaction and Establishment of Electrode Potentials

The *reversible* potential of methanol can be calculated thermodynamically to a close approximation* from the free energy, ΔG, of the reaction

$$CH_3OH + 3/2\,O_2 \rightarrow CO_2 + 2H_2O$$

in relation to the theoretical oxygen potential in the same electrolyte. From $\Delta G^0 = -166,770\,kcal\,mole^{-1}$, it follows that $E^\circ = 1210\,mV$ is the theoretical emf of the methanol–oxygen cell at 25°C. The corresponding data for formaldehyde, formic acid and hydrogen are collected together in Table III.1.

If the *rest potentials* established at platinized platinum electrodes in unimolar solutions of methanol, formaldehyde and formic acid are measured, it is found that: 1, the potentials are established very slowly and are poorly reproducible, 2, the potentials are influenced by the pH of the electrolyte, and 3, the potentials always lie positive to the theoretical values (cf. Figures III.21 and 22).

The establishment of the theoretical potentials is, however, hardly to be expected. It would be necessary (apart from the difficulty indicated in the footnote) for the electrolyte to contain, in addition to the fuel, intermediate and end-products together, at concentrations appropriate to equilibrium. Further, all the reaction steps would have to be reversible. But the first reaction step from methanol to formaldehyde is not reversible at any known electrode metal; it is therefore useless to add to the electrolyte any of the successive oxidation products.

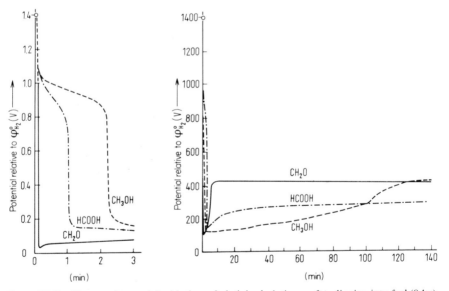

Figure III.21. Change of potential with time of platinized platinum after dipping into fuel (0.1M)-electrolyte (5N H_2SO_4) mixture. Left: reduction of oxygen charge on electrode (after Oxley, Johnson and Buzalski[88]); right: gradual poisoning of the electrode (after Vielstich and Vogel[98]).

* The ΔG value relates to pure methanol. The activity of methanol in aqueous solution is not taken into account.

Table III.1. Theoretical emf values at 25°C.

Fuel	Reaction	ΔG° (kcal mole^{-1})	E° (mV)	Potential rel. to rev. H_2 (mV)
Methanol	$CH_3OH + \frac{3}{2}O_2 \rightarrow CO_2 + H_2O$	-166.770	1210	$+20$
Formaldehyde	$CH_2O + O_2 \rightarrow CO_2 + H_2O$	-124.7	1350	-120
Formic Acid	$HCOOH + \frac{1}{2}O_2 \rightarrow CO_2 + H_2O$	-68.2	1480	-250
Hydrogen	$H_2 + \frac{1}{2}O_2 \rightarrow H_2O$	-56.690	1230	0

Oxley, Johnson and Buzalski[88] have followed the change of potential of a platinized platinum electrode with time, after immersion in a fuel-electrolyte mixture ($5N$ H_2SO_4, 30°C), with the results shown in Figure III.21.

The electrode, before use, was provided with a surface charge of oxygen by potentiostatic anodization in pure electrolyte at $+1.4$ V. It was then placed in the corresponding fuel-electrolyte mixture. The course of the change of potential with time shows that the organic substance reduces the oxygen layer on the surface of the platinum. Formaldehyde is particularly active in this respect. In the case of formic acid, CO_2 is evolved in amount corresponding to the reaction

$$Pt-O + HCOOH \rightarrow CO_2 + H_2O + Pt$$

Only very minor gas evolution is observed with formaldehyde or methanol.

After the destruction of the oxygen layer, a potential between $+80$ and $+200$ mV is set up which, according to Frumkin and Podlovchenko[89], corresponds with a dehydrogenation–hydrogenation equilibrium, e.g.

$$C_2H_5OH_{ad} \rightleftharpoons CH_3CHOH_{ad} + H_{ad}$$

The adsorbed hydrogen is again in equilibrium with hydrogen ions in the electrolyte. With time, however, the electrode becomes poisoned with intermediate products* and records potentials between $+200$ and $+450$ mV (Figure III.21).

The triangular wave potential scan method (cf. section II.4) is extremely useful for obtaining qualitative comparisons of different organic fuels under reproducible conditions. Figures II.8, III.10 and III.11 show current–potential diagrams for different electrode metals in pure, indifferent electrolytes. If an easily oxidizible substance is added to the electrolyte (Figure III.22), then the curves of the surface-charging currents (due to formation and destruction of hydrogen- and oxygen-surface layers, cf. the diagrams for the pure electrolytes) have superimposed upon them a 'spectrum' which is characteristic of the added substance and of the electrode material used[91,92].

The anodic peaks in the 'spectra' correspond to the oxidation of the fuel in the course of both anodic and cathodic sweeps of the electrode potential.

---

* In an experiment[90a] with butyl alcohol in acid electrolyte at platinum, without current flow, the following gases were detected over the electrolyte: $C_4H_{10}, C_3H_8, C_2H_6, CH_4$ (A. N. Frumkin in D. H. Collins, Batteries 2, *Proc. 4th. Int. Symp.*, Brighton, 1964, Pergamon Press, 1965, p. 537).

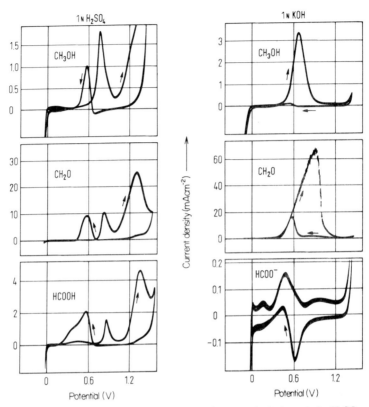

Figure III.22. Current–potential diagrams for smooth platinum in 1N H_2SO_4 or 1N KOH, with addition of methanol (1M), formaldehyde (1M) or formic acid (1M); 20°C; scan rate, 50 mV sec^{-1} (after Vielstich[17]).

III.2.1.1 *Significance of Peaks in Current–potential Diagrams in Acid Electrolytes*

If, for example, platinum is used as electrode in a set of acid electrolytes, to each of which a different organic substance has been added (methanol, formaldehyde, formic acid, ethanol, isopropyl alcohol, glycol, etc.), the triangular potential scan diagrams obtained show peaks which occur at a position along the potential axis which is independent of the kind of substance added[90,91]. From this it may be concluded that some preliminary electrochemical reaction, independent of the fuel, prefaces the reaction of the fuels.

In the anodic potential sweep, there are always two more or less well-marked current peaks (except in the case of formic acid; see below). They lie at about +900 mV and between +1200 and +1500 mV with respect to the reversible hydrogen electrode in the same electrolyte (Figure III.22).

According to the studies of Giner[96] and others, a surface layer, Pt—OH, is formed on platinum at about $+800$ mV, and is transformed in the region of $+1200$ mV into a Pt—O chemisorption layer*. This leads to the assumption[90-92,97] that in the region of the first peak, an OH layer on the platinum surface, formed in a preliminary electrochemical reaction, is destroyed by the organic reducing agent. Naturally, the formation of Pt—OH and its reaction with the fuel proceed side by side at the same time on the electrode surface, and must have a mutual influence[101]

$$H_2O + Pt \rightarrow Pt\text{—}OH + H^+ + e^-$$

$$CH_3OH + 2Pt - OH \rightarrow CH_2O + 2H_2O + 2Pt$$

$$(III.10)$$

If the reaction rates of these two steps are large, then even a very small extent of the OH coverage of the electrode surface may be adequate†. The fall of current after the first peak can be attributed to an increase in firmness of binding of the OH groups as the potential becomes more positive[92,97]. This is associated with an increase in the OH-coverage, and a corresponding decrease in that of methanol (Figure III.23).

The subsequent increase of the current to a second maximum had already been explained by Bogdanovski and Shlygin[83] (on the basis of their measurements with ethanol) in terms of the destruction of the Pt—O layer by the fuel

$$2 Pt\text{—}OH \rightarrow Pt\text{—}O + H_2O$$

$$CH_3OH + Pt\text{—}O \rightarrow CH_2O + H_2O + Pt$$

$$(III.11)$$

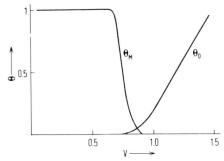

Figure III.23. Adsorption of methanol (θ_M) and oxygen (θ_0) on platinum in 1N $HClO_4$ and 1M CH_3OH at 30°C (after Breiter and Gilman[87]). Degree of coverage, θ, as function of electrode potential, from current–potential curve in the anodic direction.

---

* The formulations 'Pt—OH' and 'Pt—O' must be regarded only as working hypotheses. It is safer to assume simply that two kinds of chemisorbed oxygen are involved.
† According to Frumkin[90a] the covering of a vacuum-degassed platinum electrode with chemisorbed oxygen begins at $+400$ mV (acid electrolyte).

In the region of potential between 1200 and 1400 mV, the Pt—O bonding should still be loose enough to effect oxidation of the fuel, but with further increase of positive potential it becomes so strong that, finally, reaction with the fuel can no longer occur*. The current therefore falls, until it increases very rapidly again with the onset of oxygen evolution.

According to the reaction scheme so developed it is to be expected that catalysis of the reaction of organic fuels should depend decisively on the kind of oxygen chemisorption layer on the electrode metal. This is indeed the case. For example, if current–potential diagrams are taken for *gold* electrodes in H_2SO_4 as electrolyte, with added fuels, no anodic oxidation of any fuel is detectable under 1.2 volt[17,97]. This is consistent with the theory, since formation of an oxygen chemisorbed layer on gold begins only at 1.25 volt (Figure III.10).

At platinum, the chemisorption of oxygen from acid electrolytes is inhibited by the addition of chloride ion (Figure III.24). As a result, in presence of chloride ion concentration $> 10^{-3}$ m, the electrode is passive to organic fuels[100].

In *alkaline* media, on the other hand, the oxygen layer on the electrode is completely formed (Figure III.24d); obviously, at these much more negative real potentials, OH^- ions are more strongly adsorbed than Cl^- ions. The oxidation of organic fuels remains practically uninfluenced by chloride ions under these conditions (Figure III.24c).

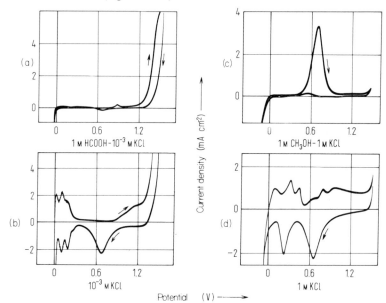

Figure III.24. Current–potential diagrams for smooth platinum at 20°C (after Vielstich[17]). (a) Electrolyte 1M HCOOH + 1N H_2SO_4 + 10^{-3}M KCl, 50 mV sec^{-1}; (b) electrolyte 1N H_2SO_4 + 10^{-3}M KCl, 1 V sec^{-1}; (c) 1M CH_2OH + 1N KOH + 1M KCl, 50 mV sec^{-1}; (d) 1N KOH + 1M KCl, 1 V sec^{-1}.

---

* The passive properties of oxygen-covered platinum electrodes have been confirmed by Giner[99], among others, in experiments with oxalic acid.

III.2.1.2   Methanol in H_2SO_4 at Increased Temperature

If the reaction velocity is increased by use of a platinized platinum electrode at 80°C [92], the potential at which the methanol reaction begins is displaced by no more than 100 mV, to the region of +400 mV with respect to the reversible hydrogen potential in the same solution. The methanol oxidation is therefore initiated under these conditions in the hydrogen-free, double-layer region of potentials.

By using electrodes of Raney-platinum at 100°C, Krupp and co-workers[102] observed a rest potential of ca. +100 mV, and at +350 to +400 mV obtained a current density of 50 mA cm^{-2}. Clearly, under these conditions methanol readily undergoes dehydrogenation.

III.2.1.3   The Current Peak on the Cathodic Sweep

On the cathodic return sweep of the triangular potential wave only one anodic peak appears, and this is attributed to renewed oxidation of the fuel (Figure III.22). Since no reaction can occur at the electrode surface completely covered with oxygen, it is clear that part of the oxygen layer must first be reduced. Renewed anodic oxidation begins on the cathodic return between 800 and 600 mV, at a potential depending on the extent of the maximum oxygen coverage attained at the extreme of the preceding anodic sweep, and on the velocity of potential change. This means that the current peak is displaced in the direction of the hydrogen potential. The current density is enhanced since the removal of oxygen has roughened and activated the surface. If the activity of the electrode is increased by platinization, the current–potential curves for anodic and cathodic sweeps come together[92]. Current–potential curves obtained by the stationary method completely fail to show this differentiation, and only two *anodic* maxima are observed (by use of a potentiostat, cf. Figure III.36).

III.2.1.4   Mechanism of Methanol Oxidation in Alkaline Electrolytes

When alkaline electrolytes are used, only one current peak appears along the anodic potential sweep (Figure III.22). Whereas in sulphuric acid at 25°C oxidation of methanol sets in at +500 mV, it is now initiated at +400 mV, i.e. at the beginning of the double-layer region.

Buck and Griffith, and Schlatter[86] have proposed the following reaction sequence:

$$CH_3OH + OH^- \rightleftharpoons CH_3O_{ad}^- + H_2O$$

$$CH_3O_{ad}^- \rightarrow CH_2O_{ad}^- + H_{ad}$$

$$H_{ad} + OH^- \rightarrow H_2O + e^-$$

$$CH_2O_{ad}^- \rightarrow CH_2O + e^-$$

(III.12)

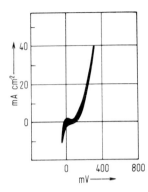

Figure III.25. Current–potential diagram for platinized platinum in 2N KOH + 1M CH_3OH at 85°C (after Vielstich[92]).
Scan rate 7 mV sec^{-1}.

However, the location of the current peak, on the potential axis for alkaline
solution (Figure III.22), corresponds with that of the previously established
charging of the platinum surface with oxygen (Figure III.11). This strongly
suggests[92] that, at least at room temperature and at a smooth platinum surface,
the methanol oxidation proceeds by reaction with the OH-layer on the electrode
surface, as in acid solutions.

In distinction from acid electrolytes, the curves relating to KOH are shifted
by slight rise of temperature or by using platinized platinum towards the
reversible hydrogen potential (Figure III.25)[92]. It is clear that under these

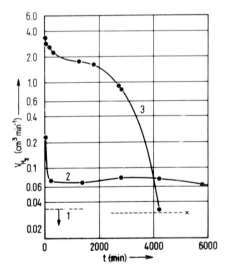

Figure III.26. Base-catalysed dehydrogenation of
formate (3.1M) in 4.3N KOH at 100°C (after
Grüneberg[92a]). 1, Without metal catalyst; 2,
Raney-copper electrode; 3, Raney-nickel electrode; x, limit of accuracy of measurement.

conditions a base-catalysed dehydrogenation occurs, as previously contended by Spengler and Grüneberg[103]. Hydrogen is split off in a preliminary chemical reaction and then undergoes electrochemical reaction:

$$CH_3OH \xrightarrow{Pt, OH^-} CH_2O + 2H_{ad}$$
$$2H_{ad} + 2OH^- \longrightarrow 2H_2O + 2e^- \qquad (III.13)$$

From pH 14 a considerable rise of oxidation current with increasing OH^- ion concentration has been established[17].

That transition to a dehydrogenation mechanism with rise of temperature occurs could be inferred from the fact that all three fuels, methanol ($> 80°C$), formaldehyde ($> 30°C$) and formate ($> 100°C$), establish the reversible hydrogen potential, and also evolve hydrogen at zero current, at Raney-nickel in concentrated KOH solution (6 to 8N).

Figure III.26 shows the quantity of hydrogen obtained as a function of time in the dehydrogenation of formate in KOH[92a]. Within the accuracy of the experiment no hydrogen evolution is observed in absence of catalyst (curve 1). Curve 2 shows the dehydrogenation at a Raney-copper electrode. With Raney-nickel as catalyst (curve 3) dehydrogenation proceeds to completion and occurs 30 times faster than at Raney-copper.

III.2.2   Galvanostatic Potential–time Curves and Stationary Current–potential Curves with Platinized Platinum

The completeness of the oxidation of methanol, through various intermediate products, to carbonate or CO_2 can be most simply investigated by means of galvanostatic potential–time curves[84]. Figure III.27 shows such curves obtained

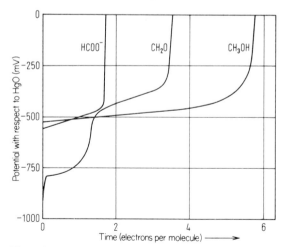

Figure III.27. Anodic oxidation of methanol, formaldehyde and formate at platinized platinum in 2N KOH at 20°C (after Vielstich[92]); methanol and formaldehyde at $1 \, mA \, cm^{-2}$, formate at $0.5 \, mA \, cm^{-2}$.

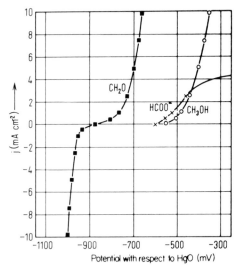

Figure III.28. Current–potential curves at platinized platinum in
2N KOH at 20°C (after Vielstich[92]).

with platinized platinum electrodes in 2N KOH, with addition in each case of
0.5 cm³ of fuel to 100 cm³ of electrolyte[92]. The measurements were carried out
in a half-cell (cf. section II.4) equipped with a magnetic stirrer. The time scale
in Figure III.27 has been normalized in terms of electron consumption per
molecule. All three curves confirm that the electrochemical reaction proceeds to
carbonate as end-product, corresponding with the consumption of 6, 4 or 2
electrons per molecule of oxidizable material. Methanol and formate came into
reaction at similar potentials, but, with formaldehyde, the oxidation curve began
at a potential more negative by some 300 mV. This result is in agreement with
the current–potential curves measured under the same conditions (Figure
III.28).

The *formaldehyde* curve shows a marked step after reaction of about two
electrons per molecule, indicating that the intermediate product, formate, does
not immediately undergo further reaction. The *methanol* curve shows no
potential step — evidently because the reaction stage from methanol to formaldehyde is hindered. A potential is then established before any oxidation can
proceed, at which formate is able to react rapidly (cf. Figure III.28). The formaldehyde formed is relatively rapidly oxidized to formate. However, a desorption
of the intermediate product, formaldehyde, during the 'adsorption time', is
possible.

In the oxidation of methanol (1.5 M) in 4.5N NaOH at platinized carbon
(1 mA cm⁻² at 24°C), Schlatter[86] found analytically a formaldehyde concentration of 10^{-4} molar. An equal formaldehyde content was observed, without
current flow, in the presence of atmospheric oxygen.

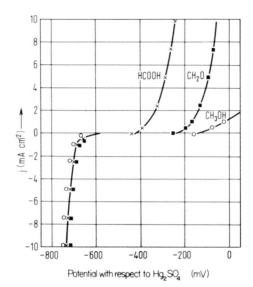

Figure III.29. Current–potential curves at platinized
platinum in $1N$ H_2SO_4 at 20°C (after Vielstich[92]).

Potential–time curves relating to $1N$ *sulphuric acid*[92] confirm that for all
three fuels oxidation proceeds to CO_2. Potential steps are not observed, since
consecutive products are all oxidized faster than the original fuel (cf. current–
potential curves of Figure III.29*). Schlatter[86] estimated stationary concentrations in respect of formaldehyde and formic acid to be ca. $5 \times 10^{-3}M$
($1.5M$ CH_3OH, $2.3M$ H_2SO_4, 23°C, 1 mA cm^{-2}).

Independently of reaction mechanisms, it is therefore true for the velocities
of partial steps in the electrochemical oxidation of methanol at a platinum
electrode through the intermediate stages of formaldehyde and formic acid or
formate that:

$$CH_3OH \xrightarrow{\ R_1\ } CH_2O \xrightarrow{\ R_2\ } \underset{HCOO^-}{HCOOH} \xrightarrow{\ R_3\ } \underset{CO_3^-}{CO_2} \qquad (III.14)$$

in acid solution, $k_1 \ll k_2 < k_3$; in alkaline solution, $k_1, k_3 \ll k_2$.

The hindrance of the first reaction steps has the consequence that the anodic
current density is reduced by strong stirring of the electrolyte, since part of the
reactive intermediate product is swept away from the electrode surface[17,88,104]

In alkaline electrolytes, at low current densities, the velocity constants k_1 and k_3 are about equal.
Since the ratio k_1/k_3 is highly dependent on conditions of experiment (temperature, current density,
concentration, activity of electrode), it is possible, with methanol as initial reactant, to secure

---

* If stationary current–potential curves are taken potentiostatically over a wide potential range,
 maxima are obtained corresponding with those of the triangular wave potential scan diagrams
 (Figures III.22 and III.36).

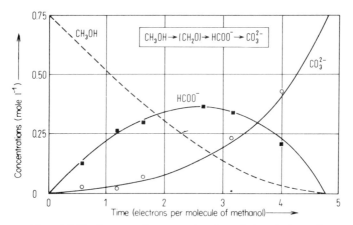

Figure III.30. Concentrations as a function of time during galvanostatic oxidation of methanol at platinized platinum in 6N KOH
at 65°C (after Bloch, Prigent and Balaceanu[85]).

considerable enrichment of formate in the solution ($k_1 > k_3$). Bloch, Prigent and Balaceanu[85] have
analysed the electrolyte at various times after the start of a methanol oxidation; for platinized
platinum in 6N KOH at 65°C, they obtained the results illustrated in Figure III.30. At Pd and
Pd–Pt catalysts, $k_1 < k_3$ (cf. Figure III.32a).

III.2.3   Oscillations at High Current Densities in the Acid Electrolytes

Müller and Tanaka[79] were the first to record a periodic oscillation of the potential of a rhodium-plated platinum electrode during the oxidation of formic
acid to carbon dioxide. When a critical current density (> 10 mA cm^{-2}) was
exceeded, the electrode potential oscillated between about $+200$ and $+600$ mV
with respect to the normal hydrogen electrode. The authors attributed this
effect to the entry of an adsorbed intermediate product, not itself electrochemically oxidizable, but decomposing to provide oxidizable hydrogen:

$$HCOO^- \rightarrow HCOO_{ad} + e^-$$

$$HCOO_{ad} \rightarrow H_{ad} + CO_2 \tag{III.15}$$

$$H_{ad} \rightarrow H^+ + e^-$$

Depending on the electrode potential, the two electrochemical reactions
contribute to the total current to varying degrees, and the coupling between the
partial reactions enforces the potential oscillations.

Oscillations of this kind, however, are not peculiar to formic acid. Quite
similar effects have been observed at smooth and platinized platinum electrodes
in sulphuric acid solutions to which additions of the following substances have

been made[17]: methanol, ethanol, isopropyl alcohol, formaldehyde[86], acetaldehyde and formic acid. Armstrong and Butler[105] observed the same phenomenon during the anodic oxidation of molecular hydrogen at platinum in sulphuric acid.

The critical current density, and the amplitude and frequency of the oscillations, are functions of the material of the electrode and its activity, the electrolyte composition and the temperature, as well as of the resistance of the outside electrical circuit. It has been shown [101a] that the electrode material is decisively significant. Studies of platinum, gold and iridium electrodes in sulphuric acid solutions of formic acid have shown that gold generally gives rise to no oscillations, and that platinum and iridium are characteristically different in behaviour. Under the same conditions, iridium gave rise to oscillations of greater amplitude but lower frequency (there is, obviously, a certain relationship between amplitude and frequency) than platinum, in a more positive range of potentials. On the other hand, platinum electrodes could carry a greater load (in mA cm^{-2}) than iridium electrodes, and maintained oscillating potentials for longer periods (5 min to 50 hr, according to conditions). Formic acid concentration had a less marked, but plainly perceptible effect; increasing concentration displaced the potential range of oscillation, positively increased the possible loading of the electrodes, and, although enhancing the amplitude, reduced the frequency of the oscillations. The amplitude (30 to 300 mV) usually increases with time until the electrode swings to such a positive potential that oxygen evolution sets in and the oscillation stops (Figures III.31).

The following explanation for this phenomenon suggests itself[17*]. During the passage of anodic current, gradual poisoning of the electrode occurs, with positive displacement of potential. Finally, the electrode develops an oxygen layer. This, however, is chemically reduced by hydrogen or the dissolved organic

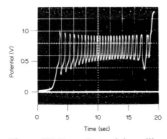

Figure III.31a. Potential oscillations at a smooth platinum electrode in 1N H_2SO_4 + 5M HCOOH, current density 1.5 mA cm^{-2} (after Vielstich[17]).

---

* A. J. Gokhstein (CITCE meeting, Moscow, 1963, No. 2.13) discussed oscillations occurring during the cathodic reduction of anions ($S_2O_3^{2-}$, $PtCl_6^{2-}$).

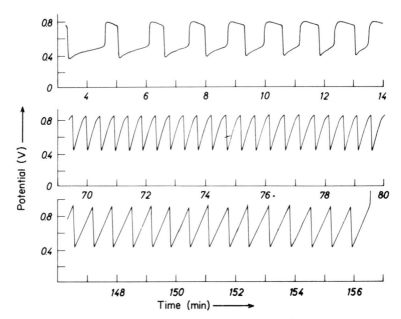

Figure III.31b. Potential oscillations at a smooth platinum electrode in $1N$ H_2SO_4 + $1M$ HCOOH, current density 0.25 mA cm^{-2} (after Vielstich[17]).

fuel. The removal of the oxygen reactivates the electrode, with the consequence that reaction of the fuel proceeds again at a more negative potential. Even the oxidation of impurities during the formation of the oxygen chemisorption layer can play a similar part. It is in harmony with this that no potential oscillations have been observed at gold electrodes; oxygen layers on gold are formed only at substantially more positive potentials (cf. Figure III.10). The labile behaviour of electrode potentials might well have been inferred from the occurrence of two or three maxima in the current–potential diagrams (Figure III.22)*.

Oscillations of this kind during the anodic oxidation of an organic fuel in *alkaline* solutions have been observed, as yet, only in the case of formaldehyde at a silver electrode (see below). Potential oscillations at high current densities occur at smooth platinum electrodes supplied with hydrogen, and also at hydrogen electrodes with catalyst suspensions (cf. Figure III.8)[17].

A possible application of such oscillations to the development of an 'alternating-current cell' has been demonstrated by Hunger[101b]. The system used consisted of a platinized platinum anode and a PbO$_2$ cathode immersed in a 2 % solution of formaldehyde in 3.75M H_2SO_4, and had a rest-voltage of 1.44 V. Oscillations occurred at current densities from 15 to 70 mA cm^{-2}, becoming more rapid with increasing current density and reaching a frequency of about

---

* For hypotheses in relation to these oscillations, as well as for their quantitative treatment, see, for example, U. F. Franck and F. Kettner, *Ber. Bunsenges. Phys. Chem.*, **68**, 875 (1964).

1 sec^{-1} at 70 mA cm^{-2}. The cell was demonstrated by means of a flashing light. Whether a system can be found of sufficiently reproducible and constant frequency to be practicable for operating, for example, signal lights on buoys, must await further investigation.

III.2.4   Catalysts

The same electrode materials are suitable for the electrochemical reaction of methanol and its oxidation products, as for the oxidation of hydrogen. An exception is provided by the anodic oxidation of formaldehyde in alkaline electrolytes, which is catalysed by silver and gold (cf. section III.2.5), as well as by metals of the platinum group and by skeleton-nickel.

The activity of sensitive platinum metals catalysing the oxidation of molecular hydrogen decreases with time of operation, particularly in the first few hours. This *passivation* is more marked for the catalysis of organic fuels dissolved in the electrolyte.

In contrast to catalytic behaviour in relation to hydrogen, skeleton-nickel in alkaline solution at lower temperatures is less active than platinum. It becomes advantageous to use nickel at temperatures at which hydrogen is formed as an intermediate product in a preliminary base-catalysed dehydrogenation reaction.

Catalyst electrodes with especially high activity relative to methanol have been developed by Krupp and co-workers[102]. By using the 'double-skeleton principle' these authors prepared Raney-metal electrodes from platinum, palladium and rhodium (construction and activation of these electrodes are discussed in section IV.1). All three electrodes, in 5N KOH at 80°C, showed limiting current densities of more than 300 mA cm^{-2}. Current–potential curves for a methanol electrode of Raney-platinum in 5N H$_2$SO$_4$ are shown in Figure III.32.

Boies and Dravnieks[106] were able to increase by nearly ten-fold the current densities for reaction of methanol or formate at Raney-nickel electrodes by means of a platinization process. They deposited nickel–aluminium alloy on to a suitable metallic support, by a special (flame spraying) process, to a thickness of 0.1 to 0.15 mm. The aluminium was then dissolved away with concentrated KOH solution. Noble metals to be studied were finally electrodeposited on the Raney-nickel electrode so obtained. The authors obtained the best activity in relation to methanol by deposition of 2 mg cm^{-2} of platinum.

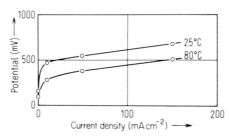

Figure III.32. Current–potential curves for a methanol electrode of Raney-platinum in 5N H$_2$SO$_4$ (after Sandstede[102]). Potential relative to hydrogen electrode in the same solution.

With these electrodes, even at 25°C, a current density of more than $100 \, mA \, cm^{-2}$ could be maintained with methanol as fuel ($6M \, CH_3OH + 5N \, KOH$), or about $20 \, mA \, cm^{-2}$ with formate. Such an electrode (methanol), passing current at $100 \, mA \, cm^{-2}$, showed a deterioration in potential of only about $30 \, mV$ after several days.

Recently, mixed catalysts have acquired particular importance; the very high activity of Pd–Ru and Pt–Ru alloys has been noted by a number of authors[106a–d]. Binder, Köhling and Sandstede[106e] have studied alloys of platinum with various metals, and Heath[106f], also using platinum alloys, has distinguished two effects of the second component. On one hand, inclusion of the foreign metal modifies the platinum surface, and on the other hand, it establishes a stable redox couple on the surface of the catalyst. Palladium–platinum mixtures have been investigated by Grimes and Spengler[106g], and platinum–gold alloys by Breiter[106h]. The general conclusion reached by these authors is that optimum catalytic activity depends not only on the nature of the second alloy constituent, but also on its relative concentration. Platinum–ruthenium catalysts (in sulphuric acid solutions of methanol), for example, show a sharp maximum of catalytic activity at a Pt:Ru ratio of 6:4.

Most of this work has been concerned with mixed catalysts in acid solutions. Special attention has recently been given[106i] to their behaviour in alkaline systems (e.g. $6M \, KOH$, $4M \, HCOOK$ or CH_3OH), which are more important in relation to fuel cell practice. This work began with the selection of the most catalytically active of many metals and alloys by an application of the triangular wave potentiodynamic method. This method provides a sensitive means of comparing catalytic activities in terms of the characteristic oxidation-current peaks, and has shown that these activities vary over many orders of magnitude. It has also been demonstrated that alloys of platinum are less suitable than those of palladium for oxidation of formate, but that the reverse is the case for the oxidation of methanol, although the difference is less marked.

Catalysts found to be the most suitable in this way have been tested in complete fuel cells, with use of carbon powder-based electrodes of 3% catalyst content, with results in conformity with the potentiodynamic measurements. Figure III.32a illustrates these results in terms of potential at a current density of $10 \, mA \, cm^{-2}$ maintained for 5 hr, plotted as a function of catalyst material. It is seen that Pd–Pt alloys are best for formate oxidation, and that Pt–Ir (3:1) and Pt–Ru (3:2) alloys are best for methanol oxidation. It is also apparent that there is no catalyst equally suitable for both. If it is desired to use fuel mixtures, which is indeed important in practice, (cf. chapters IV.3 and X) then Pd–Pt alloys provide the best compromise.

A critical comparison of these results with those of the authors cited above is impracticable because of the considerable discrepancies between them. The catalytic activities of metals and alloys depend not only on the nature of the fuel, but also on a whole series of other factors. The relationships illustrated in Figure III.32a can be disturbed, or even inverted if, for example, comparisons

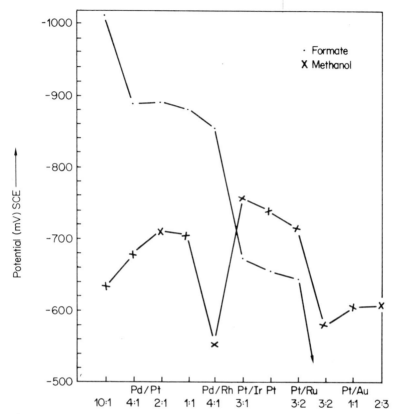

Figure III.32a. Potentials at various catalyst electrodes, loaded continuously for 5hr. at 10 mA cm^{-2}. Solution, 4M HCOOK and 2M CH$_3$OH in 6N KOH. Electrodes, carbon powder with three per cent catalyst (catalysts from Johnson, Matthey and Co.) (after Schmidt and Vielstich[106i]).

are made at other current densities or temperatures; they are critically sensitive to the conditions of preparation of the catalyst electrodes, particularly in relation to control of purity[106i].

III.2.5   Mechanism of the Anodic Oxidation of Formaldehyde in Alkaline Solution

In 1928 Müller and Takegami[80] determined stationary current–potential curves for a series of smooth metal electrodes (Pt, Pd, Rh, Au, Ag) in alkaline formalde-hyde solution (2N NaOH, 12% CH$_2$O). Surprisingly, silver was the best catalyst for the anodic oxidation of formaldehyde. Evolution of gaseous hydrogen was observed to accompany the electrochemical reaction, varying from one metal to another—particularly brisk in the case of silver but zero for platinum. The explanation given by Müller, involving the generation of hydrogen in the solution phase, and not at the electrode surface, was complex and is not now acceptable.

To investigate the mechanism of this reaction, it is of primary interest to determine whether the hydrogen that is evolved comes from the electrolyte or from the formaldehyde. By deuteration first of the formaldehyde[107], and then of the electrolyte[108], it has been shown unambiguously that the hydrogen is derived from the formaldehyde alone. The experiments in question were conducted at a silver electrode to exclude the possibility of side-reactions*.

The following reaction schemes are conceivable for the anodic reaction of formaldehyde accompanied by evolution of hydrogen:

1.

$$CH_2O + OH^- \rightarrow HCOO^- + 2H_{ad}$$

$$\begin{cases} 2H_{ad} & \rightarrow H_2\downarrow & \text{(a)} \\ 2H_{ad} + 2OH^- \rightarrow 2H_2O + 2e^- & \text{(b)} \end{cases} \qquad \text{(III.16)}$$

The anodic oxidation of adsorbed hydrogen atoms is the sole electrochemical reaction; the base-catalysed dehydrogenation of formaldehyde precedes both the formation of molecular hydrogen and the electrochemical reaction.

2.

$$M + OH^- \rightarrow M{-}OH + e^-$$

$$CH_2O + M{-}OH \rightarrow HCOOH + H_{ad}$$

$$\downarrow + OH^-$$

$$HCOO^- + H_2O \qquad \text{(III.17)}$$

$$\begin{cases} H_{ad} & \rightarrow \tfrac{1}{2}H_2\downarrow & \text{(a)} \\ H_{ad} + OH^- \rightarrow H_2O + e^- & \text{(b)} \end{cases}$$

The oxidation of formaldehyde in this case, as assumed for methanol, is initiated by the formation of a surface layer, $M{-}OH$. The anodic current is divided between the preliminary reaction and the ionization of adsorbed hydrogen atoms; not all of the hydrogen produced appears in the gaseous form.

III.2.5.1 Quantitative Estimation of Hydrogen Evolved

To test the above assumptions about reaction mechanism by experiment, the amount of gas formed anodically was determined as a function of current density, and compared with the product of current and time, as well as with the quantity of gas evolved cathodically[108].

---

* Silver does not catalyse isotope exchange between H_2 and H_2O, nor are the disproportionation products of formaldehyde, CH_3OH and $HCOO^-$, anodically oxidized at silver.

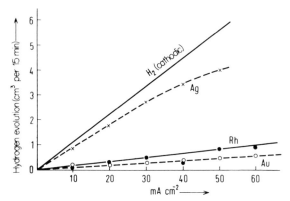

Figure III.33. Comparison of quantities of hydrogen evolved cathodically and anodically during the anodic oxidation of formaldehyde (2M) in KOH (2N). Cathode metal, Pt; anodes of Ag, Rh, Au (after Vielstich[108]).

Figure III.33 illustrates the results obtained at 20°C with smooth electrodes of Ag, Rh and Au. With smooth platinum as anode, no gas evolution was observed, even at current densities greater than $50\,mA\,cm^{-2}$. Figure III.34 shows the related current–potential curves for Ag and Pt.

The hydrogen formed at the platinum cathode corresponded with the current–time product, but, for all the metals studied, less gas was evolved in the anode compartment of the electrolysis cell than in the cathode compartment. No

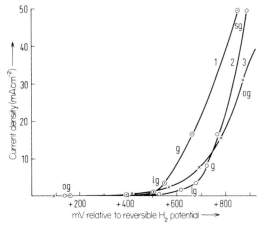

Figure III.34. Stationary current–potential curves for formaldehyde (2M) in KOH (2N) (after Vielstich[108]). 1, Silver; 2, chemically polished silver; 3, platinum—og, without gas evolution; lg, slight gas evolution, g, gas evolution; sg, strong gas evolution.

gas was produced without flow of current and hydrogen evolution increased approximately linearly with current. If the anodic current was increased excessively, the potential jumped to a value at which the electrode surface developed an oxygen layer, and eventually to the potential for oxygen evolution. In the critical transition region, *oscillations* of potential were observed (cf. section III.2.3).

The current–potential curves[108] obtained with unpolished silver and smooth platinum electrodes (Figure III.34) agree well with those published by Müller and Takegami[80]. The curve for polished silver is displaced to more positive potentials, but the increase of gas evolution with rising anodic current is the same as for the unpolished metal.

Anodic hydrogen evolution did not exceed cathodic hydrogen evolution at any of the electrodes studied. In relation to the preliminary chemical reaction of mechanism 1, it is not apparent why gas evolution should not set in at the rest-potential, and why anodic gas evolution in no case exceeded the cathodic hydrogen evolution. Since, moreover, the quantity of hydrogen evolved in unit time increases in parallel with current density, the oxidation of formaldehyde in alkaline solution must involve an initial electrochemical reaction. Mechanism 2 is consistent with the experimental results.

The fraction of atomic hydrogen formed according to mechanism 2, that subsequently undergoes combination to molecular hydrogen and escapes in the form of gas, depends on the extent to which the metal surface is covered with OH groups, or on the electrode potential, as well as on the relative velocities of the reaction steps (a) and (b). At smooth platinum, it is clear that all of the hydrogen provided undergoes immediate oxidation ($k_b \gg k_a$); at silver, recombination and desorption of molecular hydrogen is faster than oxidation ($k_b < k_a$), even at anodic overpotentials of more than 100 mV. Gold and rhodium show behaviour intermediate between platinum and silver ($k_b > k_a$); cf. Figure III.33.

The reaction of formaldehyde with the OH layer on the metal surface probably proceeds via hydroxymethoxy radical as intermediate[109]

$$\begin{array}{c} H \\ \diagdown \\ \diagup \\ H \end{array} C{=}O + M{-}OH \rightarrow \begin{array}{c} H \\ \diagdown \quad \diagup O \\ C \\ \diagup \quad \diagdown \\ H \quad OH \end{array} \rightarrow H{-}C\begin{array}{c} \diagup\!\!\diagup O \\ \diagdown \\ OH \end{array} + M{-}H$$

This assumption is supported by the investigations of Wieland and Wingler[110] on the decomposition of alkaline formaldehyde solutions by addition of H_2O_2. The formaldehyde is quantitatively converted, without a catalyst, into hydrogen and formate. Di(hydroxymethyl) peroxide has been detected as an intermediate.

Increase of temperature leads to a transition from mechanism 2 to the base-catalysed dehydrogenation, mechanism 1. Depending on the electrode material and electrode composition, more or less vigorous hydrogen evolution occurs without anodization. In this case the electrode potential is close to the reversible hydrogen potential. For Raney-nickel electrodes in 6–8N KOH, mechanism 1 begins to predominate over mechanism 2 at temperatures between 40 and 50°C.

III.2.6   Special Features of the Anodic Oxidation of Formic Acid and Formate

Formic acid and formate are dehydrogenated at the appropriate catalysts (platinum metal, Raney-nickel, Raney-copper) over the whole pH range. Base-catalysed dehydrogenation with no anodic current flow, however, first becomes appreciable at about 100°C (Figure III.26). In acid, neutral or weakly alkaline solutions hydrogen is quite briskly evolved even at room temperature.

According to Schwabe[111], the velocity of decomposition of aqueous formic acid to H_2 and CO_2 at platinum metals passes through a maximum at about pH 4 and then falls steeply with increasing pH (Figure III.35a). Measurements by Jahn[112], with Raney-nickel at 60°C, confirm the dependence on pH in the range pH 8–pH 14 found by Schwabe (Figure III.35b). Schwabe concluded from his results that the formic acid molecules, not the ions, are susceptible to catalytic decomposition.

Sachtler[113] divided the dehydrogenation into two steps:

$$HCOOH \rightarrow HCOO + H_{ad}$$

$$HCOO \rightarrow CO_2 + H_{ad}$$

(III.18)

On the basis of the fact that acetic acid undergoes no dehydrogenation under anodic conditions, it seems probable that the formic acid reaction involves fission of the hydrogen atom linked to the carbon atom.

$$HCOOH \rightarrow HCOO_{ad} + H^+ + e^-$$

$$HCOO_{ad} \rightarrow CO_2 + H^+ + e^-$$

(III.19)

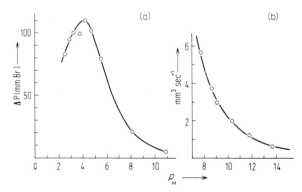

Figure III.35. Rate of decomposition of aqueous formic acid as a function of the pH of the electrolyte. (a) At platinized platinum at 25°C (after Schwabe[111]), (b) at Raney-nickel suspension at 60°C (after Jahn[112]). Bromine titration and gas volumetric methods used for the respective analyses.

Kutschker and Vielstich[97] have associated the removal and ionization of a hydrogen atom* with the first rise of current† between 0.4 and 0.6 volt in the formic acid diagram (Figure III.22‡).

The well-marked current peaks which follow at 0.9 and 1.2 volt had already been explained in the case of the methanol reaction (cf. section III.2.1) in terms of the reaction of fuel and intermediate products[94,95,98] with the oxygen chemisorption layer[97]:

$$Pt + H_2O \rightarrow Pt-OH + H^+ + e^-$$

$$Pt-OH + HCOOH \rightarrow Pt + H_2O + CO_2 + H^+ + e^- \text{ and}$$

$$Pt-OH + COOH_{ad} \text{ (or } HCOO_{ad}) \rightarrow Pt + H_2O + CO_2 \qquad \text{(III.20)}$$

or

$$2Pt-OH \rightarrow Pt-O + H_2O$$

$$Pt-O + HCOOH \rightarrow Pt + H_2O + CO_2 §$$

The anodic current densities that can be obtained are considerably greater than that corresponding with the rate of hydrogen production in the open-circuit decomposition. The catalytic decomposition is, moreover, strongly time-dependent.

Shortly after addition of formic acid to the reaction vessel, vigorous hydrogen evolution occurs which, however, decreases within an hour to about one hundredth of its initial rate. Schwabe[111] attributed this drop in decomposition velocity to hindered desorption of H_2 molecules, but detailed investigation[97,98] has shown that platinum becomes passivated by the intermediate product formed (see above). Moreover, the current in the region of the first maximum of the triangular potential scan diagram (Figure III.22) decreases, at *constant* potential, very rapidly with time[94,98]. Thus, the potentiostatic, stationary

---

* Whether, in the adsorbed intermediate, bonding to carbon $-O-\overset{\displaystyle O}{\underset{\displaystyle H}{\overset{\|}{C}}}$ or oxygen $-\overset{\displaystyle O}{\underset{\displaystyle OH}{\overset{\|}{C}}}$

is concerned has not yet been unambiguously demonstrated. Carbon monoxide or oxalic acid (by combination of adjacent formate radicals) have been excluded on experimental evidence as possible intermediate products[98a-c]. In any case the intermediate product blocks the electrode surface at potentials less positive than its oxidation potential (about $+700$ mV on platinum at 40°C). This leads to the minimum between the first and second current peaks.

† The shoulder in the single anodic peak found in the cathodic (return) traverse of potential was also correlated with fission of a hydrogen atom.

‡ By using platinum electrodes preheated to redness in vacuum, Bagotzky and Vasiljev[90] have found, for CH_3OH and CH_2O as well, a current rise at about 0.5 volt, which is similarly attributable to a preliminary dehydrogenation.

§ Hoffman and Kuhn[93] consider that this reaction can proceed by two electrochemical steps:

$$Pt-HCOOH \rightarrow Pt + CO_2 + 2H^+ + 2e^-$$

$$Pt-O + 2H^+ + 2e^- \rightarrow Pt + H_2O$$

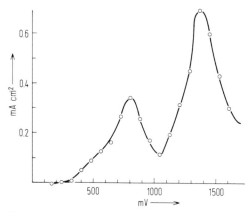

Figure III.36. Stationary current–potential curve (potentiostatic) for anodic oxidation of formic acid (1M)
at smooth platinum in sulphuric acid (1M) electrolyte,
with nitrogen bubbling (after Vielstich and Vogel[98]).
Potentials with respect to the hydrogen electrode in the
same solution.

current–potential curve (Figure III.36) shows only the two maxima attributed
to the reduction of oxygen chemisorbed on the electrode.

It is evident that the intermediate product, blocking the electrode surface,
becomes oxidized at the expense of chemisorbed oxygen at sufficiently positive
potentials[97–98]. These findings explain why anodic current densities can be
obtained which are large in comparison with the slow decomposition of formic
acid in aqueous electrolytes at room temperature.

In *alkaline* electrolytes, the currentless decomposition at room temperature
is negligible, and not until 100°C is reached does considerable formation of
hydrogen take place by base-catalysed dehydrogenation. At very active catalysts
or at elevated temperatures, however, current densities of $> 20 \, \text{mA cm}^{-2}$ can
be attained[106d,g,i].

III.3 OTHER OXYGEN-CONTAINING ORGANIC COMPOUNDS

Alcohols, aldehydes and acids with more than one carbon atom per molecule
are only of restricted suitability as fuels in fuel cells. Thus, although ethanol,
isopropyl alcohol and ethylene glycol are all comparable with methanol in
relation to reaction velocity, the reaction products formed in *alkaline* electrolytes
cause considerable technical difficulties. In acid solutions, CO_2 is the end-
product for a number of compounds, but the working potentials at current
densities of technical interest are too positive (Figure III.37).

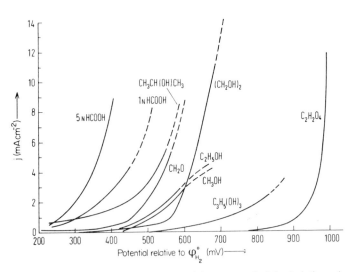

Figure III.37. Stationary current–potential curves at platinized platinum in 1N H_2SO_4, with addition of various C, H, O compounds, at 20°C (after Vielstich[17]). – – – – –, Potential oscillations (cf. Figures III.31a and b).

So far as they are known, reaction mechanisms are similar to those for compounds with one carbon atom. Reaction of fuel at room temperature occurs predominantly by reduction of the chemisorbed layer of oxygen on the electrode (at potentials > +500 mV, Figure III.38). At elevated temperatures (especially in alkaline electrolytes), a preliminary dehydrogenation is often observed.

III.3.1   Ethanol, Acetaldehyde and Acetic Acid

The current–potential diagrams in Figure III.38 show that in *acid* electrolytes, at potentials less than 0.9V (with respect to hydrogen), ethanol can be oxidized only to acetaldehyde, and that acetic acid remains inert even up to 1.6 V. A mixture of ethanol and acetaldehyde behaves in the same way as acetaldehyde alone[83], i.e., the reaction product, acetaldehyde, in time poisons the electrode.

In alkaline solution anodic oxidation of ethanol leads to an aldol condensation, which rapidly covers the electrode with resin. The presence of oxygen accelerates the process and the electrolyte turns brown.

III.3.2   Isopropyl Alcohol

Electrochemical oxidation of isopropyl alcohol yields acetone as end-product. In alkaline electrolytes, in the presence of dissolved oxygen, mesityl oxide and isophorone appear as condensation products[81]. Two electrons are obtained for

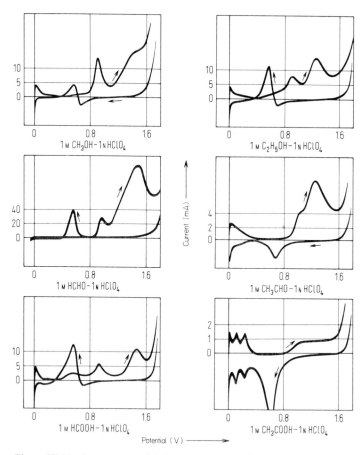

Figure III.38. Current–potential diagrams for smooth platinum in $1N\,HClO_4$ (after Kutschker and Vielstich[97]) 0.275 V sec^{-1}, 20°C, area = 0.66 cm^2.

reaction of one molecule of alcohol, but (in contrast to methanol or glycol) with no additional consumption of OH^- ion.

$$CH_3-CHOH-CH_3 \rightarrow CH_3-CO-CH_3 + 2H_{ad}$$
$$2H_{ad} + 2OH^- \rightarrow 2H_2O + 2e^-$$
(III.21)

Yeager[81], by using platinized carbon as fuel electrode in 14N KOH at 88°C and -0.85 volt (with respect to HgO), achieved a current density of 100 mA cm^{-2}. At temperatures below the boiling point of acetone (56°C), the current density decreased rapidly at constant potential (Figure III.39), but this poisoning by the acetone produced in the reaction was reversible. Evaporation of the acetone, or transfer of the electrode to a fresh solution of isopropyl alcohol, restored the initial current density[114].

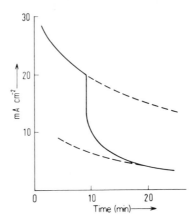

Figure III.39. Poisoning of a platinized platinum electrode during potentiostatic oxidation of isopropyl alcohol (0.67M $CH_3.CHOH.CH_3$ + 3N KOH, 20°C. Addition of 20% acetone at 10 min (after Prigent, Bloch and Balaceanu[114]).

III.3.3   Glycol, Glycerol and Oxalic Acid

Glycol, glycerol and oxalic acid are oxidized to CO_2 at platinum in *acid* solution. The working potentials, however, lie in a region far more unfavourable for use of these substances as fuels than in the case of formic acid (Figure III.37)[90].

In *alkaline* solution oxalate is but slightly soluble and is inactive. Glycol and glycerol cannot be oxidized to carbonate, but use of elevated temperatures (base-catalysed dehydrogenation) enables high current densities to be attained (up to 600 mA cm^{-2} at 80–90°C in 6N KOH with skeleton-nickel electrodes).

The base-catalysed dehydrogenation of ethylene glycol at Raney-nickel electrodes has been studied in more detail by Spengler and Grüneberg[103]. The reaction proceeds via several intermediate products (such as glycolaldehyde, glycollate, glyoxal and glyoxylate) to oxalate, with fission of eight hydrogen atoms:

$$CH_2OH.CH_2OH + 2OH^- \rightarrow C_2O_4^{2-} + 8H_{ad}$$
$$8H_{ad} + 8OH^- \rightarrow 8H_2O + 8e^-$$

(III.22)

In 6N KOH the rest potential is at -1140 mV with respect to the saturated calomel electrode (SCE).

The technical difficulty of continuous operation lies in the slight solubility of the electrochemically inactive oxalate in glycol–KOH mixtures. Crystallization of oxalate in the pores of a fuel electrode causes a very rapid fall of current density. At *room temperature*, despite oxalate formation, current densities less than $5\,\text{mA cm}^{-2}$ can be maintained for a few months (cf. section IV.3). Rest potentials and working potentials are more suited here to application as fuel[91] than in the case of methanol.

III.3.4   Carbon Monoxide

Because of its cheapness carbon monoxide is of special interest for use in fuel cells. The electrochemical reaction of carbon monoxide in acid electrolytes proceeds by a mechanism similar to that of the reaction of methanol or formaldehyde, and in alkaline solution formate is formed as an intermediate. It is therefore desirable to discuss the behaviour of carbon monoxide at the present stage.

In the period 1918–1920, Hofmann[115] carried out investigations of the electrochemical reaction of carbon monoxide. Unlike his predecessors, who used CO in fuel cells at 200 to 700°C, he conducted his experiments at room temperature. For strongly alkaline electrolytes (KOH), he used copper gauze electrodes, finding that copper could catalyse the reaction

$$CO + 2KOH \rightarrow K_2CO_3 + H_2 \qquad (III.23)$$

and that the hydrogen so formed is electrochemically active.

III.3.4.1   Mechanism in Alkaline Electrolytes

Grüneberg, in 1956, repeated Hofmann's experiments with CO in alkaline electrolytes. He found that two reaction paths were possible according to the catalyst used:

1. Catalytic conversion of CO

$$CO + 2OH^- \xrightarrow{\text{Me}} CO_3^{2-} + 2H_{ad} \qquad (III.24)$$

2. Production of formate as intermediate

$$CO + OH^- \xrightarrow{\text{Me}} HCOO^-$$

$$HCOO^- + OH^- \longrightarrow CO_3^{2-} + 2H_{ad} \qquad (III.25)$$

In either case, the atomic hydrogen formed determines the potential

$$H_{ad} + OH^- \rightarrow H_2O + e^-$$

These suggestions on the mechanism of the CO reaction have been examined by means of measurements with Raney-nickel and Raney-copper DSK electrodes (cf. section IV.1). Figure III.40 shows the anodic current–potential curves for electrodes of these two kinds, before and after transfer from operation with hydrogen to operation with carbon monoxide. The anodic current with carbon monoxide as fuel is about an order of magnitude smaller than that with hydrogen. At the same overpotential the nickel electrode passes about three times the current passed by the copper electrode.

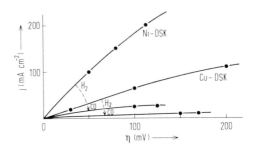

Figure III.40. Current–potential curves of Ni–DSK and Cu–DSK electrodes operating with H_2 or with CO in 5–6N KOH at about 80°C (after Grüneberg[92a]). $p_{H_2} = p_{CO} = 3$ atm (total pressure).

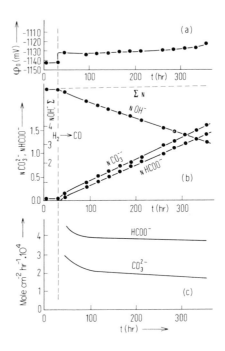

Figure III.41. (a) Potential of a Ni–DSK electrode on open circuit on transfer from operation with H_2 to operation with CO; (b) electrolyte composition; (c) rates of formation of formate and carbonate (or hydrogen). $p_{CO} = 3$ atm (total pressure), temp. = 83°C (after Grüneberg[92a]).

Over long periods of operation, carbonate formation leads to blockage of fine-pored gas-diffusion electrodes, particularly when the double-layer electrode is used. It is therefore necessary to use relatively coarsely-pored sintered electrodes, even though this leads to proportionately small electrochemical consumption of the gas supplied (less than 5%). In this way, however, a working life of more than 1000 hours can be achieved before regeneration becomes necessary.

The rest potential of a strongly-gassing Ni–DSK electrode of this kind over the course of a few hundred hours is shown in Figure III.41(a). On transfer from operation with hydrogen to operation with carbon monoxide, the potential jumps from $-1142\,\mathrm{mV}$ (relative to SCE) to $-1133\,\mathrm{mV}$. It can be inferred from Figure III.41(b), that the increases in carbonate and formate concentrations (reaction as in equation III.25 without flow of current) correspond quite closely with the consumption of OH^- ions. The quantity of carbonate formed is also equivalent to the amount of hydrogen evolved.

Finally, it can be seen from Figure III.41(c), that the rates of formation of the intermediate product (formate), and that of the end-product (carbonate), decrease from relatively high values in the first several hours, but thereafter decrease but slightly, in conformity with the consumption of OH^- ions.

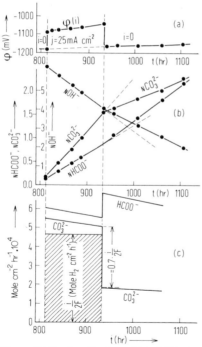

Figure III.42. Anodization of Ni–DSK electrode operating with CO (after Grüneberg[92a]). (a) Potential; (b) electrolyte composition; (c) rates of formation of formate and carbonate.

The diagrams in Figure III.42 also show the behaviour during the passage of $25 \, mA \, cm^{-2}$, and after cessation of this anodic current. The rate of carbonate formation is considerably enhanced by the anodic current. The slow positive displacement of the stationary potential corresponds with the decrease in concentration of OH^- ions. When current flow is stopped the reaction velocities of the preceding experiment are reinstated.

The experiment already illustrated in Figure III.26 is a decisive one in relation to mechanism. This figure represents the catalytic dehydrogenation of formate in KOH, in terms of hydrogen evolution, as a function of time. Three parallel experiments were carried out under similar conditions in three reaction vessels. Hydrogen evolution before the addition of any catalyst was, within experimental error, zero (curve 1). Curve 2 shows the dehydrogenation at a Raney-copper electrode. Dehydrogenation at a Raney-nickel electrode (curve 3) proceeds to completion at a rate some 30 times faster than at the copper catalyst.

The measurements conducted with anodic current flow show that reaction of CO at the copper electrode is slower than that at the nickel electrode by a factor of about 3, whereas the formate dehydrogenation is slower by a factor of 30. It is to be concluded from this that practically the whole current at the copper electrode must be determined by a base-catalysed conversion as in equation III.24, but that at Raney-nickel reaction mainly proceeds via formate. This suggests the following scheme for the CO reaction in alkaline solution at the catalysts studied by Grüneberg[92a]:

$$
\begin{array}{c}
CO_{ad} + OH^- \xrightarrow{\text{R-Cu}} CO_3^{2-} + 2H_{ad} \\
\Updownarrow \qquad\qquad\qquad\qquad\qquad \Big\downarrow \\
\boxed{CO} \\
\Updownarrow \qquad\qquad \boxed{H_2O + e^- \rightleftharpoons OH^- + H_{ad}} \cdot \\
CO_{ad} \\
+OH^- \\
\text{R-Ni} \downarrow \text{R-Cu} \qquad\qquad\qquad \Big\uparrow \\
HCOO^- + OH^- \xrightarrow{\text{R-Ni}} CO_3^{2-} + 2H_{ad}
\end{array}
\qquad (III.26)
$$

III.3.4.2 *Mechanism in Acid Electrolytes*

The anodic reaction of carbon monoxide in acid electrolytes has been studied in recent years[115a-f], but the mechanism is not yet clarified. As the triangular potential scan diagram in Figure III.43 shows, carbon monoxide, like methanol and formaldehyde, reduces the chemisorbed oxygen layer on the electrode. On the cathodic return sweep, anodic current flows over the whole potential range from 1.7 to 0.7 volt. It is obvious from this that CO reacts with a platinum surface completely covered with chemisorbed oxygen. In accordance with this

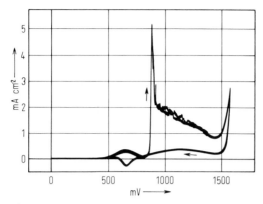

Figure III.43. Current–potential diagram for smooth platinum in 1N H_2SO_4 with CO bubbling, 50 mV sec^{-1}, 20°C (after Vielstich[17]).

the following electrochemical reaction for acid electrolytes is suggested[17,115b,115f]:

$$Me + H_2O \rightarrow Me-OH + H^+ + e^-$$

$$2Me-OH \rightarrow Me-O + H_2O \tag{III.27}$$

$$CO + Me-O \rightarrow Me + CO_2$$

Only small current densities are attainable with platinum black. Niedrach[116] obtained 2 to 5 mA cm^{-2} between $+0.7$ and 0.9 volt, relative to the hydrogen potential, at room temperature (Figure III.44).

Recently, Niedrach and Weinstock[115e] have shown the promoting effect of oxides on the platinum catalyst. At Teflon-bonded platinum-black electrodes containing 10% W_2O_5, for instance, CO was oxidized in a test cell at 85°C with an electrolyte of 5N H_2SO_4. An electrode potential of $+100$ mV at 80 mA cm^{-2} has been observed. Reformer gas has been used as a fuel in continuous operation, at a load of 50 mA cm^{-2} at constant voltage (0.8 V), over 75 hr.

The rest potential at platinized platinum lies between $+200$ and $+300$ mV, but the hydrogen potential is reached at 100°C by use of Raney-platinum electrodes. This behaviour supports the opinion of Sokolsky[118] that at platinum metals (especially palladium) the reaction

$$CO + H_2O \rightarrow CO_2 + H_2$$

proceeds even at room temperature—hydrogenation by use of CO should in this case be a possibility. It is clear that at 100°C reaction of CO occurs exclusively by way of this preliminary conversion.

III.4 HYDROCARBONS

Until a few years ago it seemed very improbable that hydrocarbons such as propane, ethane, ethylene, or even methane, could be electrochemically oxidized at room temperature. In 1962, however, Niedrach[116] demonstrated that, even at 25°C, current densities of up to 5 mA cm^{-2} could be obtained with *ethylene* at finely-divided platinum*. The overpotential relative to hydrogen was certainly considerable (Figure III.44). Measurable currents were likewise obtained with propane and natural gas.

With platinized carbon electrodes in 5N H_2SO_4 at 80°C, Schlatter[119] obtained current densities of a few mA cm^{-2} with propane, ethylene, propylene and isobutane at +0.6 volt relative to hydrogen. The best results have been recently obtained by Binder and co-workers[117], with Raney-platinum electrodes.

Raney-platinum powder (10–25 μ) was mixed with gold powder (40% by vol) and pressed into discs (diam 4 cm, 3 mm thick) at 1 ton cm^{-2}. The aluminium was dissolved out in the usual way with strong alkali warmed slowly to 90°C. More recently, carbon discs have been used as substrates, the components of the Raney-alloy being deposited in a thin layer by vacuum evaporation[120]. The electrodes were used with gas at 0.5 atm in excess of atmospheric pressure.

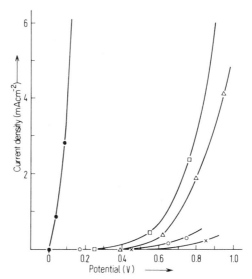

Figure III.44. Current–potential curves at a platinum electrode operating with various fuel gases, measured in a membrane cell with 18 mg cm^{-2} platinum powder at room temperature (after Niedrach[116]). ●, Hydrogen; □, carbon monoxide; △, ethylene; ○, propane; ×, natural gas.

* Measuring rest potentials of an electrode fed with ethylene in acid solution, Niedrach[116a] detected methane and ethane in the gas phase. From this, and the value of the rest potential (ca. + 200 mV), the existence of a hydrogenation–dehydrogenation equilibrium is to be inferred (cf. section III.2.1.).

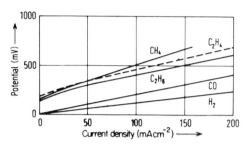

Figure III.45. Stationary current–potential curves at Raney-platinum electrodes in 3N H_2SO_4 at 100°C (after Binder, Köhling, Krupp, Richter and Sandstede[117]).

Figures III.45 and 46 show current–potential curves for the saturated hydrocarbons from methane to heptane, and for ethylene, hydrogen and carbon monoxide. Optimal working conditions are obtained with 3N H_2SO_4 and at 100°C. Reduction of sulphuric acid enters with 5N H_2SO_4 at 100°C (cf. section IV.2.3). The overpotential of the fuel electrode increases with the length of the hydrocarbon chain from ethane to heptane. Rest potentials lie between $+120\,mV\,(C_2H_6)$ and $+210\,mV\,(C_2H_4)$, and do not correspond with thermodynamic potentials.

The temperature-dependence of a reaction between 40 and 100°C is given for propane as an example in Figure III.47.

For the hydrocarbons from methane to butane, the reaction (at 90°C in 3N H_2SO_4, $\phi = +350$ to $+600\,mV$ relative to H_2) proceeds completely to CO_2[117]. No intermediate products were found (cf. section IV.2.3; reaction of propane in phosphoric acid as electrolyte).

At temperatures above 150°C electrochemical oxidation is greatly facilitated by catalytic cracking of the hydrocarbons. The reactions to be expected in

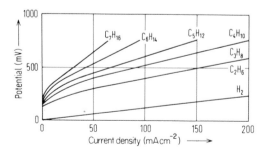

Figure III.46. Stationary current–potential curves for saturated hydrocarbons at Raney-platinum in 3N H_2SO_4 at 100°C (after Binder, Köhling, Krupp, Richter and Sandstede[117]).

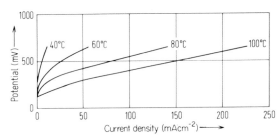

Figure III.47. Temperature dependence of the current–potential curve for propane at Raney-platinum in $3N$ H_2SO_4 (after Binder, Köhling, Krupp, Richter and Sandstede[117]). Potential relative to hydrogen electrode in the same solution.

consequence of this and the phenomena observed are discussed in the appropriate sections of chapter IV.

The '*electrochemical activity*' of hydrocarbons at *room temperature* becomes less surprising when it is considered that, in common with oxygen-containing compounds (methanol, formaldehyde, etc., cf. section III.3), they are not oxidized by a direct abstraction of electrons. The hydrocarbons, in a similar manner, reduce the oxygen chemisorbed on the noble-metal electrode in a secondary chemical reaction, or undergo a preliminary dehydrogenation.

Figure III.48 shows the current maxima at 0.9 and 1.2 volt (cf. Figure III.38), characteristic of the reduction of oxygen chemisorbed at platinum. It is interesting to note that in the H_{ad} region (< 200 mV), ethylene is hydrogenated very rapidly (Figure III.48).

Accordingly, the course of the reaction for ethylene, leading ultimately to formation of CO_2 is:

$$H_2O + Pt \rightarrow Pt{-}OH + H^+ + e^-$$

$$C_2H_4 + 4Pt{-}OH \rightarrow 2CO_2 + 8H^+ + 8e^- \quad \text{or}$$

$$H_2O + Pt \rightarrow Pt{-}O + 2H^+ + 2e^- \quad \quad \text{(III.28)}$$

$$C_2H_4 + 6Pt{-}O \rightarrow 2CO_2 + 2H_2O + 6Pt$$

over all

$$C_2H_4 + 4H_2O \rightarrow 2CO_2 + 12H^+ + 12e^-$$

Bockris and co-workers[121] have examined this reaction scheme in relation to C_2H_2 and C_2H_4 at 80°C, and have confirmed its validity over the whole pH range. Dahms and Bockris[121] have further studied the different activities of noble metals for the reaction of ethylene ($1N$ H_2SO_4, 80°C) and have established the sequence Pt > Rh > Ir, Pd > Au. With Pt, Ir and Rh, reaction proceeds completely to CO_2, but Pd and Au are said to afford aldehyde and acetone as final products.

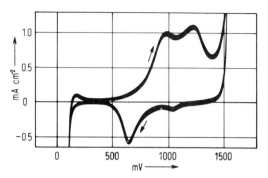

Figure III.48. Current–potential diagram obtained with smooth platinum in $1N\ H_2SO_4$ bubbled with C_2H_4 (after Vielstich[17]). $50\ mV\ sec^{-1}$, $20°C$.

Bockris and Srinavasan[121a] have recently referred to the use of mixed and alloy catalysts. Thacker and Bump[121b], have obtained current densities of $10\ mA\ cm^{-2}$ $(C_2H_2, C_2H_4, C_3H_8,$ n-butane) at $150°C$, with finely-divided platinum (on carbon as carrier) as catalyst, in an alkaline electrolyte. Further recent results[115a,121c–i] cannot be discussed.

III.5   HYDRAZINE, ALKALI METAL BOROHYDRIDES AND AMMONIA

Nitrogen and boron hydride compounds have come under consideration as possible fuels for fuel cells in the last few years. High current densities (>1 $A\ cm^{-2}$) can be attained, even at room temperature, with use of hydrazine or alkali metal borohydrides, but these substances are at present relatively expensive. Ammonia, more competitive in availability and price, is less reactive.

III.5.1   Hydrazine as Fuel

The following electrochemical reactions of hydrazine in acid and alkaline electrolytes are those cited by Latimer[122] as to be expected[123] on thermodynamic grounds:

$$N_2H_5^+ \rightarrow N_2 + 5H^+ + 4e^-$$

$$\text{with} \quad \phi_0 \rightarrow -230\ mV \tag{III.29}$$

and

$$N_2H_4 + 4OH^- \rightarrow N_2 + 4H_2O + 4e^-$$

$$\text{with}\ \phi_0 = -1160\ mV\ (-330\ mV\ \text{rel. to}\ \phi^0_{H_2/OH^-}) \tag{III.30}$$

At all metals of low hydrogen overpotential, therefore, working in an alkaline electrolyte, the establishment of a mixed potential is to be expected arising from anodic and cathodic partial currents, as follows:

$$N_2H_4 + 4OH^- \rightarrow N_2 + 4H_2O + 4e^-$$

and

$$4H_2O + 4e^- \rightarrow 4OH^- + 2H_2$$

Reactions III.29 and 30, however, are by no means reversible and the rest potentials are near to the reversible hydrogen potential. It may therefore be assumed that the rest potential is determined by hydrogen formed in a purely *chemical decomposition*, (at an activity perhaps exceeding that corresponding with 1 atm pressure), that is accelerated when the adsorbed hydrogen is electro-chemically oxidized by passing anodic current. This suggests the reaction scheme:

$$N_2H_4 \xrightarrow{\text{Me}} N_2 + 4H_{ad}$$
$$4H_{ad} + 4OH^- \longrightarrow 4H_2O + 4e^- \tag{III.31}$$

Decomposition in aqueous solution yields ammonia only to a very small extent. Grüneberg and Jung[124] detected ammonia in the gas evolved during the decomposition of hydrazine at Raney-nickel electrodes at 90°C. Decomposition of hydrazine at room temperature, with visible gas evolution, is catalysed by many metals; this applies particularly to alkaline solutions but vigorous decomposition is also observed in acid solutions in the presence of, for example, finely-divided platinum or palladium[17]. At platinized rhodium ($0.5M$ N_2H_4, $5N$ KOH, 20°C), the currentless decomposition corresponds with ca. 12 mA cm^{-2}.[125]

The *rest potential*, according to hydrazine concentration and catalyst material, lies up to some 180 mV more negative than the reversible hydrogen potential in the electrolyte concerned[126].

Figure III.49 shows current–potential curves obtained at a series of metals in alkaline solution at 20°C. At high overpotentials hydrazine is oxidized even at gold and silver. The nascent hydrogen formed must obviously reduce the chemisorbed oxygen layer at these electrodes and the preceding electrochemical reaction at these metals (Ag, Au) must be

$$M + OH^- \rightarrow M{-}OH + e^- \quad \text{or}$$
$$M + 2OH^- \rightarrow M{-}O + H_2O + 2e^- \tag{III.32}$$

At skeleton-nickel electrodes (Raney-nickel or nickel boride), the anodic current–voltage curve begins at potentials about 100 mV negative to the reversible hydrogen potential in the same solution (Figure III.49a). At Raney-nickel particularly, copious gas evolution is observed on open circuit.

The anodic current density can be raised considerably by increase of tempera-ture. Thus Grüneberg and Jung, using Raney-nickel DSK electrodes at 90°C ($0.37M$ N_2H_4, $6N$ KOH), attained 600 mA cm^{-2} even at the hydrogen poten-tial[26].

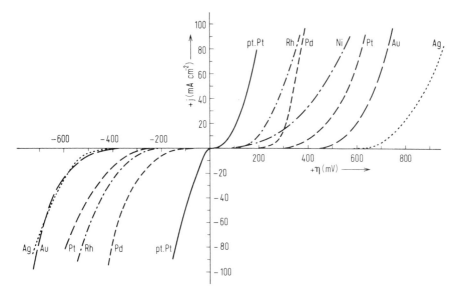

Figure III.49. Stationary current–potential curves for various metal electrodes in $0.2M$ N_2H_4 + $2N$ KOH at 20°C (after Vielstich[17]). Rest potential about -830 mV with respect to $\phi^0_{H_2,H^+}$.

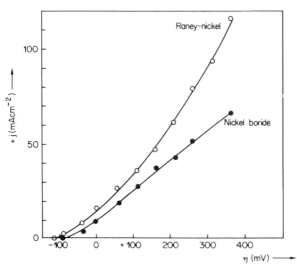

Figure III.49a. Stationary current–potential curves at Raney-nickel and nickel boride electrodes in $1M$ $N_2H_4.H_2O$ + $1M$ KOH at 20°C. Rest potentials ca. -915 mV against $\phi^0_{H_2,H^+}$ (after Gutjahr and Vielstich[177]).

Anodization leads to practically complete consumption of the hydrogen formed (equation III.31) at smooth electrodes (Pt, Rh, Au, Ag) at room temperature[17]—the gas evolved is nitrogen alone (1 mole N_2/4 faradays). 10% of hydrogen has been found in the gas from platinized rhodium at 100 mA cm^{-2} [125].

In general, current efficiency is reduced by the use of highly active electrodes, especially in the case of alkaline electrolytes at elevated temperatures, because of vigorous decomposition of hydrazine. This loss is diminished with high current densities. Table III.2 shows the maximum and average current efficiencies, in electrons per mole of hydrazine, obtained at 10 mA cm^{-2} and 20°C[17]. Hydrazine concentration was 5×10^{-2}M in this series of experiments, which extended over 50 hours.

Table III.2. Current efficiencies in terms of electrons per molecule of hydrazine decomposed

Electrode material	1N HClO$_4$		2N KOH	
	max.	average	max.	average
Platinized platinum	3.8	3.35	3.6	3.3
Palladized palladium	3.9	3.66	3.1	2.55
Palladized carbon			3.1	3.0

The table shows that even at moderate current densities and temperatures, 80 to 100 A hr per mole N_2H_4 can be expected.

Although somewhat higher current efficiencies can be attained with acid electrolytes, the position of the current–potential curves is very unfavourable (Figure III.50) and temperature dependence is accentuated (Figure III.51).

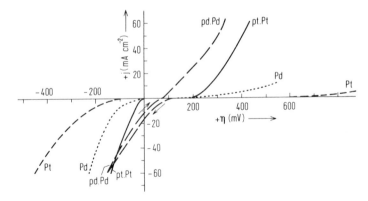

Figure III.50. Stationary current–potential curves for various metal electrodes in 0.114M N_2H_4 + 1N HClO$_4$ at 20°C (after Vielstich[17]). Overpotential(mV) relative to $\phi^0_{H_2,H^+}$.

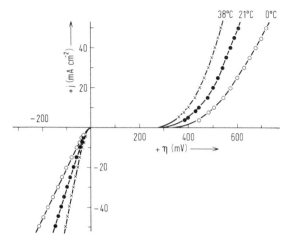

Figure III.51. Temperature dependence of the stationary current–potential curve for platinized platinum in $0.2M\ N_2H_4$ + $1N\ HClO_4$ (after Vielstich[17]).

The current–potential diagrams obtained by the triangular wave potential scan method (chapter II.4), shown in Figures III.52 and 53, indicate very clearly the difference in behaviour between alkaline and acid solutions. Whereas in KOH solution the hydrazine decomposition gives rise to a very sharp current peak, due to the nascent hydrogen formed, between the normal first and second peaks of the hydrogen region (cf. Figure III.11), yet in acid electrolytes decomposition of the hydrazine occurs mainly in the double-layer region (cf. chapter II.4 and Figure III.10).

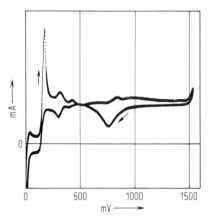

Figure III.52. Current–potential diagram for smooth platinum in $1.2 \times 10^{-2}M\ N_2H_4$ + $1N$ KOH, 20°C (after Grüneberg[127]).

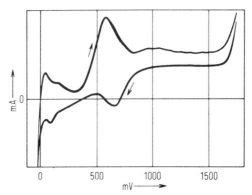

Figure III.53. Current–potential diagram for smooth platinum in 2.5×10^{-2} M N_2H_4 + 1N H_2SO_4, 20°C (after Grüneberg[127]).

III.5.2 Alkali Metal Borohydrides

Like hydrazine, the alkali-metal borohydrides are strong reducing agents[128], quite stable in alkaline solution in absence of metal catalysts[129]. In acid solution, vigorous hydrogen evolution occurs in the presence of metals. Anodic oxidation of borohydrides has been studied in some detail by Elder and Hickling[130] (KBH$_4$), Jasinski[131] (KBH$_4$) and Indig and Snyder[129] (NaBH$_4$).

As in the case of hydrazine, rest potentials at smooth electrodes are relatively ill defined, but, at platinized platinum or sintered nickel, definite potentials are established which are more negative the higher the pH value and the higher the borohydride concentration, [BH$_4^-$]. At platinized platinum the potential conforms with the relation[130]

$$\phi = \phi_0 + \frac{RT}{2F} \ln\{[H^+]/[BH_4^-]\} \tag{III.33}$$

In a fuel-electrolyte mixture of 0.1M KBH$_4$ + 0.1N KOH, $\phi = 827$ mV (relative to ϕ_{H_2,H^+}^0); the electrode potential is thus about 76 mV negative with respect to the reversible hydrogen potential in the same solution. Pecsok[132] and Stockmayer and coworkers[133], have calculated the standard potential of the reaction

$$BH_4^- + 8OH^- \rightleftharpoons H_2BO_3^- + 5H_2O + 8e^- \tag{III.34}$$

to be -1.23 to -1.24 volt.

Experimentally, however, rather less than 4e$^-$ per BH$_4^-$ are obtained. Probably, as for hydrazine, the nascent hydrogen formed as intermediate during anodic oxidation is potential-determining.

Figure III.54a shows current–potential curves obtained at a platinized platinum electrode. At about +1.6 volt relative to the hydrogen potential,

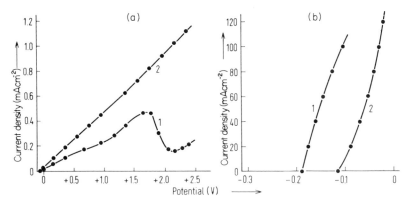

Figure III.54. Stationary current–potential curves for alkaline borohydride solutions (potentials relative to hydrogen electrode in the same solution). (a) Platinized platinum, 0.1N KOH, 20°C (after Elder and Hickling[130]). 1, 0.04M KBH$_4$; 2, 0.1M KBH$_4$. (b) KBH$_4$, 25% KOH, 25°C (after Jasinski[131]) (ohmic resistance eliminated). 1, Nickel boride, Ni$_2$B (precipitation on a sintered Nickel electrode, see chapter IV); 2, palladized sintered nickel electrode.

oxygen is simultaneously evolved. Simple sintered nickel electrodes, almost inactive to molecular hydrogen at 25°C, will pass current densities up to 400 mA cm^{-2} (6% NaBH$_4$, 20% NaOH, porosity of nickel electrode 80%)[129]. Sintered electrodes of nickel with addition of nickel boride are specially suitable (Figure III.54b[131]). The current–potential curve for an electrode activated with Ni$_2$B is more favourably placed by some 100 mV on the potential axis than the corresponding curve for a palladized electrode (3 mg Pd cm^{-2}). Even at 100 mA cm^{-2}, the electrode potential remains negative relative to a hydrogen electrode in the same solution; the molecular hydrogen evolved during the catalytic reaction cannot, of course, be electrochemically oxidized in this region of potential.

Whereas hydrogen evolution falls off quite sharply five or ten minutes after immersion of a smooth platinum electrode in a fuel-electrolyte mixture, it continues uniformly at platinized platinum. Anodization decreases it. At specially active electrodes (sintered nickel), however, hydrogen formation increases with rising anodic current density (cf. section III.2.5.).

Elder and Hickling[130] have proposed a reaction scheme according to which between 2 and 4 electrons per borohydride ion are to be expected, and between 0.5 and 1.5 mole of hydrogen per faraday. The results obtained with platinized platinum are in agreement with this. With sintered nickel, Indig and Snyder[129] obtained 3.4 to 3.9 electrons per BH$_4^-$ ion.

Pentaborane (B$_5$H$_9$) is also very reactive but, as yet, no electrolyte has been found in which it will dissolve without vigorous reaction[134].

III.5.3   Ammonia

The electrochemical oxidation of ammonia has hitherto been investigated only qualitatively[127,135a,135]. The emf expected for the ammonia–oxygen cell on thermodynamic grounds is about 1.2 volts[135a] (cf. Table II.3). Wynveen[135a] with gaseous ammonia and concentrated potassium hydroxide as electrolyte, has obtained a yield of three electrons per molecule of ammonia, as well as nitrogen and water as the sole reaction products*.

The corresponding overall reaction

$$NH_3 + 3OH^- \rightarrow \tfrac{1}{2}N_2 + 3H_2O + 3e^- \qquad (III.35)$$

is, however, certainly irreversible, and therefore cannot be potential-determining. It is not surprising that the observed open-circuit emf is less than that of the hydrogen–oxygen cell[135a], but a relatively favourable potential can be obtained with ammonia in alkaline solution (6N KOH) with a gas-diffusion electrode of skeleton-nickel (ca. -1000 mV against SCE[127]).

It can easily be seen from the current–potential diagrams (Figure III.55)[127] that reaction of ammonia occurs at bright platinum, at room temperature, only in alkaline electrolytes (pH > 9.5) and under these conditions, mainly in the double-layer region. Furthermore, Frick, Subcasky and Shaw[135] have shown that ammonia is inert in acid solutions; ammonium ion is thus not electrochemically oxidized. Comparison with Figure III.22 (oxidation of methanol in alkaline electrolyte) shows that for the region of the current maximum a reaction with the layer of OH adsorbed at the electrode must be assumed.

Elliott and Wynveen[135a] have found that the peak of anodic current (Figure III.55) is almost independent of both NH_3 and OH^- concentrations for KOH solutions between 0.1 and 1N. In 10N KOH, however, there is a pronounced maximum at about 1M NH_3.

Gas evolution has been observed at platinized carbon electrodes under open-circuit conditions[135]; indicating the existence of a dehydrogenation. Wynveen[135a] has detected no hydrogen in the gas evolved during current flow.

Current–potential curves for individual electrodes have yet to be published. According to the characteristic curves for ammonia–oxygen cells (cf. section V.7), 20 to 40 mA cm^{-2} can be obtained with platinized carbon electrodes at

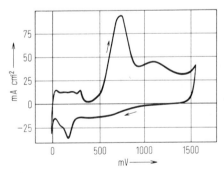

Figure III.55. Current–potential diagram for platinized platinum in 0.1M $(NH_4)_2SO_4$ + 1N KOH at 20°C, 50 mV sec^{-1} (after Vielstich[17]).

---

* Spahrbier and Wolf[176] deduced from potentiostatic current–time curves that the recombination of nitrogen atoms formed is the rate-limiting step.

30 to 80°C. This could probably be more than doubled at skeleton-nickel electrodes[127].

Aqueous solutions could be used instead of gaseous ammonia but the solubility in strongly alkaline electrolytes is small, and this would introduce mass transfer difficulties.

III.6   THE OXYGEN ELECTRODE

III.6.1   Investigations of Reaction Mechanism

The electrochemical behaviour of oxygen* in aqueous solution at electrodes of platinum, gold, silver or carbon is essentially more complicated than that of other gases such as hydrogen or chlorine. Over the whole pH range the *rest potential* of the oxygen electrode is very slowly established, poorly reproducible and more than 100 mV less than the theoretical oxygen potential. This behaviour is due to the great stability of the chemical bond in the oxygen molecule. At temperatures below 150°C breaking of the bond occurs indirectly with formation of H_2O_2. Intervention of radicals such as HO_2 and OH then further complicates the course of reaction.

In acid solution, the reversible reduction of oxygen would correspond with the equation

$$O_2 + 4H^+ + 4e^- \rightleftharpoons 2H_2O \qquad (III.36)$$

with a standard potential ($a_{H^+} = 1$, $p_{O_2} = 1$ atm, 25°C) $\phi_0 = +1.229$ volt. In alkaline solution one has

$$O_2 + 2H_2O + 4e^- \rightleftharpoons 4OH^-, \qquad \phi_0 = +0.401 \text{ V} \qquad (III.37)$$

A potential of 1.23 volt, with respect to a hydrogen electrode in the same solution at 25°C, is therefore to be expected.

The emf of hydrogen–oxygen cells of this kind had already been determined 60 years ago by numerous authors[136] using acid, neutral and alkaline solutions. Different kinds of carbon and a series of metals (mainly noble metals) have been used for the oxygen electrode. At room temperature, emf values between 1.05 and 1.15 volt have been found, and the formation of H_2O_2 has been observed.

At 200°C and 40 atm, the measured emf[138] ($+1.168$ volt in 45% KOH) is considerably closer to the thermodynamic value (under these conditions, 1.21 volt). Haber[137] experimentally confirmed the theoretical oxygen potential at gold and platinum electrodes in molten salts between 300 and 1000°C.

The *formation of hydrogen peroxide* in aqueous electrolytes appears to be a possible reason for the low value of the oxygen potential observed experimentally. Berl[139] has carried out the most thorough investigation of this problem by

* For a review see: J. P. Hoare, *The Electrochemistry of Oxygen*, Interscience Publishers, New York 1968; J. P. Hoare in P. Delahay and Ch. W. Tobias, *Advances in Electrochemistry and Electrochemical Engineering*, Interscience Publishers, New York 1967, Vol. 6, p. 201.

using the carbon–oxygen electrode in alkaline solution. From potential measurements at various hydroxyl ion and peroxide concentrations, he proposed that

$$O_2 + H_2O + 2e^- \rightleftharpoons HO_2^- + OH^- \tag{III.38}$$

is the only reversible and potential-determining reaction. According to Kordesch and Martinola[140] the dependence of potential on the partial pressure of oxygen also appeared to be consistent with equation III.38.

The thermodynamic standard potential expected for this reaction is, according to Lewis and Randall[141], $\phi_0 = -76\,\text{mV}$ relative to hydrogen. Then, under standard conditions, the potential relative to a hydrogen electrode in the same solution would be only 0.75 volt instead of 1.23 volt (Figure III.56).

Extrapolation of Berl's measured potentials for various OH^- and H_2O_2 concentrations gives a value for the standard potential at *carbon* which deviates by 35 to 40 mV, in the positive direction, from the reversible O_2, H_2O_2 potential. The potential at *silver* is displaced in the same direction by about 150 mV (Figure III.57). Platinized carbon in alkaline H_2O_2 solutions assumes a rest potential intermediate between those found for carbon and silver electrodes[142].

Results obtained for silver and platinized carbon, together with Berl's curves for carbon, are shown in Figure III.57. Under the condition that $[H_2O_2] \ll [KOH]$, the potential becomes about 30 mV more negative for a ten-fold increase in H_2O_2 concentration; this conforms with the concentration dependence expected for an O_2, H_2O_2 electrode based on equation III.38, namely,

$$\phi_{O_2|HO_2^-}^{25°} = -76 - 29.5 \log \frac{a_{OH^-} \cdot a_{HO_2^-}}{p_{O_2} \cdot a_{H_2O}} [\text{mV}] \tag{III.39}$$

With decreasing activity of OH^- ion, the potential becomes less negative—at carbon electrodes, by about 30 mV for tenfold decrease in agreement with equation III.39, but at silver electrodes by 40 mV or more.

For $[H_2O_2] \sim [KOH]$, the potential passes through a maximum negative value, and then becomes less negative again with continued addition of hydrogen peroxide. This is attributed by Berl[139] to the partial neutralization of OH^- ions by the weak acid, hydrogen peroxide.

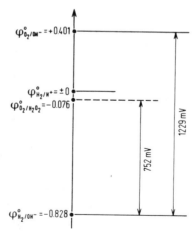

Figure III.56. Location on the potential scale of the O_2–H_2O_2 electrode in alkaline electrolytes.

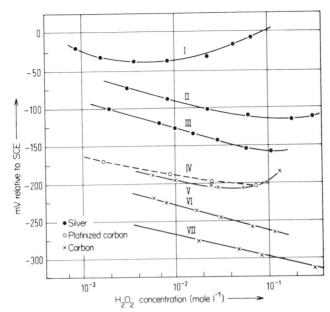

Figure III.57. Oxygen potentials plotted as functions of H_2O_2 concentration at 27°C, at various concentrations of KOH (after Vielstich[142]), as follows: I, 2.0×10^{-2}N; II, 1.6N; III, 6.0N; IV, 1.6N; V, 1.2×10^{-1}N; VI, 1.6N; VII, 11.4N.

The stationary concentration of HO_2^- established during the reduction of oxygen is between 10^{-5} and 10^{-3} molar. According to equation III.39, this corresponds with a positive displacement of the O_2, HO_2^- potential by 100 to 150 mV with respect to the standard value. The stationary concentration of hydrogen peroxide is small because it takes part in a secondary electrochemical reaction, and is susceptible to catalytic decomposition at the metal surface or in the electrolyte (especially in alkaline solution)[143]. Chemisorbed oxygen is produced at the metal surface and its electrochemical reduction has an additional influence on the electrode potential (see below). Consequently, the overall reactions for the cathodic reduction of oxygen in acid and alkaline solutions, represented by equations III.36 and III.37, must be split up as follows:

Acid solution

I
$$O_2 + 4H^+ + 4e^- \rightleftharpoons 2H_2O, \quad \phi_0 = +1.229 \text{ V}$$

$$O_2 + 2H^+ + 2e^- \rightleftharpoons H_2O_2, \quad \phi_0 = +0.682 \text{ V} \qquad \text{(III.40a)}$$

II $\left\{ \begin{array}{l} \text{(a) } H_2O_2 + 2H^+ + 2e^- \rightleftharpoons 2H_2O, \quad \phi_0 = +1.77 \text{ V} \\[2mm] \text{(b)} \quad \left\{ \begin{array}{l} \quad\quad H_2O_2 \rightarrow H_2O + \frac{1}{2}O_2 \\ M{-}H_2O_2 \rightarrow H_2O + M{-}O \end{array} \right. \end{array} \right.$

Alkaline solution

$$I \qquad\qquad O_2 + 2H_2O + 4e^- \rightleftharpoons 4OH^-, \qquad \phi_0 = +0.401 \text{ V}$$

$$O_2 + H_2O + 2e^- \rightleftharpoons HO_2^- + OH^-, \quad \phi_0 = -0.076 \text{ V}$$

$$\text{II} \quad \begin{cases} \text{(a)} \quad HO_2^- + H_2O + 2e^- \rightleftharpoons 3OH^-, \qquad \phi_0 = +0.88 \text{ V} \\[2mm] \text{(b)} \quad \begin{cases} HO_2^- \rightarrow OH^- + \frac{1}{2}O_2 \\ M\text{—}HO_2^- \rightarrow OH^- + M\text{—}O \end{cases} \end{cases}$$

$$\text{(III.40b)}$$

In stage I there is a 2-electron reduction of the oxygen molecule to hydrogen peroxide without cleavage of the O—O bond*. Further reaction occurs either electrochemically (IIa) or chemically (IIb). The electrochemical reduction of hydrogen peroxide (IIa) is irreversible and usually strongly hindered (see below). In general, however, there is still a consumption of 4 electrons per oxygen molecule reduced, because the molecular oxygen formed by homogeneous decomposition of hydrogen peroxide and the chemisorbed oxygen on the metal surface formed in the heterogeneous decomposition are both susceptible to reduction[145-147].

The reduction of chemisorbed oxygen[142] proceeds according to

$$M\text{—}O + 2H^+ + 2e^- \rightarrow M + H_2O \quad \text{or} \qquad \text{(III.41a)}$$

$$M\text{—}O + H_2O + 2e^- \rightarrow M + 2OH^- \qquad\qquad \text{(III.41b)}$$

The loss of energy (and hence of potential) is the greater the stronger the M—O bonding.

At *silver*, reaction III.41b is strictly reversible, and the equilibrium potential is only about 50 mV less than the oxygen potential.

Silver is an exceptionally good catalyst for the decomposition of hydrogen peroxide and facilitates the establishment of a stationary oxygen potential. A comparison[147a] of the rates of electrode potential adjustment at $40 \, \mu\text{A cm}^{-2}$ between gas-activated and silvered carbon electrodes is shown in Figure III.57a.

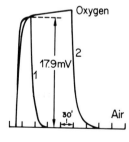

Figure III.57a. Potential adjustment at porous carbon diffusion electrodes on change of gas composition (air–oxygen–air); 12N KOH, 20°C, $40 \, \mu\text{A cm}^{-2}$. 1, Silvered carbon; 2, gas-activated carbon (after Mrha and Vielstich[147a]).

* Yeager and co-workers[148], in experiments with active carbon and ^{18}O, have shown that all of the peroxide oxygen comes from gaseous oxygen without breakage of the O—O bond.

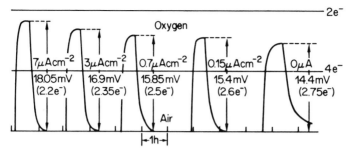

Figure III.57b. Potential response of a silvered carbon electrode to changes from air to oxygen and back again at various current densities; 12N KOH, 20°C. The electrons consumed per O_2 molecule and reaction step are shown in brackets. The reference lines indicate electron consumption in the Berl reaction ($2e^-$, equation 2) and in the overall reduction ($4e^-$, equation 1) (after Mrha and Vielstich[147a]).

The time for potential adjustment decreases with increasing current density (Figure III.57b).

Transfer from oxygen to air, or vice versa, should, according to equation III.39, lead to a potential change of 19.8 mV for the transfer of two electrons per molecule. For the silvered carbon electrode illustrated in Figure III.57b, however, the change under open-circuit conditions is only 14.4 mV, corresponding

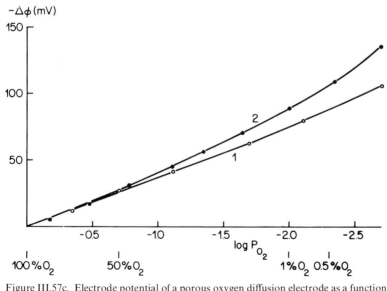

Figure III.57c. Electrode potential of a porous oxygen diffusion electrode as a function of partial pressure of oxygen at constant flow of gas mixture; 12N KOH, 20°C, current density 0.4 mA cm⁻². 1, 450 cm³ H_2–O_2 mixture min⁻¹; 2, 225 cm³ H_2–O_2 mixture min⁻¹ (after Mrha and Vielstich[147a]).

with consumption of 2.75 e⁻ per reaction step. In spite of this deviation from the Berl mechanism, silvered carbon electrodes are very serviceable for the determination of very low partial pressures of oxygen[147a]. Figure III.57c shows the dependence of electrode potential on oxygen partial pressure (down to 0.2%) at constant flow of gas mixture. Although the current density is low (0.4 mA cm⁻²), mass transfer hindrance has a noticeable effect at 1% of oxygen in the gas mixture, in the form of a diffusion potential (cf. curves 1 and 2 in Figure III.57c).

Other investigations of oxygen reduction have been carried out with carbon[147b], palladium[147c], rhodium[147d], gold[147e] and silver[147f,g] electrodes.

III.6.2   Discussion of Reaction Mechanism in Alkaline Electrolytes

Experimental results have shown that, at temperatures up to at least 100°C, reduction proceeds via HO_2^- ions as intermediates. The first 2-electron step is reversible in alkaline electrolytes:

$$O_2 + H_2O + 2e^- \rightleftharpoons HO_2^- + OH^- \tag{III.38}$$

The succeeding electrochemical step

$$HO_2^- + H_2O + 2e^- \rightarrow 3OH^- \tag{III.42b}$$

is, according to the electrode material and the working conditions, more or less strongly hindered and irreversible. This is particularly the case for mercury[149,150] and carbon electrodes and it has been shown by Müller and Nekrassov[151] that step II (equation III.40b) is also hindered at smooth platinum. An increase in HO_2^- ion concentration in the electrolyte is therefore observed during the reduction of oxygen (cf. section III.7.1).

The molecular oxygen formed in the electrolyte by decomposition of HO_2^- ions diffuses to the electrode surface and enters the reaction cycle again, according to equation III.38. The heterogeneous decomposition of H_2O_2 can lead to

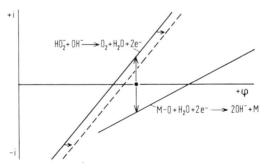

Figure III.58. The establishment of a mixed potential at
the oxygen electrode in alkaline solution.

the formation of a surface oxide or layer of chemisorbed oxygen of undefined stoichiometry. The reduction of this bound oxygen adds a further increment to the cathodic current. Finally, at high overpotentials, reaction III.42b comes into effective operation.

These considerations lead to the following scheme of reaction for alkaline electrolytes:

$$O_2 + H_2O + 2e^- \rightleftharpoons HO_2^- + OH^-$$

$$M-HO_2^- \leftrightarrow OH^- + M-O$$

$$M-O + H_2O + 2e^- \rightarrow M + 2OH^-$$

$$2M-O \rightarrow 2M + O_2$$

$$HO_2^- \leftrightarrow OH^- + \tfrac{1}{2}O_2$$

$$HO_2^- + H_2O + 2e^- \nleftrightarrow 3OH^-$$

$\leftrightarrow$ hindered $\nleftrightarrow$ strongly hindered

To what extent the electrochemical reduction of HO_2^- ions plays a part in this scheme is discussed in section III.7.1.

The results illustrated in Figure III.57, as well as the consumption of 2.75 electrons per molecule at silver-activated carbon electrodes (cf. Figure III.57b), show clearly that the Berl reaction is accompanied by another reaction involving more than two electrons. The reduction of chemisorbed oxygen, $M-O$, answers this description[147a].

The scheme allows the time-dependence of the rest potential of the oxygen electrode in alkaline solution to be explained. As long as the potential for the reduction of chemisorbed oxygen ($M-O$) is positive with respect to the equilibrium potential of reaction III.38, a mixed potential[152] is established at which H_2O_2 is anodically oxidized (Figure III.58):

$$HO_2^- + OH^- \rightarrow O_2 + H_2O + 2e^-$$

The associated cathodic current is provided by reduction of $M-O$

$$M-O + H_2O + 2e^- \rightarrow M + 2OH^-$$

and the chemisorbed oxygen layer is constantly renewed by the heterogeneous decomposition of HO_2^- ions

$$M-HO_2^- \rightarrow OH^- + M-O$$

Decrease of the concentration of HO_2^- causes displacement of the equilibrium potential for reaction III.38 and consequently a shift of the mixed potential in a positive sense (Figure III.58, broken line).

III.6.3   The Oxygen Electrode in Acid Electrolytes

As for the electrode in alkaline solution, many investigations have been directed to processes occurring at the oxygen electrode in acid electrolytes, but the details have not been clarified[153–155]*. It has become generally assumed that, under acid conditions, the mechanism involves the formation of H_2O_2 and that both HO_2 and OH radicals are concerned.

The rest potential is established very slowly, and, in solutions of a normal degree of purity, often falls very short (more than 200 mV) of the reversible oxygen potential.

According to Bockris and Huq[156] however, the thermodynamic potential of $+1.23$ volt with respect to the hydrogen electrode is established at bright platinum in sulphuric acid solutions in very specially purified systems. The impurities must not exceed 10^{-11} mole l^{-1} otherwise the exchange current due to them becomes greater than that of the oxygen reaction; this leads to concurrent electrode processes which establish a mixed potential, with a corresponding shift of the rest potential. The very high degree of purity required to avoid this was attained by cathodic (24 hr, 10^{-2} A cm^{-2}) and anodic (48–72 hr, 10^{-2} A cm^{-2}) preelectrolysis of solution.

In acid electrolytes, not only the second, but also the *first* step in the reduction of oxygen is hindered. Müller and Nekrassov[151] consider the influence of anion adsorption at positive electrode potentials to be responsible for hindrance of the first step [$O_2 \to H_2O_2$, cf. equation III.40a]. The characteristic potential–current plot for oxygen is displaced towards more negative potentials in the sequence $HClO_4 < H_2SO_4 < HCl < HBr$, i.e., in the order of increasing susceptibility to adsorption of the anions (Figure III.59).

The yield of H_2O_2 produced in the reduction of oxygen at smooth platinum in 1N H_2SO_4 has been estimated by Müller and Nekrassov[151] by use of a rotating disc electrode associated with a platinized ring electrode. The H_2O_2 formed by reduction of O_2 at the smooth platinum disc was determined by oxidation at the ring. Figure III.60 shows the percentage yield of H_2O_2 as a function of electrode potential.

Frumkin[144], on the basis of studies by Bagotzky and Yablokova[149] with mercury electrodes, divided the first step of O_2 reduction in acid electrolytes as

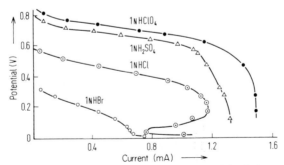

Figure III.59. Current–potential curves (potentials relative to $\phi_{H_2}^0$) for the reduction of oxygen at a platinum rotating disc electrode (2650 rpm) in different acids (after Müller and Nekrassov[151]).

* For a review, see K. J. Vetter, *Elektrochemische Kinetik*, Springer Verlag, 1961, or K. J. Vetter, *Electrochemical Kinetics*, Academic Press, New York, 1967.

follows:

$$O_2 + e^- \rightleftharpoons O_2^-$$

$$O_2^- + H^+ \rightleftharpoons HO_2$$

$$HO_2 + e^- \rightleftharpoons HO_2$$

$$HO_2^- + H^+ \rightleftharpoons H_2O_2$$

According to this, it is the acceptance of the first electron by the O_2 molecule that is rate-determining and irreversible. Comparison with the first step in the corresponding analysis of the reaction in an *alkaline* electrolyte indicates why, in the latter case, the reaction may be reversible. Because of the low proton concentration a comparatively high O_2^- concentration can be attained and this will facilitate the back-reaction.

For the *second* step the same reaction path as for alkaline solution is possible; the homogeneous and heterogeneous decomposition of H_2O_2 and the electrochemical reduction

$$H_2O_2 + 2H^+ + 2e^- \rightarrow 2H_2O \qquad\qquad (III.42a)$$

Subdivision of this electrochemical reaction has been proposed[143,149,153,154]:

$$H_2O_2 + e^- \rightarrow OH^- + OH$$
$$OH + e^- \rightarrow OH^- \qquad\qquad (III.44)$$

The existence of OH radicals as dissociation products of heated water vapour has been demonstrated spectroscopically by Bonhoeffer[154a].

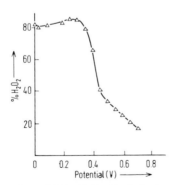

Figure III.60. Yield of H_2O_2 in the reduction of oxygen at a rotating disc platinum electrode in $1 \text{N } H_2SO_4$ as a function of electrode potential (relative to $\phi_{H_2}^0$), referred to total oxygen consumption (after Müller and Nekrassov[151]).

The following reaction scheme for the reduction of oxygen in acid solution can be suggested on the evidence described:

$$O_2 + e^- \rightleftharpoons O_2^- + 2H^+ + e^- \rightleftharpoons H_2O_2$$

$$M-H_2O_2 \rightarrowtail H_2O + M-O$$

$$M-O + 2H^+ + 2e^- \rightarrow H_2O \qquad\qquad (III.45)$$

$$2M-O \rightarrow 2M + O_2$$

$$H_2O_2 \rightarrowtail H_2O + \tfrac{1}{2}O_2$$

$$H_2O_2 + 2H^+ + 2e^- \rightarrowtail\!\!\!\!\rightarrow 2H_2O$$

$$\rightarrowtail \text{hindered} \qquad \rightarrowtail\!\!\!\!\rightarrow \text{strongly hindered}$$

In the reduction of oxygen at *mercury cathodes* there is a considerable differentiation of the two steps $O_2 \rightarrow H_2O_2$ and $H_2O_2 \rightarrow H_2O$ in the reduction of oxygen. If the electrode potential is made more negative at a constant rate, two current steps appear, corresponding with 2- and 4-electron reactions. This resolution of the steps is possible only because of the high hydrogen overpotential at mercury, and because of its very low catalytic activity for the decomposition of H_2O_2. According to working conditions (e.g. composition of electrolyte), the two steps may vary in relative height; oxygen arising from peroxide decomposition can accentuate the first step[157].

Recent investigations by Hamann[157a] have raised doubt as to the correctness of mechanism III.45. Hamann considered the initial step of oxygen reduction, represented by

$$O_2 + 2H^+ + 2e^- \rightarrow H_2O_2$$

or

$$O_2 + H_2O + 2e^- \rightarrow HO_2^- + OH^-$$

as a duplex redox-electrode reaction ('zweifache Elektrode'). In this case, as shown by Vetter[157b], the electrode process may be analysed into two coordinated charge-transfer steps, each involving one electron, with associated oxidation-exchange current density, $j_{0,ox}$, and reduction-exchange current density, $j_{0,red}$.

By means of galvanostatic current–potential curves at low and high overpotentials at both smooth and platinized platinum electrodes, it was established that, in acid or alkaline media, it is the first charge-transfer step that is rate-determining, i.e., it is always the case that $j_{0,ox} \ll j_{0,red}$. The exchange current density which is consequently of principal interest is $j_{0,ox}$, and its dependence on the pH of the electrolyte is illustrated in Figure III.60a. In contrast to these experimental curves, calculations in terms of equation III.45, using the theory

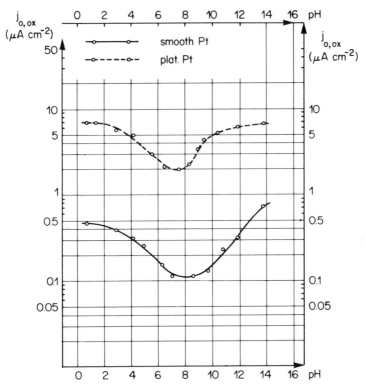

Figure III.60a. Dependence of the oxygen exchange current density, $j_{o,ox}$, at platinum on the pH of the electrolyte (H_2SO_4, K_2SO_4, KOH) at 25°C.

of duplex electrodes, predict an exchange current density increasing exponentially with pH across the whole pH range. It must be concluded that mechanism III.45 is incorrect for the acid region. On the other hand, all the measurements are well explained by the following reaction sequence proposed by Hamann[157a].

$$O_2 + H^+ \rightarrow HO_2^+ ; \qquad HO_2^+ + e^- \rightarrow HO_2$$
$$HO_2 + e^- \rightarrow HO_2^- ; \qquad HO_2^- + H^+ \rightarrow H_2O_2 \qquad \text{(III.45a)}$$

III.6.4   Catalysts

Catalysts for the evolution and reduction of oxygen have not yet been so extensively studied as the corresponding catalysts for hydrogen. Platinum metals and chromium are suitable as activators for oxygen reduction over the whole pH range. In alkaline conditions, iron, cobalt and nickel, in addition to silver and various kinds of carbon, have proved serviceable. The special properties of silver and platinum in this connexion depend in large part on their catalytic activity for the decomposition of H_2O_2.

Kordesch has specified the following *impregnating solution* for the activation of porous *carbon* electrodes for use as oxygen electrodes in alkaline solution[158]: 10g silver nitrate, 5g ferric nitrate, 1g cupric nitrate, 0.1g ammonium vanadate, 100g water. The use of mixtures of heavy-metal salts has proved to be advantageous—principally nitrates or nitrites, since these are easily decomposed on heating (300°C) the impregnated electrode.

For use at room temperature in alkaline electrolytes, sintered electrodes of *silver* or Raney-silver are very suitable. According to Friese, Justi and Winsel[159], stable Raney-silver electrodes can be prepared by mixing no more than 80 wt.% of Raney-silver with at least 20 wt% of supporting skeleton material (preferably silver). It is advantageous to include in the Raney-alloy up to 10 wt.% of manganese and/or 5 wt.% of chromium and/or 5 wt.% of molybdenum.

Electrodes which contain silver as catalytically active material can, according to Jung and Kröger[160], be considerably improved in properties by treatment with alkaline KCNO solution— up to 200 mV increase in potential at 1 to 50 mA cm^{-2}. After such activation the rest potential is only about 30 mV negative to the theoretical O_2, OH^- potential.

The International Nickel Company (Mond) has recently studied porous electrodes of sintered nickel containing 1% of palladium; it appears that these electrodes are very suitable for oxygen reduction in alkaline solution. Yeager and co-workers[161] have reported good results with a *lithium*-doped NiO electrode for oxygen, also in alkaline solution.

Certain kinds of active carbon, with no additions of metal, are intrinsically excellent catalysts for hydrogen peroxide decomposition and oxygen reduction. With adequate supply of oxygen, hydrophobic diffusion electrodes of active carbon can be loaded at 200 mA cm^{-2} at room temperature[161a]. According to Mrha[161b], the activity of active carbon is improved by the presence of nitrogen complexes bound to its surface; these are more stable than oxygen complexes and are formed by heating active carbon in ammonia at 850°C. Active carbon treated in this way, with addition of carbon black, polystyrene, crude rubber and grease, was formed into laminar electrodes of area 7 cm^2 [161c] that, at 25°C in 12N KOH, could be loaded at 40 mA cm^{-2} for more than 1000 hr (potential, -350 mV relative to SCE). An additional impregnation with silver led to attainment of current densities up to 200 mA cm^{-2} under the same working conditions.

For the catalysis of oxygen reduction in *acid* electrolytes, platinum metals take the lead, and, of these, platinum itself is preeminent.

The catalyst metal in finely divided form is used on an appropriate porous support (e.g. carbon). The introduction of porous membranes or diaphragms has already led to good results.

Platinized smooth metal electrodes are unsuitable, since their activity falls away rapidly and they are particularly vulnerable to poisoning. Trachuk and Voronin[162] have developed a sintered electrode of non-noble material for oxygen reduction over the whole pH range. These authors have made porous discs from chromium powder with 10–15% of ammonium bicarbonate added to increase porosity. The mixture was pressed at 1.2 ton cm^{-2} and sintered in hydrogen at 800°C. The porosity was 55–62%. The electrodes so made could pass current densities of up to 80 mA cm^{-2} in 1N NaOH at room temperature, but considerably less than this in sulphuric acid solutions.

For use at temperatures over 150°C, sintered electrodes of carbonyl nickel or nickel–lithium oxide are sufficiently active for the reduction of oxygen. A nickel–lithium oxide electrode was developed by Bacon[138]. The inclusion of lithium atoms in the semi-conducting nickel oxide lattice, on account of the influence of Li^+ on the nickel ions, leads to the formation of Ni^{3+} ions, with a consequent improvement of electrical properties. These electrodes will deal with current densities higher than $1\ A\ cm^{-2}$ in concentrated alkali solutions at 200–240°C.

The reaction of oxygen, at least at platinum, is not very susceptible to disturbance by other processes occurring at the electrode, provided that no liquid fuel finds its way into the electrolyte (the influence of hydrogen admixture has already been described in section III.1.5).

Figure III.61 shows a current–potential diagram, obtained by the triangular wave potential scan method (cf. section II.4), which clearly indicates that the formation and dissipation of the M—H layer on smooth platinum foil occur when oxygen is bubbled through the solution in very much the same way as when it is not (cf. the upper part of the diagram relating to the case of nitrogen-bubbling). The effect of oxygen appears to be simply a downward displacement of the diagram. In the hydrogen region the O_2-reduction current has assumed the value of the limiting diffusion current. This becomes evident if the rate of the potential sweep of the applied emf is greatly reduced (Figure III.62). Processes concerned with surface layers on the electrode then contribute only a small part to the total current. The current–potential curve for oxygen reduction becomes clearly predominant. Comparison with the corresponding diagram relating to the case of nitrogen-bubbling (upper curves in Figure III.62a), shows that part of the oxygen layer on the electrode must be destroyed during the cathodic return potential sweep before oxygen reduction can be reinstated (hysteresis in this part of the diagram).

There is an additional aspect of the catalysis of oxygen reduction which has to be considered in the case of fuel cells using *liquid fuels*. It is required that the catalyst to be used for oxygen reduction should catalyse this reduction alone, and not the oxidation of the fuel (e.g. alcohols, acids, aldehydes). If, for example, a platinum–carbon–oxygen electrode is used in an alkaline medium containing fuel, oxygen reduction and fuel reaction will occur simultaneously at the electrode, with the establishment of a mixed potential (see Figure III.63 curve II.b).

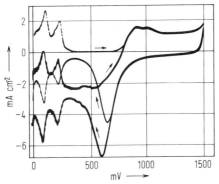

Figure III.61. Current–potential diagram for smooth platinum in 1 N H_2SO_4 (after Vielstich[91]). Upper: N_2-bubbling, lower: O_2-bubbling.

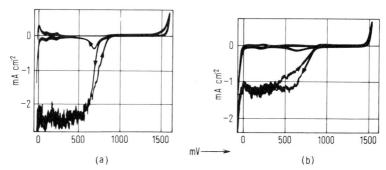

Figure III.62. Oxygen limiting diffusion currents (lower curves) and currents due to generation and removal of surface layers alone (upper curves, N_2-bubbling) at equal scan rates (30 mV sec^{-1}) (after Grüneberg[127]). (a) 2N H_2SO_4; (b) 2N KOH.

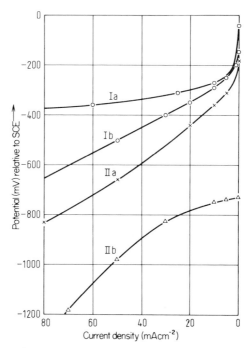

Figure III.63. Cathodic current–potential curves for a porous carbon-oxygen electrode at 20°C and p_{O_2} (total) = 1.9 atm (after Vielstich[17]). I, Pure 6N KOH; II, electrolyte 6N KOH + 3M glycol: a, metal catalyst after Kordesch (see p. 130); b, electrolytically platinized carbon.

The electrode activated with silver, iron and copper also shows rather unfavourable behaviour in the glycol–KOH mixture (compare curves Ia and IIa). If no suitably selective catalyst for oxygen reduction can be found, the problem can be met by *separation* of the oxygen electrode domain from that of the electrolyte-fuel mixture (see chapter IV). This can be done either by separating the electrode compartments from each other with a membrane impervious to the fuel, or by using as an electrolyte for the oxygen electrode an ion-exchange or asbestos membrane that simultaneously serves the purpose of keeping the fuel-electrolyte mixture away from the electrode[163].

In recent years experiments have been undertaken with the object of improving the catalytic properties of electrodes by incorporating *radioactive isotopes* in their construction. The experimental results obtained by Salcedo and Lang[46] with the hydrogen electrode have already been described (section III.1.6). The same authors have also impregnated sintered electrodes of silver with salts containing ^{63}Ni and ^{14}C, but these electrodes showed the normal characteristics of silver–oxygen electrodes—on the other hand, similar unimpregnated electrodes were unusually poor in performance. It is therefore yet unproven whether the reduction of overpotential is to be attributed to the influence of radiation, or to a difference of surface structure arising from the impregnation procedure. The results of Schwabe[164], involving use of ^{204}Tl (3 mc cm^{-2}) and ^{106}Ru (110 mc cm^{-2}), are similarly indecisive.

The rest potentials of isotope-containing electrodes are more positive, and therefore closer to the thermodynamic oxygen potential, than those of platinized platinum electrodes to which no such treatment has been given. These potential displacements can, however, be attributed to the formation of oxygen atoms or ozone by radioactive irradiation. Similarly, the displacement of potential–time curves (Figure III.64) at low current densities may be due to a radiolysis effect. It is suggestive, for example, that at 3.4 mA cm^{-2} curves for active and inactive electrodes come very nearly into coincidence.

More recent experiments by Schwabe and co-workers[165], in contrast to those mentioned above, have shown unambiguously that the activity of carbon–

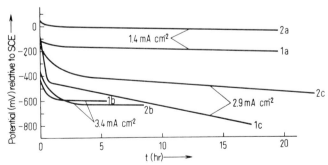

Figure III.64. Galvanostatic potential–time curves of a platinized platinum electrode, bubbled with oxygen, in 1N KOH at 20°C (after Schwabe[164]). 1, Inactive electrode; 2, 110 mc ^{106}Ru cm^{-2}.

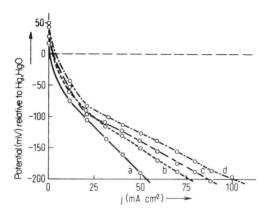

Figure III.65. The influence of reactor irradiation on current density–potential curves of carbon–oxygen electrodes (after Schwabe, Köpsel, Wiesener and Winkler[165]). 6N KOH, 20°C, p_{O_2} = 300 mm. a, Untreated electrode; b, electrode activated by heating from 300 to 1100°C in a gas stream (air, O_2, N_2, CO_2, CO or NH_3); c, as a, but irradiated for 100 hr; d, as b, but irradiated for 100 hr.

oxygen electrodes can be improved by *neutron irradiation*. Porous carbon electrodes, variously pretreated (Figure III.65, curves a and b) were subjected to irradiation for 100 hr in a reactor (neutron flux 9 . 10^{12} cm^{-2} sec^{-1}). The electrodes treated in this way showed, at 20 cm distance, an activity of 1 Roentgen hr^{-1}. At a given potential, the irradiated electrodes passed some 30% higher current density (curves c and d). The authors assumed that the irradiation produced especially energetic surface sites. Nothing is yet known about the permanence of the effect.

III.7   HYDROGEN PEROXIDE, NITRIC ACID, BROMINE AND CHLORINE AS OXIDANTS

Although, normally, an oxygen (air) electrode is used for the positive pole of a fuel cell in the interests of economy and of simple and safe operation, there are special cases in which use of an oxidizing agent in the liquid state, such as hydrogen peroxide, nitric acid or bromine solution, is advantageous. This, of course, introduces technical difficulties (see chapter IV), and this also applies to the use of chlorine gas owing to its corrosive and poisonous nature.

III.7.1   Hydrogen Peroxide

It is well known that hydrogen peroxide is unstable in aqueous solution (especially in alkaline electrolytes) decomposing into water and oxygen even at room temperature

$$H_2O_2 \rightarrow H_2O + \tfrac{1}{2}O_2$$

This decomposition is accelerated* at the surfaces of metals (platinum metals, silver) and also by active carbon. The formation of oxygen has the result that a mixed O_2–H_2O_2 potential is always established at electrodes (cf. section III.6).

In section III.6 it was shown that the cathodic reduction of O_2 to HO_2^- ions and the corresponding reverse reaction

$$O_2 + H_2O + 2e^- \rightleftharpoons HO_2^- + OH^- \tag{III.38}$$

are highly reversible. The electrochemical reduction of HO_2^- or H_2O_2 to OH^- or water is, however, irreversible and proceeds only at high overpotentials

$$H_2O_2 + 2H^+ + 2e^- \rightleftharpoons 2H_2O, \qquad \phi_0 = +1.77\,V \tag{III.42a}$$

$$HO_2^- + H_2O + 2e^- \rightleftharpoons 3OH^-, \qquad \phi_0 = +880\,mV \tag{III.42b}$$

Reaction of H_2O_2, in the sense of the second stage of the reduction of oxygen, therefore takes place mainly by the decomposition of H_2O_2, and the electrochemical reduction of the molecular or chemisorbed oxygen so formed (see section III.6). Thus, in alkaline solution

$$HO_2^- \rightarrow OH^- + \tfrac{1}{2}O_2$$

or

$$M-HO_2^- \rightarrow OH^- + M-O$$

According to the investigations of Jacq and Bloch[166], at platinized platinum in 1N KOH (25°C), at the relatively high overpotential of $-500\,mV$ (relative to $\phi^0_{H_2,H^+}$), (III.42b) is the preferred reaction. On the other hand, with smooth platinum under the same conditions, the cathodic current is almost exclusively due to the oxygen formed by decomposition of peroxide.

Figures III.66 and 67 show current–potential diagrams obtained by the triangular wave potential scan method at smooth platinum in electrolytes to which H_2O_2 (10^{-3}M) has been added. The diagrams indicate that:

1. The formation and removal of the hydrogen layer on the electrode occur independently of H_2O_2 addition (cf. Figure III.61).
2. At the equilibrium potential the current–potential curves intersect the zero potential axis very steeply. There is also a displacement upward of the anodic partial reaction in the oxygen region. It is clear that, even at potentials not far from the equilibrium potential, both anodic and cathodic currents become limited by diffusion. The magnitude of the limiting current is indicated by the effect of intensity of stirring (by N_2 bubbling)—for comparison, diagrams obtained with and without stirring are included in Figure III.66. As to be expected the contribution to the current attributable to oxidation or reduction of surface layers is not dependent on stirring.

---

* The decomposition of H_2O_2 is also catalysed by metal ions in homogeneous solution and by irradiation. A good summary is to be found in reference 38; see also chapter VII.

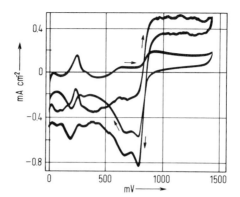

Figure III.66. Pt/2$_N$ KOH, current–potential diagram after addition of H_2O_2 (10^{-3}M), 60 mV sec^{-1}, 20°C, with and without stirring (after Vielstich[91]).

3. If a comparison is made of the location along the potential axis of the current maximum for the cathodic reduction of the chemisorbed oxygen layer on the platinum surface (Figure III.61 and 62), with the rise of the reduction current observed when H_2O_2 has been added, it is seen that in *alkaline* electrolytes (Figure III.66) reaction of H_2O_2 starts when the metal surface still carries a chemisorbed oxygen layer. In *acid* electrolytes, however, the oxygen layer must first be partially destroyed. This leads to a crossing of the current–potential curves for the anodic and cathodic potential sweeps. The same applies in the case of the reduction of molecular oxygen (Figure III.62).

A conclusion of practical importance can be drawn from these results. It is that a platinum surface covered with chemisorbed oxygen, inactive to a large number of fuels, can be used for the selective reduction of H_2O_2 in alkaline

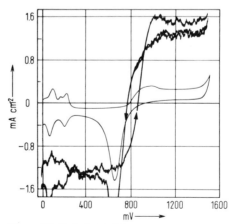

Figure III.67. Pt/2$_N$ H_2SO_4, current–potential diagram after addition of H_2O_2 (10^{-3}M), 60 mV sec^{-1}, 20°C, with and without stirring (after Grüneberg[127])

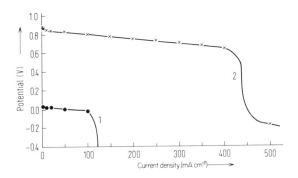

Figure III.68. Cathodic reduction of H_2O_2 at platinized platinum at 30°C (after Monsanto Research Company[134]). 1, 5N KOH, 1M H_2O_2; 2, 5M H_2SO_4, 1M H_2O_2. Potentials relative to standard hydrogen electrode.

solution—even when fuel is present in the electrolyte (see chapter IV, alcohol–H_2O_2 fuel cell).

In Figure III.68, current–potential curves for platinized platinum are plotted. It is striking that the limiting current density is greater in the case of 5M H_2SO_4 than that for 5N KOH. In alkaline solution the rest potential is more positive than expected from reaction III.38; this is due to the establishment of a mixed O_2–HO_2^- potential.

III.7.1.1   Analytical Estimation of H_2O_2 Concentration Near an Electrode

The concentration of H_2O_2 can be determined from the magnitude of the limiting diffusion currents, such as represented in Figures III.66 and 67. Use of the triangular wave potential scan method has a special advantage if there is a pronounced concentration gradient prevailing near the electrode, and it is only the concentration of H_2O_2 in the immediate vicinity of the electrode which is of interest (for example, in the formation of H_2O_2 as an intermediate in oxygen reduction, or in the radiolysis of water, see chapter VII).

The anodic part of the diagram serves best for quantitative estimation. In the upper part of Figure III.69, the anodic sections of curves obtained by adding 0, 3, 6, 9, 12 and 15 drops of H_2O_2 to 2N H_2SO_4 are shown. The current density of about 4 mA cm^{-2} for zero addition of H_2O_2 corresponds to that required for the formation of the chemisorbed oxygen layer. For the concentration range studied, there is a linear relation between diffusion current and H_2O_2 concentration (lower part of Figure III.69).

III.7.1.2   Enrichment of H_2O_2 in Electrolytes by Reduction of Oxygen

In the reduction of oxygen via the intermediate product H_2O_2 to H_2O or OH^- ions, the second reaction step is always slower than the first (see section III.6). The consequent enrichment of H_2O_2 which occurs depends on very many factors*, since H_2O_2 undergoes reaction not only by the electrochemical

* The current efficiency of H_2O_2 production depends not only on the electrode material but also on the concentrations of oxygen and H_2O_2 in the electrolyte, on temperature, current density, pH and composition of the electrolyte.

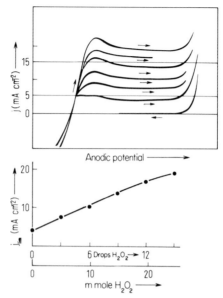

Figure III.69. Pt/H_2SO_4;   anodic   limiting
currents on addition of various quantities of
H_2O_2 (after Vielstich[91]).

routes of equations III.42a and III.42b, but also by homogeneous decomposition
in the electrolyte, or by heterogeneous reaction at the electrode (e.g. equations
III.45, p. 128).

Strongly alkaline conditions in solution, and raised temperature, favour the
homogeneous decomposition; platinum metals, silver and active carbon
catalyse the heterogeneous reaction especially strongly. In a hydrogen–oxygen
fuel cell with an alkaline electrolyte the H_2O_2 concentration is therefore always
less than 10^{-3} molar.

In experiments on the reduction of O_2 at various kinds of carbon in 1N KOH at 5°C, at 10 to
20 mA cm^{-2}, the following current efficiencies of H_2O_2 have been obtained (over 20 to 50 hours)[17]:

Ringsdorff carbon	70–40%*
Active carbon Hydraffin B	35–30%
Elorit DA	10– 7%
Elorit TD	3– 2%

Hydrogen peroxide concentrations of up to 0.5% were attained in this way.

The possibility of applying this effect to the *technical production* of H_2O_2 has
been investigated, principally by Berl[167]. By using specially developed carbon
electrodes in potassium hydroxide as electrolyte, H_2O_2 concentrations of 2 to
5% were achieved with current efficiency of 90%, and up to about 25% with

* First figure = initial current efficiency; second figure = efficiency at end of the experiment.

efficiency of 65%. Current density was about 20 mA cm^{-2}, and energy consumption was 7.5 kW hr per kg H_2O_2. This 'Berl Process' has not, however, been applied on the technical scale. The difficulties lie in neutralizing the electrolyte and in repressing the activity of the carbon electrodes for hydrogen peroxide decomposition (see above).

With *acid* electrolytes (1% H_2SO_4), Fischer and Priess[168] increased the yield of H_2O_2 by application of 100 atm pressure of O_2. Gold foil was used as cathode, platinum as anode. With 2 V applied voltage and 23 mA cm^{-2}, they attained 2.7% H_2O_2 concentration at a current efficiency of 83%.

Kozawa[169] has recently reported that presence of Ca^{2+}, Sr^{2+} or Ba^{2+} ions in 1N NaOH as electrolyte, markedly decreases the yield of H_2O_2; Cl^- and Br^- ions in 1N H_2SO_4 as electrolyte, increase the yield (cf. Figure III.59).

III.7.2   Nitric Acid as Oxidizing Agent

Nitric acid is a powerful oxidizing agent and because of the numerous possible chemical and electrochemical modes of reaction, equilibria and reaction products, its electrochemical behaviour may be very complex[122,170]. In the present context, therefore, only the facts most important in relation to operating conditions of fuel cells will be stated, and only a few conceivable reaction routes will be discussed.

With carbon or platinum electrodes, rest potentials lie between +1.0 and 1.2 volts. Even at room temperature, current densities of more than 100 mA cm^{-2} can be attained (Figures III.70 and 71; Table III.3).

The porous carbon cited in Table III.3*, used in the electrolyte indicated, operated for 57 days[134] at 20 mA cm^{-2}. In the first 10 days the electrode potential fell from 1.04 to 1.00 volt and then remained approximately constant. In the

Table III.3.   Nitric acid as oxidant

Electrode	Electrolyte	Temp. (°C)	Rest-potential (Volt)	mA cm^{-2} at $\eta = -300$ mV	Ref.
platin. Pt	2.5 M HNO_3 5.0 M NH_4HSO_4	30	1.17	10	134
platin. Pt	2.5 M HNO_3 5.0 M H_2SO_4	30	1.14	50	134
platin. Pt	2.5 M HNO_3 5.0 M $NaNO_3$	30	1.09	10	134
platin. Pt	10.0 M HNO_3 0.9 M $NaNO_3$	30	1.20	500	134
porous C.	10.0 M HNO_3 0.8 M $NaNO_3$	30	1.22	30	134
platin. C	1.0 M HNO_3 3.7 M H_2SO_4	82	1.17	140	171

* Porous carbon 'National Carbon grade 20', pore width 150 μ, heated to 700°C and quenched in distilled water.

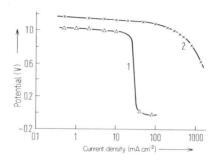

Figure III.70. Cathodic reduction of HNO_3 at platinized platinum at 30°C (after Monsanto Research Company[134]). 1, 2.5M HNO_3 + 5M $NaNO_3$; 2, 10M HNO_3 + 0.9M $NaNO_3$. Potentials relative to standard hydrogen electrode.

latter half of the period of experiment the gaseous reaction product had the following composition:

$$NO, 98.8\%; \qquad NO_2, 1.1\%; \qquad N_2O, 0.1\%.$$

In experiments conducted with platinized carbon at 82°C[171], NO also appeared as the main product of reaction. The experimental rest potentials and reaction products are to be compared with the following equilibrium potential derived by Latimer[122].

$$HNO_3 + 2H^+ + 2e^- \rightleftharpoons HNO_2 + H_2O, \qquad \phi_0 = +0.94 \text{ V}$$

$$HNO_3 + 3H^+ + 3e^- \rightleftharpoons NO + 2H_2O, \qquad \phi_0 = +0.96 \text{ V}$$

$$HNO_2 + H^+ + e^- \rightleftharpoons NO + H_2O, \qquad \phi_0 = +1.00 \text{ V} \qquad (III.46)$$

$$HNO_2 + 2H^+ + 2e^- \rightleftharpoons \tfrac{1}{2}N_2O + \tfrac{3}{2}H_2O, \qquad \phi_0 = +1.29 \text{ V}$$

HNO_2 can also be formed in a purely chemical reaction:

$$HNO_3 + 2NO + H_2O \rightarrow 3HNO_2 \qquad (III.47)$$

and the following equilibria are also to be considered:

$$HNO_3 + HNO_2 \rightleftharpoons N_2O_4 + H_2O$$

$$N_2O_4 \rightleftharpoons 2NO_2 \qquad (III.48)$$

$$2HNO_2 \rightleftharpoons NO + NO_2 + H_2O$$

Shropshire and Tarmy[171] have pointed out that HNO_3 can be regenerated by supplying the electrode with a stream of oxygen or air. It is assumed that NO

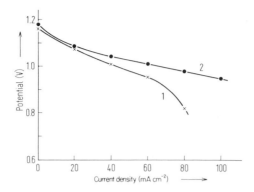

Figure III.71. Reduction of HNO_3 at platinized carbon at 80°C (after Shropshire and Tarmy[171]). 1, 0.2M HNO_3 + 3.7M H_2SO_4; 2, 1.0M HNO_3 + 3.7M H_2SO_4. Potentials relative to standard hydrogen electrode.

formed at the electrode reacts with molecular oxygen according to

$$2\,NO + O_2 \rightarrow 2NO_2$$

$$3NO_2 + H_2O \rightarrow 2HNO_3 + NO \quad \text{or} \qquad\qquad \text{(III.49)}$$

$$NO + NO_2 + H_2O \rightarrow 2HNO_2$$

It is possible to increase the faradays obtained per mole of HNO_3 in this way; under optimum conditions the factor attained with oxygen was 225 (with air 30)[171].

III.7.3   The Bromine–Bromide Electrode

Bromine is also a strong oxidizing agent. Since molecular bromine is readily soluble in water and establishes with bromide ion a highly reversible potential, the two provide, even at room temperature, a redox electrode of low polarizability, i.e., one which will pass high faradaic current densities with relatively small deviations of potential from the equilibrium value:

$$Br_2\,(aq) + 2e^- \rightleftharpoons 2Br^-, \qquad \phi_0 = +1.066 \text{ volt}$$

The value of 1.066 volt relates to the saturation concentration $c = 0.2125$ mole of bromine per litre, with a partial vapour pressure of bromine, $p_{Br_2} = 211$ Torr at 25°C.

The *platinum* metals are specially suitable as catalyst materials. Chang and Wick[172] have measured anodic and cathodic current–potential curves for smooth platinum and iridium electrodes in bromine-saturated 1M KBr solution.

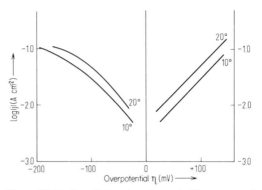

Figure III.72. Br_2, Br^- redox electrode (Br_2 saturated
1M KBr), logarithm of current density, j, as function of
overpotential, η, at a rotating electrode of smooth Ir at
10° and 20°C (after Chang and Wick[172]).

At platinum, except for the ohmic potential drop in the electrolyte, there is no
detectable overpotential at current densities up to 100 mA cm^{-2}. Measurements with an iridium electrode gave a log (current density)–overpotential
diagram (Figure III.72) which includes an anodic Tafel line consistent with
$\alpha = 0.58$. On the cathodic side of the diagram there are small deviations from
Tafel behaviour which Loshkarev and Esin[172] have attributed to the anodic
back-reaction. The reaction mechanism has not yet been investigated in detail.

If the oxygen electrode of an acid hydrogen–oxygen cell is replaced by a
bromine–bromide redox electrode, there is an improvement in characteristics.
It is, however, essential to separate the cathode and anode compartments of
the cell effectively. Membrane cells are specially suitable for this purpose (see
chapter IV).

III.7.4   The Chlorine–Chloride Electrode

A platinum foil electrode, bubbled with chlorine gas in a chloride solution,
records a reversible, thermodynamic potential as satisfactorily as the hydrogen
electrode (Figure III.73)

$$Cl_2 + 2e^- \rightleftharpoons 2\,Cl^-, \qquad \phi_0 = +1.359 \text{ volt*}$$

The reaction does not proceed in *alkaline* electrolytes because the dissolved
Cl_2 is hydrolysed to hypochlorite:

$$Cl_2 + H_2O \rightarrow Cl^- + H^+ + HOCl$$

Investigations of the *mechanism* of this redox electrode (in which the electrode
metal is involved only in exchange of electrons) have been carried out by

* Cl_2 partial pressure of 1 atm corresponds with 0.09M Cl_2 in solution.

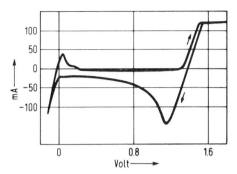

Figure III.73. Current–potential diagram at smooth platinum in 1_N HCl (after Kutschker, Diplomarbeit, Bonn, 1962), 0.57 volt sec^{-1}; area, $0.66 \, cm^2$, anodic 'limiting current' is an artefact.

Frumkin and Tedoradze[173], by using a platinum rotating disc electrode, with the following results:

1. At sufficient cathodic overpotential ($\eta > 60 \, mV$), the charge transfer current is proportional to the Cl_2 concentration in the electrolyte, i.e. there is a reaction of first order with respect to Cl_2.
2. The cathodic current–potential curve (*not*, however, the rest potential and the overpotential) is practically independent of the Cl^- concentration (Figure III.74). The Tafel slope corresponds with a transfer coefficient of $(1 - \alpha) = 0.69$.

Thus, sufficiently far from the rest potential, Cl_2 reduction proceeds irreversibly in accordance with the relation

$$j = 2k_1[Cl_2](1 - \theta) e^{-(1 - \alpha)\phi F/RT} \qquad (III.50)$$

where k_1 is a rate constant expressed in electrical units, and θ is the fraction of the electrode surface covered with adsorbed chlorine atoms. It must be assumed that θ is so small that variation in θ does not disturb the first-order relation of reaction rate to Cl_2 concentration.

From their measurements with platinum electrodes, Frumkin and Tedoradze concluded that

$$Cl_2 + e^- \rightleftharpoons Cl_{ad} + Cl^- \qquad (III.51a)$$

is the first and rate-controlling step. The following reaction step is assumed to be

$$Cl_{ad} + e^- \rightleftharpoons Cl^- \qquad (III.51b)$$

The cathodic reduction of chlorine, however, is not confined to noble metals used as electrode material. Figure III.75 shows current–potential curves obtained with porous carbon electrodes copiously supplied with chlorine gas[174]. The overpotential is least in 5_N HCl.

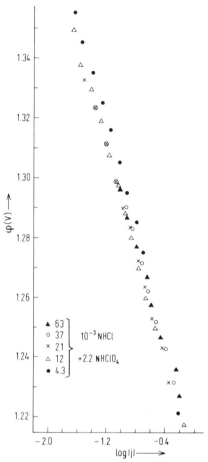

Figure III.74. Dependence of the logarithm of current density on electrode potential, ϕ, for cathodic Cl_2 reduction at bright platinum, in presence of various Cl^- concentrations, at 25°C (after Frumkin and Tedoradze[173]). Potentials relative to standard hydrogen electrode.

III.7.5   Halogen Compounds as Oxidants in Solution for Cells with Alkaline Electrolytes

Aqueous solutions of a series of halogen compounds have been investigated by Boies and Dravnieks[175] in relation to their suitability as electrochemical oxidizing agents. Table III.4 displays the equations for the reactions and the related standard potentials according to Latimer:

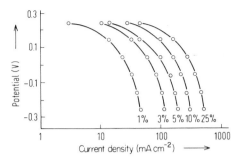

Figure III.76. $NaClO_2$ as oxidant; current–potential curves with a porous silver electrode in $5N$ KOH at 25°C, in relation to concentration of $NaClO_2$ (after Boies and Dravnieks[175]). Potentials relative to standard hydrogen electrode.

Table III.6. Rest potential and current–potential data for chlorite electrodes at 23°C, 10% $NaClO_2$ in $5N$ KOH (after Boies and Dravneiks[175]). Potentials relative to the standard hydrogen electrode.

Kind of electrode	Flame-sprayed Raney-Ag–Ni	Flame-sprayed silver	Platinized flame-sprayed Raney-Ni	Platinized platinum	Carbon
Rest potential (V)	0.26	0.27	0.30	0.30	0.10
Potential (V)	Current density (mA cm^{-2})				
+0.20	35	48	1.1	3	
+0.10	92	118	6	13	
0.00	170	172	25	50	
−0.10	280	215	90	104	0.03
−0.20		245		200	0.08

REFERENCES

1. J. Bockris and E. Potter, *J. Chem. Phys.*, **20**, 614 (1952); *J. Electrochem. Soc.*, **99**, 169 (1952).
2. K. J. Vetter, *Z. Elektrochem.*, **59**, 435 (1955); see also K. J. Vetter, *Electrochemical Kinetics*, Academic Press, New York, 1967.
3. M. Breiter and R. Clamroth, *Z. Elektrochem.*, **58**, 493 (1954).
4. K. J. Vetter and D. Otto, *Z. Elektrochem.*, **60**, 1072 (1956).
5. M. Breiter, C. A. Knorr and R. Meggle, *Z. Elektrochem.*, **59**, 153 (1955).
6. A. N. Frumkin and E. A. Aikazyan, *Izvest. Akad. Nauk SSSR, Otdel. Khim. Nauk*, p. 202, (1959).
7. W. Vielstich, *Chem.-Ing.-Techn.*, **33**, 75 (1961).
8. M. D. Zholudev and V. V. Stender, *Zh. priklad. Khim.*, **31**, 719 (1958).

9. E. Joachim, B. Struck and W. Vielstich, unpublished; H. Lauer, Diplomarbeit, Bonn, 1961.
10. A. N. Frumkin, *Z. Physik. Chem.* (*Leipzig*), **207**, 321 (1957); also cf. A. N. Frumkin, *Advances in Electrochemistry and Electrochemical Engineering*, Vol. 3, John Wiley, New York, 1963.
11. F. G. Will, *J. Electrochem. Soc.*, **110**, 145 (1963).
12. G. Grüneberg, Diplomarbeit, Braunschweig, 1956.
13. A. N. Frumkin and A. Shlygin, *Acta Physicochim. URSS*, **3**, 791 (1935); **5**, 819 (1936).
14. A. Obrucheva, unpublished experiment.
15. P. Dolin and B. Ershler, *Acta Physiocochim. URSS*, **13**, 747 (1940); P. Dolin, B. Ershler and A. Frumkin, *Acta Physiocochim. URSS*, **13**, 779 (1940); K. Rosental, P. Dolin and B. Ershler, *Zh. Fiz. Khim.*, **19**, 601 (1945).
16. M. Tahlinger and M. Volmer, *Z. Physik. Chem.*, Pt. A, **150**, 401 (1930).
17. W. Vielstich, unpublished.
18. A. Eucken, *Lehrbuch der Chemischen Physik*, Vol. II, 2, p. 1234, 1243ff.
18a. C. E. Heath and W. I. Sweeney, in W. Mitchell, *Fuel Cells*, Academic Press, New York, 1963, p. 102.
19. A. Podvyazkin and A. I. Shlygin, *Zh. Fiz. Khim.*, **31**, 1305 (1957).
20. M. Pilkuhn and A. Winsel, *Z. Elektrochem.*, **63**, 1056 (1959).
21. F. C. Tompkins, private communication.
22. A. Travers and J. Aubry, *Atti del X° Congresso Internationale di Chimica*, Rome, 1938.
23. W. Vielstich, *Z. Instrumentenk.*, **67**, 154 (1959).
24. D. V. Sokolsky and S. R. Omarova, *Dokl. Akad. Nauk SSSR*, **102**, 977 (1955).
25. D. V. Sokolsky and K. K. Dzhardamalieva, *Dokl. Akad. Nauk SSSR*, **113**, 860 (1957), Engl. Transl. p. 231.
26. E. Justi and A. Winsel, *Ger. Pat.* 1019361 (23.10.1954).
27. L. Horner, *Dechema-Arbeitsausschuss*, 'Elektrochemische Prozesse', Frankfurt, November, 1960.
28. R. Paul, P. Buisson and N. Joseph, *Ind. Eng. Chem.*, **44**, 100 (1952); H. I. Schlesinger, *U.S. Pat.* 2461611 (9.1.1945).
29. I. P. Tverdovski and I. F. Tupitsyn, *Problemy Kinetiki i Kataliza, Akad. Nauk, SSSR, Inst. Fiz. Khim.* Soveshch., Moscow, 1956, **9**, 84–90 (Publ. 1957).
30. R. Jasinski, *Amer. Chem. Soc.*, Fuel Cell Symp., New York, 1963, p. 95 (Publ. 1965).
31. M. Jung and H. Kröger, *Ger. Pat. Appl.* 1156768 (17.12.1960).
32. F. G. Will and C. A. Knorr, *Z. Elektrochem.*, **64**, 270 (1960).
33. F. S. Feates, *Trans. Faraday Soc.*, **56**, 167 (1960).
33a. P. G. Grimes, J. N. Murray and H. M. Spengler, *The Electrochem. Soc. Meeting*, Washington, 1964, Ext. Abstr. No. 22.
34. K. Bennewitz and W. Neumann, *Z. Physik. Chem.* Pt. A, **164**, 277 (1933).
35. K. I. Rozental and V. I. Veselovsky, *Zh. Fiz. Khim.*, **31**, 1555 (1957).
36. K. I. Rozental and V. I. Veselovsky, *Zh. Fiz. Khim.*, **35**, 2256 (1961).
37. K. I. Rozental and V. I. Veselovsky, *Zh. Fiz. Khim.*, **35**, 2481 (1961).
37a. B. D. Struck, Dissertation, Bonn, 1966.
38. W. C. Schumb, C. N. Satterfield and R. L. Wentworth, *Hydrogen Peroxide*, Reinhold Publishers Co., New York, 1955.
39. G. Braun, *Ger. Pat.* 232005 (1909).
40. C. E. Crossley, *Brit. Pat.* 220422 (1923).
41. G. L. Clark, P. C. McGrath and M. C. Johnson, *Nat. Acad. Sci., Nat. Res. Council, Publ.*, **11**, 646 (1925).
42. G. J. Harker, *Soc. Chem. Ind. Victoria* (*Proc.*), **51**, 314 (1932).
43. P. H. Emmet and E. J. Jones, *J. Phys. Chem.*, **34**, 1102 (1930).

44. G. M. Schwab, R. Sizmann and N. Todo, *Z. Naturforsch.*, **16a**, 985 (1961).
45. R. W. Clarke and E. J. Gibson, *Nature (London)*, **180**, 140 (1957).
46. R. Salcedo and M. Lang, *The Electrochem. Soc. Meeting*, Detroit, 1961, Ext. Abstr. No. 30.
47. cf. E. Justi and A. Winsel, *Fuel Cells—Kalte Verbrennung*, Steiner, Wiesbaden, 1962, p. 50ff.
48. W. M. H. Sachtler and L. L. van Reijen, *Shokubai (Tokyo)*, **4**, 147 (1962).
49. G. C. Bond, *Catalysis by Metals*, Academic Press, 1962, p. 149ff; O. Beeck, J. W. Givens and A. W. Ritchie, *J. Colloid Sci.*, **5**, 141 (1950).
50. B. M. W. Trapnell, *Chemisorption*, Butterworth, London, 1955, p. 216.
51. F. Sweet and Sir Eric Rideal, *Actes du Deuxième Congrès International de Catalyse*, Paris, p. 175 (1960).
52. H. Zimmermann, *Angew. Chem.*, **76**, 1 (1964).
53. W. Eichenauer, *Mém. Sci. Rev. Mét.*, **57**, No. 12, 943 (1960).
54. D. P. Smith, *Hydrogen in Metals*, University of Chicago Press, 1948.
55. G. G. Libowitz, *J. Nucl. Mater. (Amsterdam)*, 2, 1 (1960).
56. J. B. Hunter, *Platinum Metals Rev.*, **4**, 130 (1960).
56a. H. Connor, *Platinum Metals Rev.*, **6**, 130 (1960).
56b. A. Küssner, *Z. Naturforsch.*, **21a**, 515 (1966).
57. T. J. Tiedema, C. Kooy and W. G. Burgers, *Proc. Koninkl. Ned. Akad. Wetenschap.*, Ser. B, **62**, 34 (1959).
58. M. v. Stackelberg and P. Ludwig, *Z. Naturforsch.*, **19a**, 93 (1964).
58a. E. Wicke and K. Otto, *Z. Physik. Chem.*, *(Frankfurt)*, **31**, 222 (1962).
58b. N. A. Scholtus and W. Keith Hall, *J. Chem. Phys.*, **39**, 868 (1963).
59. G. Rosenhall, *Ann. Physik.* (5), **24**, 297 (1935).
60. H. Brodowsky and E. Poeschel, *Z. Phys. Chem. (Frankfurt)*, **44**, 143 (1965).
61. E. Wicke and G. J. Nerst, *Ber. Bunsenges. Phys. Chemie*, **68**, 224 (1964).
62. W. Jost and A. Widmann, *Z. Physik. Chem.*, Pt. B, **29**, 247 (1935); Pt. B, **45**, 285 (1940). P. S. Perminow, A. A. Orlow and A. N. Frumkin, *Dokl. Akad. Nauk SSSR*, **84**, 749 (1952).
63. T. B. Flanagan and F. A. Lewis, *Trans. Faraday Soc.*, **55**, 1400 (1959); *J. Electrochem. Soc.*, **108**, 473 (1961).
63a. L. Lederer and N. D. Greene, *Elektrochim. Acta (London)*, **8**, 883 (1963).
64. R. Ulbrich, *Z. Physik*, **121**, 351 (1943).
65. F. A. Lewis and A. Ulbrich, *Das Zweistoffsystem Gas-Metal*, Munich, 1950, p. 110.
66. A. Küssner, *Z. Elektrochem.*, **66**, 675 (1962).
67. H. O. v. Samson-Himmelstjerna, *Z. Anorg. Chem.*, **186**, 351 (1930).
68. T. J. Tiedema, B. C. de Jong and W. G. Burgers, *Proc. Koninkl. Ned. Akad. Wetenschap.*, Ser. B, **63**, 422 (1960).
69. E. Wicke, A. Küssner and K. Otto, *Actes du Deuxième Congrès International de Catalyse*, Paris, 1960, 1035.
E. Wicke and A. Küssner, *Ger. Pat. Appl.* 1148981 (29.3.1961).
71. H. G. Oswin and S. M. Chodosh, *Amer. Chem. Soc.*, Fuel Cell Symp., New York, 1963, p. 61 (Publ. 1965).
71a. F. v. Sturm and H. Kohlmüller, unpublished.
72. H. B. Wahlin, *J. Appl. Phys.*, **22**, 1503 (1951).
73. A. S. Darling, *Platinum Metals Rev.*, **7**, 126 (1963).
74. A. Küssner, *Z. Physik. Chem.*, *(Frankfurt)*, **36**, 383 (1963).
75. E. Wicke and G. Bohmholdt, *Z. Phys. Chem.*, *(Frankfurt)*, **42**, 715 (1964).
76. G. V. Elmore and H. A. Tanner, *J. Electrochem. Soc.*, **108**, 669 (1961).
77. A. Küssner and E. Wicke, *Z. Phys. Chem.*, *(Frankfurt)*, **24**, 152 (1960).
77a. A. Küssner, *Journées Intern. Étud. Piles Combust.*, Brussels, 1965, Report 2, p. 116.

77b. S. D. Axelrod and A. C. Makrides, *J. Phys. Chem.*, **68**, 2154 (1964).
77c. Y. Yamamoto and R. Goto, *French Pat.* 1 320 481 (20.4.1962).
78. E. Müller, *Z. Elektrochem.*, **28**, 101 (1928).
79. E. Müller and S. Tanaka, *Z. Elektrochem.*, **34**, 704 (1928).
80. E. Müller and S. Takegami, *Z. Elektrochem.*, **34**, 704 (1928).
80a. T. O. Pavela, *Ann. Acad. Sci. Fennical.*, *A*, *2*, **59**, 7 (1954).
81. J. F. Yeager, *The Electrochem. Soc. Meeting*, Indianapolis, 1961, Ext. Abstr. No. 109.
82. P. G. Grimes, B. Fielder and J. Adam, 15*th Power Sources Conf.*, Atlantic City, 1961.
83. C. A. Bogdanovski and A. I. Shlygin, *Zh. Fiz. Khim.*, **33**, 1769 (1959); **34**, 57 (1960).
84. W. Vielstich, *The Electrochem. Soc. Meeting*, Indianapolis, 1961, Ext. Abstr. No. 113.
85. O. Bloch, M. Prigent and J. C. Balaceanu, *The Electrochem. Soc. Meeting*, Indianapolis, 1961, Ext. Abstr. No. 116.
86. R. P. Buck and I. R. Griffith, *J. Electrochem. Soc.*, **109**, 1005 (1962); M. J. Schlatter, *Symp. Amer. Chem. Soc.*, Chicago, 1961, B 149.
87. M. Breiter and S. Gilman, *J. Electrochem. Soc.*, **109**, 622 and 1099 (1962).
88. J. E. Oxley, G. K. Johnson and B. T. Buzalski, *Elektrochim. Acta* (*London*), **9**, 897 (1964): *CITCE-Meeting Moscow*, 1963, No. 2, 17 (also cf. 89).
89. A. N. Frumkin and B. I. Podlovchenko, *Dokl. Akad. Nauk SSSR*, **150**, 349 (1963).
90. V. S. Bagotzky and Y. B. Vasiljev, *Electrochim. Acta* (*London*), **9**, 869 (1964).
90a. A. N. Frumkin, in D. H. Collins, *Batteries 2, Proc. 4th. Int. Symp.*, Brighton, 1964; Pergamon Press, 1965, p. 537.
91. W. Vielstich, *Z. Instrumentenk.*, **71**, 29 (1963).
92. W. Vielstich, *Chem.-Ing.-Techn.*, **35**, 362 (1963).
92a. G. Grüneberg, Dissertation, Braunschweig, 1958.
93. A. Hoffmann and A. T. Kuhn, *Electrochim. Acta* (*London*), **9**, 835 (1964).
94. S. B. Brummer and A. C. Markides, *J. Phys. Chem.*, **68**, 1448 (1964).
95. J. Giner, *Electrochim. Acta* (*London*), **8**, 857 (1963); **9**, 63 (1964).
96. J. Giner, *Z. Elektrochem.*, **63**, 386 (1959).
97. A. Kutschker and W. Vielstich, *Electrochim. Acta* (*London*), **8**, 985 (1963); cf. J. Giner, *Electrochim. Acta* (*London*), **9**, 63 (1964).
98. W. Vielstich and U. Vogel, *Ber. Bunsenges. Physik. Chem.*, **68**, 688 (1964).
98a. P. R. Johnson and A. T. Kuhn, *J. Electrochem. Soc.*, **112**, 599 (1965).
98b. D. R. Rhodes and E. F. Steigelmann, *J. Electrochem. Soc.*, **112**, 16 (1965).
98c. W. Vielstich and U. Vogel, *J. Electrochem. Soc.*, **113**, 628 (1966).
99. J. Giner, *Electrochim. Acta* (*London*), **4**, 42 (1961).
100. M. Breiter, *Electrochim. Acta* (*London*), **9**, 827 (1964).
101. S. Gilman, *Gen. Electr. Res. Lab. Rep.*, No. 63-RL-3400C.
101a. W. Vielstich and D. Kudlorz, unpublished.
101b. H. F. Hunger, *Proc. Journ. Intern. Étude Piles Combust.*, Brussels, 1965, Ed. Serai, Brussels, 1966, Vol. 4, p. 87.
102. H. Krupp, R. Mc. Jones, H. Rabenhorst, G. Sandstede and G. Walter, *J. Electrochem. Soc.*, **109**, 553 (1962); G. Sandstede, *Brennstoffzellen für die elektrochemische Verbrennung*, Battelle Institut, Frankfurt, 1964, p. 55.
103. H. Spengler and G. Grüneberg, *Dechema Monogr.*, **38**, 579 (1960).
104. E. A. Aikasjan and Y. V. Plesko, *Zh. Fiz. Khim.*, **31**, 205 (1957).
105. G. Armstrong and J. A. V. Butler, *Discuss. Faraday Soc.*, **1**, 122 (1947).
106. D. B. Boies and A. Dravnieks, *Electrochem. Techn.*, **2**, 351 (1964).
106a. J. O'M. Bockris and H. Wroblowa, *J. Electroanal. Chem.*, **7**, 428 (1964).

106b.   A. N. Frumkin in D. H. Collins, *Batteries, Proc. 4th Intern. Symp. Batteries*, Brighton, Pergamon Press, London, 1965, p. 537.

106c.   O. A. Petry, B. I. Podlovchenko, A. N. Frumkin and H. Lal, *J. Electroanal. Chem.*, **10**, 253 (1965).

106d.   W. Vielstich in B. S. Baker, *Hydrocarbon Fuel Cell Technology*, Academic Press, New York, London, 1965, p. 79.

106e.   H. Binder, A. Köhling and G. Sandstede in B. S. Baker, *Hydrocarbon Fuel Cell Technology*, Academic Press, New York, London, 1965, p. 91.

106f.   C. E. Heath, *Proc. Journ. Étud. Piles Combust.*, Brussels, Ed. Serai, Brussels, 1966, Vol. 3, p. 99.

106g.   P. G. Grimes and H. M. Spengler, *Electrochem. Soc. Meeting*, Washington, 1965, Battery Div. Ext. Abstr. No. 27; in full in B. S. Baker, *Hydrocarbon Fuel Cell Technology*, Academic Press, New York, London, 1965, p. 121.

106h.   M. W. Breiter, *Gen. Electr. Res. Lab. Rep.* No. 65-RL-3927M (1965).

106i.   H. Schmidt and W. Vielstich, *Z. Anal. Chem.*, **224**, 84 (1967).

107.   H. Hoyer, *Z. Naturforsch.*, **4a**, 355 (1949).

108.   W. Vielstich, *Habilitationsschrift*, Bonn, 1962; *The Electrochem. Soc. Meeting*, Los Angeles, 1962.

109.   P. Hersch, personal communication.

110.   H. Wieland and A. Wingler, *Liebigs Ann. Chem.*, **1923**, 431.

111.   K. Schwabe, *Z. Elektrochem.*, **61**, 744 (1957).

112.   D. Jahn, Diplomarbeit, Bonn, 1958.

113.   W. M. H. Sachtler, *Meeting Bunsenges*, Bonn 1960, Paper No. 76.

114.   M. Prigent, O. Bloch and J. Balaceanu, *Bull. Institut Francais du Pétrole*, February 1962; *Bull. Soc. Chim. France*, **368** (1963).

115.   K. A. Hofmann, *Ber. Dtsch. Chem. Ges.*, **51**, 1562 (1918); **52**, 1185 (1919); **53**, 914 (1920).

115a.   H. Binder, A. Köhling and G. Sandstede, *Electrochem. Soc. Meeting*, Washington, 1964, Ext. Abstr. No. 11; G. Sandstede, *Chem. Ing. Techn.*, **37**, 632 (1965).

115b.   S. Gilman, *J. Phys. Chem.*, **66**, 2657 (1962); **67**, 78 (1963); **67**, 1898 (1963); **68**, 70 (1964).

115c.   A. B. Fasman, G. L. Padyakova and D. V. Sokolsky, *Dokl. Akad. Nauk SSSR*, **150**, 856 (1963).

115d.   H. S. Taylor and P. V. McKinney, *J. Amer. Chem. Soc.*, **53**, 3604 (1931).

115e.   L. W. Niedrach and I. B. Weinstock, *Electrochem. Techn.*, **3**, 270 (1965).

115f.   P. Stonehart, *Proc. 5th Intern. Symp. Batteries*, Brighton, 1966, paper No. 24.

116.   L. W. Niedrach, *J. Electrochem. Soc.*, **109**, 1093 (1962).

116a.   L. W. Niedrach, *J. Electrochem. Soc.*, **111**, 1309 (1964).

117.   H. Binder, A. Köhling, H. Krupp, K. Richter and G. Sandstede, *The Electrochem. Soc. Meeting*, Toronto, 1964.

118.   A. B. Fasman, G. L. Padynkova and D. V. Sokolsky, *Dokl. Akad. Nauk SSSR*, **150**, 856 (1963).

119.   M. J. Schlatter, *Amer. Chem. Soc.*, Fuel Cell Symp., New York, 1963, p. 234, cf. R. R. Paxton, J. F. Demendi, G. J. Young and R. B. Rozelle, *J. Electrochem. Soc.*, **110**, 932 (1963).

120.   A. Köhling, H. Krupp, K. Richter, G. Sandstede and G. Walter, *DAS* 1 172 650 (7.9.1962).

121.   J. O'M Bockris, J. W. Johnson and H. Wroblova, *J. Electrochem. Soc.*, **111**, 863 (1964); H. Dahms and J. O'M. Bockris, *J. Electrochem. Soc.*, **111**, 728 (1964); M. Green, J. Weber and V. Drazic, *J. Electrochem. Soc.*, **111**, 721 (1964).

121a.   J. O'M. Bockris and S. Srinivasan, *Proc. Journ. Intern. Étud. Piles Combust.* Brussels, 1965, Ed. Serai, Brussels, 1966, Vol. 2, p. 68.

121b. R. Thacker and D. D. Bump, *Electrochem. Techn.*, **3**, 9 (1965).

121c. M. Fukuda, C. L. Rulfs and P. J. Eloing, *Electrochim. Acta*, **9**, 1551 (1964); *Electrochim. Acta*, **9**, 1563 (1964); *Electrochim. Acta*, **9**, 1581 (1964).

121d. L. W. Niedrach, *J. Electrochem. Soc.*, **111**, 1309 (1964).

121e. V. R. José and V. D. Long, *J. Appl. Chem. (London)*, **14**, 64 (1964).

121f. H. Binder, A. Köhling and G. Sandstede, *Proc. Journ. Intern. Étud. Piles Combust.*, Brussels, 1965, Ed. Serai, Brussels, 1966, Vol. 1, p. 74.

121g. S. Gilman, in B. S. Baker, *Hydrocarbon Fuel Cell Technology*, Academic Press, New York, London, 1965, p. 349.

121h. L. W. Niedrach, in B. S. Baker, *Hydrocarbon Fuel Cell Technology*, Academic Press, New York, London, 1965, p. 377.

121i. R. Thacker, in B. S. Baker, *Hydrocarbon Fuel Cell Technology*, Academic Press, New York, London, 1965, p. 525.

122. W. M. Latimer, *Oxidation Potentials*, Prentice-Hall, New York, 1956.

123. R. Glicksman, *J. Electrochem. Soc.*, **108**, 922 (1961).

124. G. Grüneberg, private communication.

125. G. Susbielles and O. Bloch, *Bull. Inst. Français du Pétrole*, Ref. 7512 (1962).

126. cf. Ref. 47, p. 41–42.

127. G. Grüneberg, unpublished.

128. H. I. Schlesinger and others, *J. Amer. Chem. Soc.*, **75**, 215 (1953); H. C. Brown and A. C. Boyd, *Anal. Chem.*, **27**, 156 (1955); G. W. Schaefer and Sister M. Emilius, *J. Amer. Chem. Soc.*, **76**, 1203 (1954).

129. M. E. Indig and R. N. Snyder, *J. Electrochem. Soc.*, **109**, 1104 (1962); **110**, 591 (1963).

130. J. P. Elder and A. Hickling, *Trans. Faraday Soc.*, **58**, 1852 (1962).

131. R. Jasinski, *The Electrochem. Soc. Meeting*, Pittsburg, 1963, Abstr. No. 29.

132. R. L. Pecsok, *J. Amer. Chem. Soc.*, **76**, 1203 (1954).

133. W. H. Stockmayer, D. W. Rice and C. C. Stephenson, *J. Amer. Chem. Soc.*, **77**, 1980 (1955).

134. *Monsanto Res. Co.*, USAF Contract No. AF 33(616)-7735 Report ASD-TOR-62-42 (1962).

135. G. Frick, W. J. Subcasky and M. Shaw, *The Electrochem. Soc. Meeting*, Pittsburg, 1963, Ext. Abstr. No. 183.

135a. R. W. Wynveen, *Symp. Amer. Chem. Soc.*, Chicago, 1961, B 49; W. E. Elliott and R. A. Wynveen, *The Electrochem. Soc. Meeting*, Los Angeles, 1962.

136. R. Lorenz and H. Hauser, *Z. Anorg. Allgem. Chem.*, **51**, 81 (1906).

137. F. Haber and others, *Z. Anorg. Allgem. Chem.*, **51**, 245 (1906).

138. F. T. Bacon, *Ind. Eng. Chem.*, **52**, 301 (1960); *Brit. Pat.* 667 298 (8.6.1949) 725 661 (1.1.1954).

139. W. G. Berl, *Trans. Electrochem. Soc.*, **83**, 253 (1943).

140. K. Kordesch and F. Martinola, *Mh. Chem.*, **84**, 39 (1943).

141. G. N. Lewi and M. Randall, *Thermodynamics*, New York, 1923.

142. W. Vielsticn, *Z. Physik. Chem., (Frankfurt)*, 15, 409 (1958).

143. J. Weiss, *Trans. Faraday Soc.*, **31**, 1547 (1935).

144. A. N. Frumkin, *Vosprosy Khim. Kinetiki, Kataliza i Reaktsionoi Sposobnosti*, (Akad. Nauk SSSR, 1955, p. 402–19.

145. E. Yeager and co-workers, Office of Naval Res., Techn. Rep. No. 4 (1954).

146. F. Kornfeil, Diss. University, Vienna, 1952; H. Hunger, Diss. University, Vienna, 1954.

147. M. Dittmann, E. Justi and A. Winsel, *Symp. Amer. Chem. Soc.*, Chicago, 1961, B 139. E. Justi and A. Winsel, *Kalte Verbrennung—Fuel Cells*, Steiner Verlag, Wiesbaden, 1962.

147a. J. Mrha and W. Vielstich, *Z. Instrumentenk.*, **74**, 235 (1966).
147b. E. Yeager, P. Krouse and K. V. Rao, *Electrochim. Acta*, **9**, 1057 (1966).
147c. Y. A. Mazitov, K. I. Rosental and V. I. Veselovski, *Dokl. Akad. Nauk SSSR*, **148**, 152 (1963).
147d. E. I. Khrushcheva, N. A. Shumilova and M. R. Tarasevich, *Electrokhimia (Akad. Nauk SSSR)*, **2**, 363 (1966).
147e. G. Bianchi, F. Mazza and T. Mussini, *Electrochim. Acta*, **11**, 1509 (1966).
147f. N. A. Shumilova, G. V. Zhutaeva and M. P. Tarasevich, *Electrochim. Acta.* **11**, 967 (1966).
147g. T. Hurlen, Y. L. Sandler and E. A. Pantier, *Electrochim. Acta*, **11**, 1463 (1966).
148. M. O. Davies, M. Clark, E. Yeager and F. Hovorka, *J. Electrochem. Soc.*, **106**, 56 (1959).
149. W. S. Bagotzky and N. E. Yablokova, *Zh. Fiz. Khim.*, **27**, 1663 (1953).
150. J. Koryta, *Coll. Czch. Chem. Comm.*, **18**, 21 (1953); D. M. H. Kern, *J. Amer. Chem. Soc.*, **76**, 4208 (1954).
151. L. Müller and L. Nekrassov, *Electrochim. Acta (London)*, **9**, 1015 (1964).
152. C. Wagner and W. Traud, *Z. Elektrochem.*, **44**, 391 (1938).
153. R. and H. Gerischer, *Z. Physik. Chem.*, *(Frankfurt)*, **6**, 178 (1956).
154. D. Winkelmann, *Z. Elektrochem.*, **60**, 731 (1956).
154a. K. F. Bonhoeffer, *Z. Physik. Chem.*, **131**, 363 (1928).
155. J. O'M. Bockris, *Modern Aspects of Electrochemistry*, London, 1954.
156. J. O'M. Bockris and A. K. M. Sh. Huq, *Proc. Roy. Soc. (London) Ser. A* **237**, 277 (1956).
157. R. Cornelissen and L. Gierst, *J. Electroanal. Chem. (Amsterdam)*, **3**, 219 (1962).
157a.. K. Hamann, *Ber. Bunsenges. Phys. Chem.*, **71**, 612 (1967).
157b. K. J. Vetter, *Electrochemische Kinetik*, Springer Verlag, 1961; *Electrochemical Kinetics*, Academic Press, New York, 1967.
158. K. Kordesch, *Austrian Pat.* 168 040 (10.4.1951).
159. K. H. Friese, Diplomarbeit, Braunschweig, 1956. E. Justi and A. Winsel, *Kalte Verbrennung—Fuel Cells*, Steiner Verlag, Wiesbaden, 1962.
160. M. Jung and H. Kroger, *Ger. Pat. Appl.* 1 137 779/21b (23.2.1961).
161. E. Yeager, A. Kozawa and M. Savy, *The Elektrochem. Soc. Meeting*, Pittsburg, 1963, Ext. Abstr. No. 164.
161a. W. Güther and W. Vielstich, F. Goebel and U. Vogel, unpublished.
161b. J. Mrha, *Coll. Czech. Chem. Comm.*, **31**, 715 (1966).
161c. J. Mrha, *Coll. Czech. Chem. Comm.*, **32**, 708 (1967).
162. S. V. Trachuck and N. N. Voronin, *Zh. Fiz. Khim.*, **32**, 201 (1958).
163. W. Vielstich and G. Grüneberg, *Ger. Pat. Appl.* 1150 420 (6.5.1960).
164. K. Schwabe, *Symp. Amer. Chem. Soc.*, Chicago, 1961, B 61.
165. K. Schwabe, R. Köpsel, K. Wiesener and E. Winkler, *Electrochim. Acta (London)*, **9**, 413 (1964).
167. J. Jacq and O. Bloch, *Electrochim. Acta (London)*, **9**, 551 (1964).
167. E. Berl, *Trans. Electrochem. Soc.*, **76**, 359 (1939); *Ger. Pat.* 648 964 (13.8.1937).
168. F. Fischer and O. Priess, *Ber. dtsch. Chem. Ges.*, **46**, 698 (1913).
169. A. Kozawa, Battery Division, *The Electrochem. Soc. Meeting*, New York, 1963, Ext. Abstr. No. 36.
170. K. J. Vetter, *Z. Physik. Chem. (Leipzig)*, **194**, 199 (1950); *Z. Elektrochem.*, **55**, 121 (1951).
171. J. A. Shropshire and B. L. Tarmy, *Amer. Chem. Soc.*, Fuel Cell Symp., New York, 1963, p. 153, (Publ. 1965).
172. F. T. Chang and H. Wick, *Z. Physik. Chem.* Pt. A **172**, 448 (1935); M. Loshkarev and O. Esin, *Acta Physiochim. USSR*, **8**, 189 (1938).

173. A. N. Frumkin and G. A. Tedoradze, *Z. Elektrochem.*, **62**, 251 (1958); *Dokl. Akad. Nauk SSSR*, **118**, 530 (1958).
174. S. Yoshizawa, F. Mine, Z. Takehara and Y. Kanaya, *J. Electrochem. Soc., Japan*, **38**, E10 (1962).
175. D. B. Boies and A. Dravnieks, *Amer. Chem. Soc.*, Fuel Cell Symp., New York, 1963, p. 262 (Publ. 1965).
176. D. Spahrbier and G. Wolf, *Z. Naturforsch.*, **19a**, 614 (1964).
177. M. Gutjahr and W. Vielstich, *Chem. Ing. Techn.*, **40**, 180 (1968).

CHAPTER IV

Classical Fuel Cells

The description 'classical' is applied to cells that by long historical association are regarded as the prototypes of galvanic fuel cells. Oxidizing agents are confined to oxygen and hydrogen peroxide.

This group of cells includes a number that have been associated with developments important to speedy commercial application (cf. chapter X). The principal representatives are:

1. *The $H_2/KOH/O_2$ cell with free electrolyte and diffusion electrodes of carbon or sintered metal (predominantly skeleton-nickel or silver as catalysts) (see section IV.1.1),*
2. *The H_2KOH/O_2 cell with an asbestos diaphragm as electrolyte carrier (see section IV.1.2.2),*
3. *The N_2H_4/O_2 cell with asbestos diaphragm (see section IV.3.2) and*
4. *CH_3OH/O_2 and $HCOOH/O_2$ cells, especially those with alkaline electrolytes and water-repellent, air-diffusion electrodes (see section IV.3.1).*

IV.1 H_2/O_2 CELLS (FOR TEMPERATURES BELOW 100°C)

The electrochemical reaction of hydrogen and oxygen in a galvanic cell, forming water, was studied as long ago as 1839 by Grove[1]. This cell has remained a centre of interest ever since, because hydrogen reacts readily at temperatures below 100°C. It was not until 1959, however, that Bacon[2] was able to produce a hydrogen–oxygen battery* of technological construction, with a nominal output rating of 5 kW (10 kW for short periods).

* The Bacon cell works at 200°C and 20 atm and is therefore appropriately discussed in section IV.2.

Modern hydrogen–oxygen cells operate almost exclusively with *alkaline* electrolytes—usually with 6 to 10N KOH, but cells with *acid* ion-exchange membranes form an exception. Cells with H_2SO_4 as electrolyte have not yet been developed in a form economical to construct and operate, because only the platinum metals* adequately catalyse oxygen reduction under these conditions and are stable to corrosion at temperatures of 50 to 80°C.

The current output of electrodes lies between 50 and 150 mA cm^{-2}; i.e. for 10 cm^2 of surface area at 0.7 to 0.8 volt terminal voltage, joulean heat is evolved corresponding with about $\frac{1}{4}$ watt output, and in one hour 0.15 to 0.45 cm^3 of water is formed. *Temperature control* and *regulation of water content* are therefore factors of extreme importance in operation of these cells. For this reason, in the case of cells with *free electrolyte*, both gas and electrolyte are circulated; for cells with diaphragms as *electrolyte carriers*, however, gas circulation alone must suffice.

The *ideal* hydrogen–oxygen cell should be capable of operation with hydrogen derived from hydrocarbons, ammonia or methanol, and with air instead of oxygen. Appropriate generators have been constructed for the supply of hydrogen from these readily available and cheap raw materials. The use of air in place of oxygen for cells with *alkaline* electrolytes requires the removal of *carbon dioxide*; for a battery rated at 1 kW, at least 1 l (measured at 1 atm and 20°C) per hour of CO_2 must be reckoned with.

IV.1.1 Cells with Free Electrolyte

The *principles* of hydrogen–oxygen cells with free electrolyte and two gas-diffusion electrodes consisting of porous plates, discs or cylinders have already been treated in chapters I and II. The gas is supplied under pressure (0.1 to 3 atm above atmospheric) to the porous electrodes (Figure I.3), which are at least partly furnished with catalysts for the desired electrochemical reactions (Schmid, 1923[3]). For alkaline electrolytes, carbon and nickel are mainly used for supporting materials. Catalysts for hydrogen are platinum metals and skeleton-nickel (see section III.1.4); for *oxygen*, silver, platinum, palladium and active carbon.

To secure as complete reaction of the gas supplied as possible, Bacon[2] constructed the *double-layer* electrode (see Figure II.12, section II.6.2.1). The three-phase zone (electrode/gas/electrolyte) is localized with a combination of a *homogeneously* fine-pored surface layer with a coarse-pored supporting layer.

If single-layer electrodes are used, entry of gas bubbles to the electrolyte can be prevented either by hydrophobic treatment of the porous electrodes, or by attaching a suitable membrane (e.g., of asbestos) to them as a surface layer.

* Recently H. Jahnke and M. Schönborn (CITCE Meeting, Detroit, 1968; Fuel Cell Meeting, Brussels, 1969) have shown that iron phthalocyanine polymers deposited on graphite catalyse the oxygen reduction in acid electrolyte extremely well even at room temperature. It should be mentioned at this point that also for hydrogen in the same medium a catalyst without noble metal content has been found in the form of tungsten carbide (F. A. Pohl and H. Böhm, Fuel Cell Meeting, Brussels, 1969).

The preparation of the electrodes and their properties, as well as the construction and methods of operation of complete cells, will be described in this chapter.

IV.1.1.1 Cells with Metallic Diffusion Electrodes for Use with Alkaline Electrolytes

The double-layer electrodes of sintered carbonyl nickel developed by Bacon[2] are not of sufficient catalytic activity for use in hydrogen–oxygen cells working at temperatures below 100°C.

Skeleton-nickel has proved to be a suitable catalyst for the *hydrogen* electrode. Travers and Aubry[4] had discovered in 1938 that the hydrogen potential is established reversibly at Raney-nickel (alloy of aluminium and nickel) in alkaline solution. A mechanically stable electrode cannot, however, be made from the alloy powder by pressing, sintering, and finally dissolving out the aluminium in warm alkali.

IV.1.1.1.1 The DSK electrode

Justi, Scheibe and Winsel[6] were the first to solve this problem by development of the so-called DSK (Doppelskelett-Katalysator) electrode. A mechanically stable and highly active pyrophoric metallic material is obtained by pressing and sintering a powder mixture consisting of about 2 parts of carbonyl nickel and 1 part of Raney-alloy, and finally dissolving out the aluminium by means of hot alkali. The active skeleton of Raney-nickel is held together by a supporting skeleton of carbonyl nickel. Substitution of carbonyl iron for carbonyl nickel leads to a deterioration of catalytic properties.

Justi and co-workers[7,8] have given a detailed description of the preparation, properties and applications of these hydrogen electrodes. *Optimal* conditions for the preparation are[8a] as follows:
(a) Ratio of alloy components: 50 wt % Al to 50 wt % Ni.
(b) Grain size of Raney-alloy: diam. = 6 to 10 μ.
(c) Grain size of carbonyl nickel (supporting skeleton): diam. = 5 μ.
(d) Ratio of Raney-alloy to carbonyl nickel powder: one part by weight of Raney-alloy powder to 2 parts by weight of carbonyl nickel powder.
(e) Pressing pressure (for a 20 g weight, 2–3 mm thick electrode): 3.8 ton cm^{-2}.
(f) Sintering temperature: 680°C.
(g) Sintering time: 30 min (exclusive of heating and cooling).

The dissolution of the aluminium can be accelerated by anodization of the electrode (cf. Jap. Pat. 6611/56). This 'controlled' activation leads to an improvement (at least initially) of the catalytic activity of the electrode[8b].

Figure IV.1 shows current–potential curves obtained by Grüneberg[9] with a strongly gassing 'monolayer'–Ni–DSK electrode, supplied with hydrogen under pressure, in 6N KOH at various temperatures.

The circular disc electrodes (diam. = 4 cm, thickness 2–3 mm) were activated without anodization. Whilst measurements for curves 1 to 3 were being taken, it is clear that some residual aluminium was dissolving out, with an increase in the number of active centres. This is why curves 4 to 8 are more satisfactory (lower overpotentials).

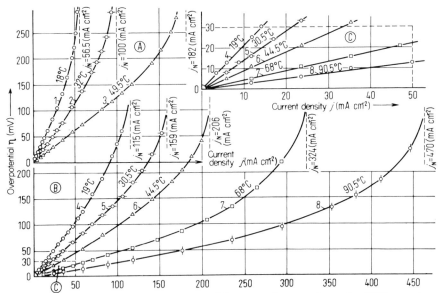

Figure IV.1. Stationary anodic current–potential curves of a strongly-gassing DSK monolayer electrode, diam. = 4 cm, p_{H_2} = 3.5 atm total pressure (after Grüneberg[9]), j_N = maximum continuous current density.

Grüneberg has also studied the *working life* of several monolayer electrodes in half cells (cf. section II.4). Optimal results were 200 mA cm^{-2} for 500 hr at 80°C (overpotential $\eta \sim 150$ mV).

The following reasons for limitation of working life are suggested:
1. blocking of pores by deposition of carbonates and hydroxides,
2. recrystallization of the active skeleton-nickel,
3. poisoning of catalytic centres by catalyst poisons from the fuel gas or the electrolyte.

Electrodes with blocked pores can be *regenerated* by the following procedures[8,10]:
1. dissolution of any adhering layer of K_2CO_3 in distilled water.
2. removal of aluminium hydroxide deposits by boiling with repeatedly renewed, concentrated KOH solution,
3. dissolution of nickel hydroxide by boiling with a slightly alkaline solution of ethylene diamine-tetra-acetic acid (EDTA),
4. renewal of hydrogen in the nickel lattice by strong cathodization.

Ten regenerations carried out by these methods gave a total time of operation of 3500 hr at 200 mA cm^{-2} ($\eta \sim 150$ mV) at 80°C. This corresponds to 700 A hr cm^{-2}.

Only 5 to 10% of the gas supplied is brought into electrochemical reaction by sintered monolayer electrodes of this type, in contrast with more than 90% attainable with double-layer electrodes.

Example of preparation[8]:

Raney-alloy, Al/Ni: 50/50	Pressing pressure: 3.8 ton cm^{-2}
Grain size for the coarse layer: 6 to 12 μ	Sintering temperature: 700°C
Surface layer: 6 μ	Sintering time: 30 min.
Mixture ratio, alloy/carbonyl nickel: 1/2	

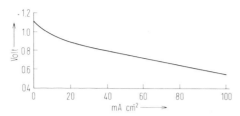

Figure IV.4. Characteristic curve for a hydrogen–oxygen cell with double-layer DSK electrodes (after Justi and Winsel[8]); 6N KOH, 35°C, electrode separation, 4 mm.

electrodes was demonstrated at Essen on 31.1.64, operating two 5-watt lamps (1A, 5V); nominal rating about 20 W; 40 W for short periods.

Clemm[15], General Electric, has also described H_2–DSK electrodes made to the specification of Justi and co-workers[8]. The best non-gassing double-layer electrode in 6N KOH at 23°C showed a performance similar to that indicated in Figure IV.1. The overpotential of monolayer electrodes was about 50 to 100 mV less; the difference was attributed to the ohmic potential drop in the surface layer, which could not, however, be made thinner than 0.25 mm.

Four double-layer hydrogen electrodes (diam. 5 cm) were combined with air electrodes in complete cells and tested for working life, without change of electrolyte. The first pair became inoperative after 24 days; the remaining pair failed after 77 days.

Electrodes of 15 cm diameter* were considerably more difficult to prepare, and showed inferior electrochemical properties.

IV.1.1.1.2   Variations of the DSK electrode

Krupp and co-workers[16] made DSK electrodes from *platinum metals* (Pt, Pd, Rh). The mixed Raney-alloy (Pt–Al, Pd–Al, etc., grain-size 25–40 μ) and noble metal (in 75 volume % excess) powder, with addition of sodium chloride to form macropores, was pressed at 10 ton cm^{-2} into discs of 12 mm diameter. To avoid the formation of intermetallic phases, the discs were not sintered†, in contrast to the technique of Justi and Winsel[8]. After activation by anodization at 60 to 100°C in KOH, these electrodes provided current densities of more than 300 mA cm^{-2} with hydrogen or methanol as fuel, either in KOH or H_2SO_4 as electrolyte.

Burshtein and co-workers[17] developed a *triple-layer* electrode for the anodic oxidation of hydrogen. In one of the methods used, two inactive layers of carbonyl nickel (coarse-pored supporting layer and fine-pored surface layer) enclosed between them an active, third layer of skeleton-nickel. This third layer was made from a Raney-nickel–titanium alloy (Al:Ni:Ti = 49:49:2) by pulverization and dissolution of the aluminium. The titanium addition stabilizes the skeleton structure. The catalytic properties of the active powder resemble those of normal Raney-nickel, but the powder is not pyrophoric, and is insensi-

* A 1000 ton press is required for making electrodes of this size.
† The discs, on sintering, became inactive for methanol oxidation and, fed with hydrogen, passed much less current per cm^2 than the unsintered discs.

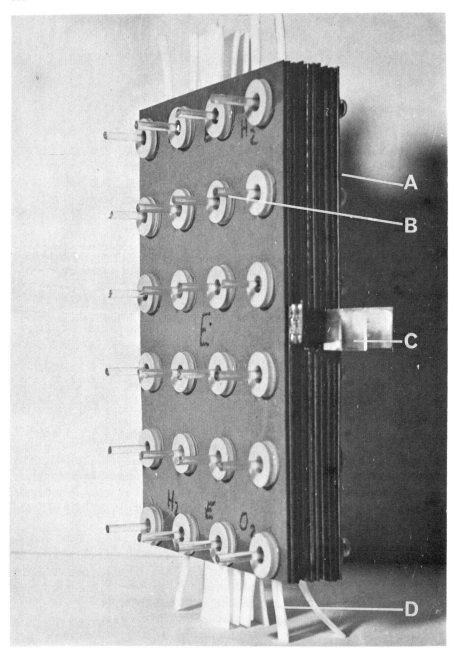

Figure IV.4a. Electrode assembly with the electrodes supported by plastic tie-rods (after Weidinger[17c]). (A) Hydrogen or oxygen electrodes; (B) plastic tie-rods; (C) electrical contacts; (D) section for provision of electrolyte and gas channels.

Figure IV.4b. Eight-cell battery with thin electrodes assembled with plastic tie-rods (after Weidinger[17c]).

tive to oxygen; this facilitates the preparation and handling of the electrode. Hydrogen–oxygen cells with these hydrogen electrodes (diam. = 12 cm, thickness 5 mm) in 6N KOH at 90°C provide a terminal voltage of 0.7 volt at 200 mA cm^{-2}. The overpotential of the hydrogen electrode amounts to 80 mV.

Dousek, Jansta and Riha[17a] have devised an interesting method of preparing non-pyrophoric nickel catalysts. Raney-nickel is first activated in KOH. The essential feature of the method is then to remove adsorbed hydrogen before the catalyst is dried. This is carried out under mild conditions either by neutralization of the alkali with tartaric acid, followed by desorption of the hydrogen in distilled water at 90°C, or by reaction of the hydrogen with O_2 or H_2O_2, preferably with potentiometric control. Reactivation of the catalyst after drying can only be relied upon if the drying temperature does not exceed 200°C, since otherwise the nickel oxide formed in the drying is not easily reducible.

The non-pyrophoric powder so obtained has a specific surface area of 60 to $80 \, m^2 g^{-1}$. It has been used to make double-layer electrodes of the Janus type, 30 mm diameter, 2.7 mm total thickness, with two surface layers (0.7 mm) on either side of a 'working layer' (1.3 mm), by pressing at $1.5 \, ton \, cm^{-2}$ and sintering at 500°C. These, working as hydrogen electrodes at 90°C, could handle $500 \, mA \, cm^{-2}$ at an overpotential of 50 mV. Complete cells using such electrodes have been run at 95°C for more than 10,000 hours at $80 \, mA \, cm^{-2}$. The following table shows the terminal voltages obtained, with oxygen cathodes based on a silver catalyst.

Current density $(mA \, cm^{-2})$	Terminal voltage (mV)	Temperature (°C)	Working life (hr)
30	750	25	15000[a]
90	720	95	17000[a]
120	640	95	4000[b]
160	610	95	1400

[a] Tests still in progress.
[b] A further 3200 hr after a special regeneration process.

Another preparation of non-pyrophoric skeleton-nickel electrodes has recently been described by Jung and van Döhren[17b]. Raney-nickel, after activation in KOH solution containing added potassium tartrate, is washed first with alkali until free from aluminium, and then with water until neutral. Adsorbed hydrogen is quantitatively oxidized by an aqueous solution of, for example, KIO_3, $KBrO_3$ or $KClO_3$, and the catalyst washed again. Electrodes are made in three layers—a fine-pored layer of Mond-nickel powder with a very low apparent density ($0.35 \, g \, cm^{-3}$); a 'working layer' consisting of a mixture of preactivated Raney-nickel and Mond-nickel powders with a sodium carbonate filler; a gas distribution layer of Mond-nickel powder with a potassium chloride filler. The electrodes are hot-pressed (0.5 ton cm^{-2}, 450°C); no special protection against air-oxidation (inert gas or reducing agent) is needed during pressing. After the fillers have been leached out by boiling in water for a short time, the electrodes are dried in vacuum at 40°C, and can then be stored in air for a month without deterioration. Reactivation of the electrodes takes place in 2–3 hours at 80°C, with hydrogen passing, after they are mounted in the battery cells.

The activity of Raney-nickel can be improved by addition of 0.1 to 0.5 % of a promoter metal; platinum and copper are predominantly used. The addition is carried out by introducing a salt of the metal during the activation process, most effectively at an early stage of the activation when the first violent evolution of hydrogen has ceased; the promoter is then almost instantly precipitated on the nickel powder. Electrodes promoted with copper have performed well at 60°C and 100 mA cm^{-2} for more than 2000 hours.

For trials of hydrogen–oxygen modules, electrodes of postcard size (active area 160 cm^2, thickness ca. 2 mm) were combined in the electrode assembly shown in Figure IV.4a. Plastic bolts take the mechanical stress set up by gas pressure, and eliminate the need for thick end-plates, with their adverse effect on performance-weight and volume ratios. The complete assembly is set in rotation and filled with plastic (Figure IV.4b).

The use of two-faced Janus electrodes, with electrolyte on either side, obviates the necessity of plastic tie-rods for support. The Janus electrodes of the Varta A.G. are about 2.5 mm thick and consist of five layers; 2 surface layers, 2 working layers, and a central gas distribution layer (see above). Modules incorporating these electrodes are made in the form of rectangular blocks (12 × 7 × 19.5 cm^3; weight, 4.0 kg) by a plastic moulding process. Rubber spacers are used for electrode positioning and insulation, and a module contains 18 hydrogen and 19 oxygen electrodes. Current density in continuous operation at 60°C (gas pressure 2.5 atm) amounts to about 50 mA cm^{-2} for each electrode face, at a terminal voltage of 0.65–0.85 V, gas pressure being 2 atm. A battery of 2.5 kW continuous rating (or 4.0 kW for 1 hr, 10 kW peak) has been assembled from 32 modules and has been subjected to practical trials in operating a fork-lift. The weight- and volume-load ratios are 33 kg kW^{-1} and 14 l kW^{-1}, relative to conditions of 50 mA cm^{-2}, 0.85 V terminal voltage and a working temperature of 60°C.

The electrolyte (initially 6N KOH) is circulated through the cells and an electrolyte tank by means of a pump with a magnetic drive (Figure IV.4c), involving a ceramic magnet and a polythene pump shaft. The alkaline electrolyte serves as lubricant, the impeller being suitably bored for the purpose. Characteristics of the pump are: 3 l min^{-1} pumping speed; lift, 1 m of water; 9 V, 150–200 mA; weight of pump, 75 g. It is driven by a dc motor without commutator, weighing, according to type, 100–200 g.

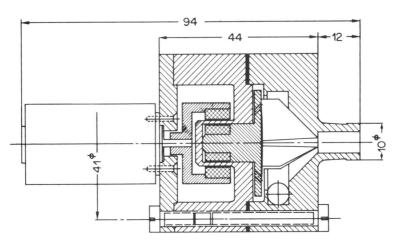

Figure IV.4c. Section through Hahn's alkali pump[17d]. Left, motor; centre, magnetic coupling; right, pump.

IV.1.1.1.3   MSK skeleton-nickel electrodes[18]

The catalytic activity of a skeleton-nickel electrode depends critically on its disperse surface structure rather than on its method of preparation (e.g. by way of Raney-nickel; see above). The electrode must have both electronic conductivity and mechanical stability.

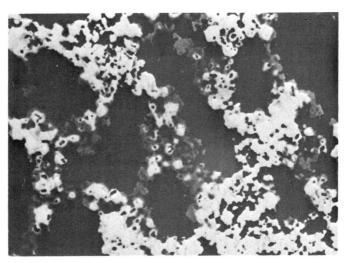

Figure IV.5. Pore structure of an active nickel-MSK electrode, magnification × 1000 (after Plust[19]). The light patches (nickel framework) are grey at the edges (catalytically active skeleton-nickel). The dark areas are pores.

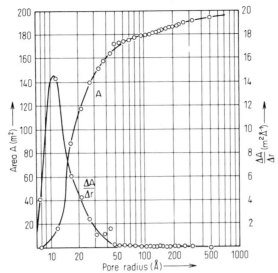

Figure IV.6. Surface distribution of a nickel–MSK electrode[21].

Skeleton-nickel electrodes with the desired properties are obtained by mixing powdered nickel and aluminium, pressing and sintering, or hot pressing, and finally dissolving out the aluminium with alkali. This gives monoskeleton-nickel (MSK, in contrast to the double-skeleton electrode, DSK), with catalytically active pore walls (Figure IV.5). Figure IV.6 gives information about the pore distribution in the microstructure of the internal surfaces of an MSK electrode. It is seen that the surface structure is characterized by pores of about 12 Å radius.

The optimal conditions of preparation are:
(a) Ni powder and Al powder (grain size 5 to 10 μ) mixed in 50:50 ratio,
(b) pressing pressure 5 ton cm^{-2}, electrode thickness, 1 to 2 mm,
(c) sintering temperature, 450°C,
(d) sintering time, 30 min (excluding heating and cooling), or, *hot pressing* with
(b') pressing pressure 1 to 2 ton cm^{-2}, electrode thickness, 1 to 2 mm,
(c') temperature, 250 to 300°C,
(d') time, 20 min (excluding heating and cooling).
These monoskeleton catalyst MSK electrodes of nickel can be prepared in larger sizes by a rolling process developed by the International Nickel Company (Mond)[20].

An oxygen MSK electrode of silver can be prepared in a similar way from a mixture of silver and aluminium powders. Since their catalytic properties in prolonged use do not differ significantly from those of normal, porous silver electrodes, they need not be discussed here.

Figure IV.7 shows the current–voltage characteristic of a hydrogen–oxygen cell with heteroporous, gassing MSK electrodes[19].

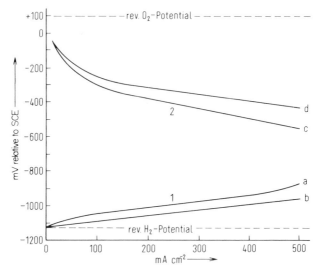

Figure IV.7. Current–potential characteristic of a hydrogen–oxygen cell with MSK electrodes (after Plust[19]); 6N KOH, 80°C. Curve 1: Ni–MSK–H_2 electrode (impregnated with Pt); p_{H_2} = 0.5 atm above atmospheric; (a) before, (b) after activation by anodic oxygen evolution. Curve 2: Ag–MSK–O_2 electrode; p_{O_2} = 0.3 atm above atmospheric; (c) pure silver, (d) addition of cadmium.

IV.1.1.1.4   Incorporation of hydrogen in electrodes by use of strong reducing agents

Jung and Kröger[23] have subjected porous sintered materials of nickel, cobalt, and mixtures of the two, to treatment with approximately 5 % sodium borohydride solution in 4N KOH, for about $\frac{1}{2}$ hr at room temperature. The sintered metals then showed pyrophoric properties, an increase in volume, and a tendency to form amalgams on contact with mercury. The authors assumed that the nascent hydrogen liberated on contact of the metals with the reducing solution very largely enters the metal lattice and is held there partly as interstitial alloy and partly in the form of a true hydride. Electrodes activated in borohydride solutions are capable of passing several hundred mA per cm^2. In addition to alkali borohydrides, lithium aluminium hydride, hydrazine and hydroxylamine have been proposed as reducing agents. Nothing is known about the behaviour of such electrodes in prolonged use.

IV.1.1.1.5   Nickel boride as catalyst

The skeleton-nickel properties of *nickel boride* have recently been exploited for the preparation of active hydrogen electrodes for use in alkaline electrolytes (cf. section III.1.4).

A solution of sodium or potassium borohydride mixed with a solution of a nickel salt forms a voluminous black precipitate of nickel boride of composition Ni_2B, with simultaneous evolution of hydrogen[24]. The product is neither magnetic nor pyrophoric, and dissolves less rapidly in acid or alkaline solution than does Raney-nickel. Its catalytic activity is comparable with that of Raney-nickel, and its sensitivity to poisoning or loss of activity with time is considerably less. Its catalytic activity can be further improved by small additions (ca. 2%) of chromium, molybdenum, tungsten or vanadium[24].

Whilst the reaction of alkali borohydride with nickel or cobalt salts produces borides, other metal salts (e.g. silver, mercury or bismuth salts) are reduced to the metals.

Other methods of preparation that have been tried are the reaction of nickel chloride with boron trichloride and hydrogen at 700 to 1000°C[22e], and the reaction of elementary boron with metallic nickel[22a,22b].

Properties of nickel boride

(a) Composition: Nickel boride is black and, when precipitated, spongy and voluminous; the other methods of preparation give a considerably denser powder, or sintered cakes or pressings. All the methods give mixtures of four borides[22a], NiB, Ni$_4$B$_3$, Ni$_2$B, Ni$_3$B, and Ni, of composition varying from one method to another. The last process named above allows the stoichiometry of the nickel borides to be established within certain limits[22b]. The precipitation method gives Ni$_2$B by the action of potassium borohydride on solutions of nickel acetate, sulphate or chloride[22c], or Ni$_3$B from the reaction of sodium borohydride on nickel chloride solution[22d]. Under special conditions, catalysts containing NiB$_{12}$ and elementary boron can be prepared[22b].

(b) Corrosion, specific area and electrocatalytic activity: All the nickel borides are unstable, and therefore unusable, in sulphuric acid electrolytes[22b]. In potassium hydroxide solutions there is a slight, long-continued corrosion process, decreasing in the sequence NiB > Ni$_4$B$_3$ > Ni$_2$B > Ni$_3$B. Both nickel and boron can be detected in solution, but boron is lost preferentially to an extent which depends, among other things, on the presence of oxidizing or reducing agents, e.g., hydrogen or oxygen[22d]. As loss of boron from a catalyst proceeds, its internal surface area increases. Freshly precipitated Ni$_3$B has a specific surface of about 30 m^2 g^{-1} (B.E.T.), which corrosion can increase to more than 60 m^2 g^{-1}. Nickel boride formed by sintering for use as an electrode has a specific area of 0.5 to 2.0 m^2 g^{-1}, but this may be multiplied by a factor of 3 or more by corrosion[22d].

According to Lindholm[22d], electrocatalytic activity increases linearly with surface area to a limiting value reached at 2 to 4 m^2 g^{-1} (B.E.T.). This is illustrated in Figure IV.8; relative catalytic activity was assessed in terms of current density at given electrode potential and hydrogen pressure. It can be suggested that dissolution of the comparatively small boron atoms (radius 0.89 Å) from the surfaces of the orthorhombic crystals of Ni$_3$B produces a catalytic surface particularly effective for the chemisorption of hydrogen (atomic radius 0.79 Å).

Electrodes from nickel boride

Electrodes can be made from nickel boride alone by pressing and sintering, but as yet the resistance to corrosion of such electrodes is quite inadequate. They partially decrepitate even in water. It is therefore necessary to add to the boride, before pressing and sintering, a metal such as nickel or gold that will

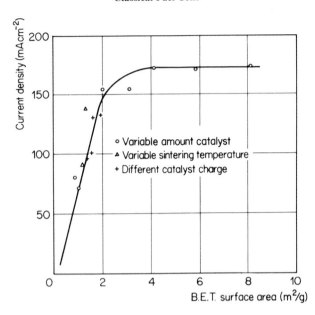

Figure IV.8. Current density at 100 mV overpotential for nickel boride electrodes of varying surface area: 50°C, 7N KOH, hydrogen pressure 2.5 atm (after Lindholm[22d]).

form, during the sintering, a coherent, supporting framework in which the porous nickel boride is embedded. Electrodes thus prepared contain between 40 and 60% of supporting metal[22b] (cf. DSK electrodes, section IV.1.1.1.1).

Another possible method of preparation consists of introducing the nickel boride into an existing porous sintered nickel matrix; thus, a porous, sintered nickel plate may be flooded with an aqueous solution of a nickel salt, and internal precipitation of boride carried out by means of alkali borohydride[22c,22d]. Alternatively again, nickel boride incorporated with Teflon as binder can be attached to a nickel support[22,22c].

The activity of such electrodes depends primarily on the catalytic activity of the nickel boride used, but other factors which may have a large effect are pressing pressure (of the order 4 ton cm^{-2}), sintering temperature (600 to 1000°C), the gas atmospheres in which pressing and sintering are carried out, and whatever binding material is used[22b]. The porosity of electrodes can be controlled by addition of an inert substance that decomposes with gas evolution (e.g., $(NH_4)_2CO_3$).

The internal surface area and activity of nickel boride electrodes can further be increased by an application of the Raney process, in which nickel boride and aluminium are sintered together. The catalyst so obtained is combined with the supporting material and formed into electrodes in the usual way, but

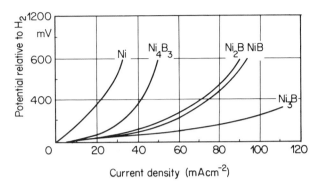

Figure IV.8a. Potential–current density curves for the anodic oxidation of hydrogen at 80°C with different nickel boride catalysts (after Jahnke[22b]).

dissolution of the aluminium must be carried out cautiously at a controlled temperature. Electrodes so prepared are specially active in catalysing the oxidation of methanol, which probably occurs preferentially at centres of NiB$_{12}$ and elementary boron[22b].

Performance of nickel boride electrodes

Sintered electrodes of supporting nickel and nickel boride made from boron and nickel by the sintering process give serviceable results with hydrogen and methanol.

With hydrogen, Ni$_3$B will provide a current density of 100 mA cm^{-2} at 80°C with an overpotential of 200 mV (cf. Figure IV.8a)[22b]: additional activation with aluminium reduces the overpotential to only 110 mV under the same conditions of temperature and current density.

According to Jasinski[25,22c] and Thacker[22], Teflon-bonded Ni$_2$B will provide higher current densities with hydrogen. Jasinski has produced a cell with Ni$_2$B for hydrogen and Pt (9 mg cm^{-2}) for oxygen electrodes, with an asbestos membrane as electrolyte (30% KOH) carrier, that provides 20 mA cm^{-2} at 0.7 V at room temperature, increased to 0.8 V and 180 mA cm^{-2} at 90°C (Figure IV.9). At this temperature there is a limiting current density of about 240 mA cm^{-2}. Teflon-bonded Ni$_2$B electrodes have an average catalyst content of 20 mg cm^{-2}.

Lindholm[22d] has described a hydrogen anode, made by a sintering process from nickel and precipitated Ni$_3$B, that will carry 200 mA cm^{-2} at 50°C, at a potential of 150 mV relative to a hydrogen electrode in the same solution. Complete cells furnished 0.7 V at this current density at 80°C; in endurance trials cell voltage fell by about 0.1 V at a constant load of 50 mA cm^{-2} at 80°C in 3600 hours of continuous operation. Batteries of 200 kW rating have been built using nickel boride as catalyst for the hydrogen electrodes; they are

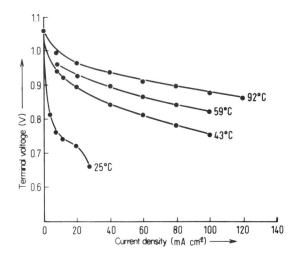

Figure IV.9. Current–voltage curves of an alkaline hydrogen–oxygen cell with Ni_2B anode and asbestos diaphragm as electrolyte carrier (after Jasinski[25]). Ohmic resistance eliminated, H_2O control by the hydrogen circulation system, gas pressure 1.2 atm above atmospheric.

assembled from square-shaped modules. At a load density of 140 mW cm^{-2}, volume and weight per kW of 50 l kW^{-1} and 10 kg kW^{-1} are attained. Figure IV.9a shows a test-bench for endurance tests of AQD modules.

Up to the present, a temperature of 80 to 90°C has been taken as optimal for current densities up to 250 mA cm^{-2}. Thacker[22] has claimed that a cell with Teflon-bonded Ni_2B hydrogen anodes can provide 720 mA cm^{-2} at 0.59 V at 150°C; the limiting current at this temperature is about 800 mA cm^{-2}.

IV.1.1.1.6   The skeleton-nickel electrode as an accumulator plate

It has already been mentioned in section III.1.4 that Raney-nickel can be charged with hydrogen to an extent of 1 atom of hydrogen per atom of nickel. This corresponds with an electrical charge of 450 A hr kg^{-1} of Raney-nickel. Although a skeleton-nickel electrode does not wholly consist of active material, about 100 to 200 A hr kg^{-1} in the form of hydrogen can be reversibly taken up and released.

As when platinum is charged with hydrogen (Figure III.9), the available hydrogen capacity is a function of the electrode potential. In 6N KOH, the hydrogen capacity of Raney-nickel is reduced to about a half at $\eta = +100$ mV (relative to the hydrogen potential in the same solution), and at $+400$ mV practically all the hydrogen is expelled from the metal[26].

At anodic overpotentials that are not too high, therefore, a skeleton-nickel electrode can be *overloaded* for a short time, because the hydrogen bound to the nickel contributes to the current. At 60 to 80°C, several hundreds of mA cm^{-2}

Figure IV.9a. Test bench for endurance tests of AQD modules of the ASEA Company (Sweden)[22f].

can be drawn from the electrode without any supply of gaseous hydrogen to it[27].

A skeleton nickel electrode can be cathodically charged at low overpotentials, with current densities of magnitude normal in the charging of accumulators (cf. Figures IV.10a and b). An alkaline accumulator can therefore be constructed by combining, for example, a Raney-nickel electrode with a nickel oxide or silver oxide electrode as the positive plate[28] (Figure IV.10c).

If the thickness of the Raney-nickel electrode is reduced, the characteristic curves (Figure IV.10b and c) become 'harder', i.e. the current–potential curve becomes flatter, and the energy efficiency, η (kW hr), is thereby somewhat improved.

The performance data of accumulators of this new type are compared in Table IV.1 with those of the usual commercial lead and nickel/cadmium accumulators.

Table IV.1. Comparison of performance data of a Raney-nickel–nickel oxide accumulator (after Justi, Vielstich and Winsel[27,28]) with lead and nickel-cadmium accumulators.

Type	η(A hr)	η(kW hr)	A hr kg^{-1}	kg kW^{-1} hr^{-1}
Lead	90	75–85	10–20	30–50
Ni/Cd	70	60	30	35–40
Ra-Ni/Nioxide	65–85	55–75	25–50	25–45

It can be seen from Figure IV.10 that the working emf of the new accumulator is 1.1 volt, and the average charging voltage 1.3 to 1.4 volt. The fully-charged emf is 1.5 volt. In comparison with the Ni/Cd accumulator, the self-discharge is faster and more extensive, but the material used is cheaper, and the manufacture simpler. Furthermore, in contrast to that of cadmium, the world production of nickel is such as to allow these accumulators to be produced in large quantities. The greater self-discharge rate would not be of great significance in their use, for example, as a power source for vehicles.

To try out this new type of accumulator for use in vehicles, we have recently built[27] a 6 V, 40 A hr battery consisting of five cells, each with ten pairs of electrodes. The total weight of each electrode, of active area 164 cm^2 is 75g, and the terminal voltage of the battery is 5 V with a load of 15 A.

Accumulators of this kind can also be made with skeleton nickel plates of MSK type (see above); 50 A hr kg^{-1} can be obtained at a discharge rate of 5 mA cm^{-2}, with terminal voltage of 1 volt (Figure IV.11).

Guth and Plust[29] have found that small quantities of platinum remove the undesirable pyrophoric properties of skeleton-nickel, without altering activity and hydrogen capacity.

The hydrogen capacity of nickel boride has not yet been investigated.

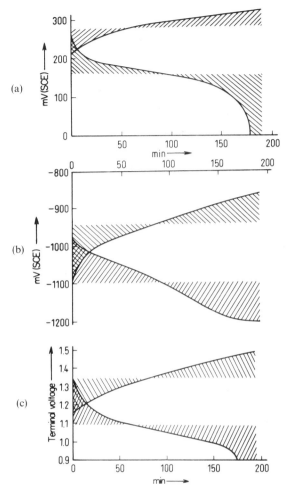

Figure IV.10. Potential–time curves for the charge and discharge of a nickel oxide electrode, a Raney-DSK electrode (relative to a saturated caiomel electrode), and an alkaline accumulator consisting of both these electrodes (after Vielstich[27]). Electrode surface area 12 cm², electrolyte 6N KOH, 20°C, current density ~10 mA cm⁻². (a) Nickel oxide electrode (VARTA 3.5 g, thickness = 0.8 mm, 33 mA hr cm⁻², 115 mA hr g⁻¹, $\eta_{A\,hr} = 93.5\%$. (b) Raney-nickel DSK electrode, 7 g, thickness 3 mm, 58 mA hr cm⁻², 100 mA hr g⁻¹, $\eta_{A\,hr} > 95\%$. (c) Charge and discharge characteristics of an accumulator with electrodes (a) and (b) for a 3 hr charge,

$$\eta_{kW\,hr} = \frac{109 \times 177}{135 \times 190} = 75\%.$$

Figure IV.11. MSK nickel/nickel oxide accumulator after Plust[19]. Left, the complete accumulator; right, electrodes and connexions.

IV.1.1.1.7   Cells with non-pyrophoric hydrogen electrodes of sintered metal with palladium additions

Ruetschi and co-workers[30] have developed a hydrogen–oxygen cell with an alkaline electrolyte and a non-pyrophoric, monolayer hydrogen electrode. The hydrogen electrode was prepared[31] from a powder mixture of 10% of nickel (particle size 1 μ), 7.5% of palladium (5 μ) and 82.5% of silver (1 μ). Considerable mechanical stability was attained by embedding the mixed powder in a supporting, perforated screen. The powder was pressed at 50 to 70 kg cm^{-2} and finally sintered for about 20 min at 450°C. More recently, electrodes of the same type have been made containing 12% Ag, 3% Pd and 85% Ni, sintered at an appropriately higher temperature. The oxygen electrodes were made in a similar way[32] from a mixture of silver and nickel powders.

The following potentials were measured in endurance tests with half-cells containing 6N KOH at 40°C, run at 70 mA cm^{-2}:

H_2 electrode after   10 hr + 135 mV (relative to H_2)
                after 3500 hr + 210 mV (relative to H_2)
O_2 electrode after   10 hr + 835 mV (relative to H_2)
                after 1600 hr + 810 mV (relative to H_2)

Laboratory cells with 10 cm^2 electrodes (Figure IV.12) could be operated at 2 to 5 mA cm^{-2} for up to 24,000 hours.

In larger, semi-technical scale cells, cylindrical electrodes are arranged concentrically. The surface area of the hydrogen electrode amounts to about

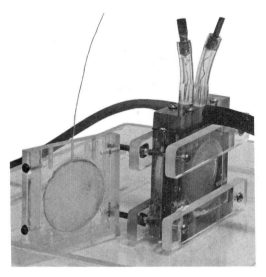

Figure IV.12. Hydrogen–oxygen cell for the study of the working life of sintered electrodes at low current densities (after Ruetschi, Duddy and Ferell[30]).

100 cm^2; that of the oxygen electrode, 160 cm^2. At 60°C and 150 mA cm^{-2}, the terminal voltage of this cell is approximately 0.7 volt.

For a study of its working life, the oxygen electrode was operated at 200 mA cm^{-2} in 15% NaOH at 60°C. For more than 3500 hr the electrode potential was sensibly constant between +510 and +460 mV (relative to H$_2$). Extension of this experiment indicated a working life of more than 7500 hours.

Electrodes based on Raney-nickel with added palladium have recently been used[154] for hydrogen consumption in alkaline cells at the Battelle-Institut e.V., Frankfurt am Main. Porous nickel discs (diam. 4 cm; thickness 3 mm), impregnated with silver were used as oxygen electrodes. A battery of 3 cells (diam. 65 mm; length 110 mm) provided 1.6 A at 1.3 volt, and 1.8 A at 1.2 volt terminal voltage (electrolyte, 6N KOH; gas pressure, 0.5 atm in excess of atmospheric pressure). In operation, a temperature of 52°C was maintained.

Auclair, as also Dubois and Biro[228] have reported on an alkaline cell with identical electrodes, of nickel and silver, functioning equally well as hydrogen anodes or oxygen cathodes. The sintered electrodes contain about 30% of silver. The identity of cathodes and anodes carries certain constructional advantages; exchangeability, replaceability and economy in manufacture. Laboratory experiments have been carried out with Janus electrodes, but the batteries so far made have involved single-layer electrodes 1.6 to 1.7 mm thick, requiring a total gas pressure of 2.8 to 3.0 atm. Each module contains 12 double cells, heat is abstracted by circulation of electrolyte. Batteries of various ratings have been built—10 V, 50 W; 4 V, 250 W; 24 V, 1.5 kW—and, according to the particular application, a 1 kW unit weighs between 40 and 80 kg. Current densities of 50 mA cm^{-2} at 20°C, or 250 mA cm^{-2} at 80°C have been attained at 0.75 V terminal voltage.

IV.1.1.2   Carbon Diffusion Electrodes

Since 1955, Kordesch, with the Union Carbide Corporation[33–35], has worked on the development of porous carbon diffusion electrodes for hydrogen and oxygen, and has successfully produced practically non-gassing, porous diffusion electrodes that will deal with current densities up to 100 mA cm^{-2} at room temperature.

Adequate activity for the *oxygen* electrode was attained by the incorporation of metal- and metal oxide-catalysts into the carbon structure.

An example of the impregnating solutions used by Kordesch has already been quoted in section III.6.5. Heating of the impregnated carbon to 900°C leads to the formation of the catalytically active spinel (Al$_2$O$_3$.CoO); the activity can be further increased by addition of noble metal. After introduction of the catalyst, the carbon electrode is made water-repellent by treatment with a solution of wax, or other high molecular weight organic compound.

The activity of the *hydrogen* electrode was established by deposition of small quantities of finely-divided platinum ($<$ 1 mg cm^{-2}).

The *older* types of electrode consisted of carbon alone and were 6.3 mm thick. The layout of an individual cell is shown schematically in Figure IV.13. The

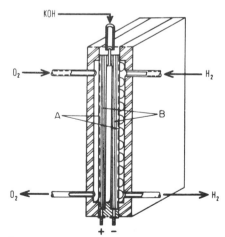

Figure IV.13. Construction of an individual cell with 6.3 mm carbon plate electrodes and free electrolyte (after Kordesch[33]). (B) carbon plate electrodes; (A) perforated nickel plates.

electrode separation is 2 to 3 mm. The reverse sides of the carbon plates are nickel-plated, for better contact with the nickel plates that simultaneously serve the purpose of distribution of gas behind the electrodes. The nickel plating is done by a flame-spraying process to ensure the production of a thoroughly porous surface. Gases and electrolyte pass round a circulatory system (cf. section IV.1.1.3).

The terminal voltage of a cell with the 6.3 mm electrodes, at 50 mA cm^{-2} is 0.84 V. These cells have attained a working life of 7000 hours at 70–100 mA cm^{-2}. A 500 W battery incorporating these electrodes is illustrated in Figure IV.14.

Characteristic data for various batteries, with electrodes up to 30.5 × 27.6 cm^2 in size[35a] are collected in the following table:

Material for electrode mounting	Electrode area (cm^2)	No. of cells	kg kW^{-1} without electrolyte	l kW^{-1}	kg l^{-1}
Rubber	30.5 × 35.6	32	80	65	1.23
Polystyrene	15.2 × 15.2	8	158	82	1.93
Epoxy-resin	7.6 × 12.7	8	68.2	74	0.92

Better figures are obtainable for short-term loads, e.g., 41.7 W kg^{-1} (24 kg kW^{-1}) at 215 mA cm^{-2} from a 1.25 kW battery with large, rubber-mounted

Figure IV.14. Five hundred watt hydrogen–oxygen battery, 9 V, electrodes, 30.5 × 35.6 × 0.63 cm^3 (after Kordesch[34]).

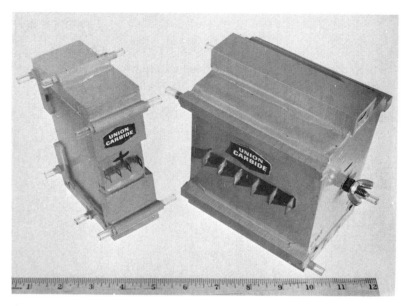

Figure IV.14a. Comparison of seven-cell experimental batteries (after Clark, Darland and Kordesch[35b]). (1) Using 0.9 mm electrodes; (2) using 6.3 mm electrodes.

electrodes, improved to 13.6 kg kW^{-1} for motor-starting at 400 A. A smaller 20 W battery, 7.3 × 12.2 × 14.7 cm^3, 1.5 kg, with 6.3 mm electrodes supported by epoxy-resin, is illustrated in Figure IV.14a.

Further reduction in the weight and volume of cells calls for use of thinner electrodes, but this was impracticable with electrode materials hitherto available without loss of adequate mechanical stability. The difficulty has, however, been overcome in the new composite-metal electrodes of Clark, Darland and Kordesch[35b,d]. Mechanical strength is provided by a nickel grid which carries a sintered carbonyl nickel layer (Figure IV.14b). On the side of this layer facing the electrolyte, a double layer of carbon is deposited by a technique which is detailed in Figure IV.14c. This provides electrodes which are only 0.9 mm thick which can, in addition, be spaced no more than 1.0 mm apart.

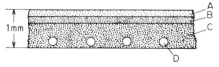

Figure IV.14b. Carbon diffusion electrode with metal support (after Kordesch[36]). (A) Carbon layer with catalyst; (B) carbon contact layer; (C) sintered carbonyl nickel; (D) nickel grid. In practice the layers are not as well separated as this, but rather penetrate each other.

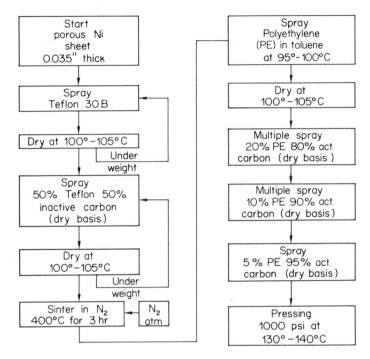

Figure IV.14c. Process flow-sheet for preparation of thin electrodes (after Clark, Darland and Kordesch[35b]).

Since greater module reliability can be attained by use of a larger number of smaller electrodes rather than fewer large ones, there is a trend to build batteries with small electrodes connected in parallel. Thus the Clark, Darland and Kordesch battery uses electrodes of $7.6 \times 12.7 \text{ cm}^2$; Figure IV.14a shows a 7-cell battery with the new thin electrodes, in comparison with the older type, with 6.3 mm electrodes—both of 20 W rating. Weight- and volume-power ratios of 10 kg kW^{-1} and 141 kW^{-1} have now been obtained at 50 mA cm^{-2} for a 16-cell, 50 W battery. Features of construction of these batteries are shown in Figure IV.14d.

Difference in operational characteristics caused by use of air instead of oxygen is considerably reduced because of the short diffusion path of the thin electrodes. This is illustrated by the current densities obtained[35b,d] at 70°C and 0.8 V relative to the reversible hydrogen potential (ohmic loss eliminated), as follows:

Gas	O_2	O_2	Air	Air	Air	Air
Pressure (atm)	2.0	1.0	2.0	1.7	1.35	1.0
mA cm^{-2}	450	350	200	170	130	110

Figure IV.14d. Eight cell (6 V) hydrogen-⌐ ᵢ ᵢel cell battery (after Clark, Darland and Kordesch[35d]).

Kordesch[35e,f] has recently still further reduced electrode thickness (Figure IV.14e), and has made electrodes 10×12.5 cm^2 0.7 mm thick (0.19–0.22 g cm^{-2}); their properties are illustrated in the following table—all the data refer to a current density of 300 mA cm^{-2}, the electrolyte being 9N KOH and the temperature 65°C; ohmic voltage loss is not included:

Gas	O_2	O_2	Air	Air	Air	Air
Pressure (atm)	2.0	1.0	2.0	1.7	1.35	1.0
Terminal voltage of H_2—O_2 cell (V)	0.82	0.80	0.77	0.73	0.65	0.42

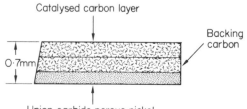

Figure IV.14e. Cut through a thin, fixed-zone electrode (after Kordesch[35e]).

Table IV.2. Comparison of 30-kW fuel cell power plant weights (H$_2$–O$_2$)(after Kordesch[35e]).

	Battery weight (wet) (lb)	External KOH (lb)	Auxiliaries* (lb)	Total (lb)
1965 Baked carbon system (50 ASF)	4860	390	750	6000
Early 1965, fixed zone system (50 ASF)	1290	180	540	2010
Late 1965, fixed zone system (100 ASF)	630	60	450	1140
Projected Improvements: Fixed-zone system (150 ASF)	507	60	300	867
Fixed-zone system (200 ASF)	420	60	300	780
Fixed-zone system (200 ASF) Fully optimized	360	50	240	650

* Fuel and fuel tankage not included.

Finally, Table IV.2 illustrates the 1965 stage of development of carbon electrodes for hydrogen–oxygen cells, and the improvements to be anticipated. The expression 'fixed-zone system' relates to the recent composite electrodes with the catalyst confined to the outermost carbon layer. In these, it is arranged, by adjustment of hydrophobic treatment to suit working conditions, to confine the three-phase zone, and therefore the location of the electrochemical reaction, to the catalyst-bearing layer. Further reference to the data in Table IV.2 will be made in chapter X in relation to applications.

Barak, Gillibrand and Gray[37] have reported good results with silver-impregnated carbon electrodes for oxygen in alkaline electrolytes. Fine, amorphous carbon powder is soaked in silver nitrate solution, and the silver is reduced at 500°C. The powder thus impregnated is mixed with a hydrophobic binder and pressed into a silver gauze (area 20 cm^2). Results of measurements with these electrodes are shown in Table IV.3.

Electrodes 15 × 10 cm^2 for oxygen and hydrogen (probably provided with platinum catalyst) were made[37a,b] on the basis of these results, a single cell consisting of four hydrogen and five oxygen electrodes[37b,c] (cf. Figure IV.87). A battery of 1 kW rating (ca. 50 A at 20 V, 50–60°C)[37c] was built and subjected to endurance tests (although without automatic water content control) over the course of a year. It would provide 1 kW output even from a cold start (Figure IV.15).

The optimum operating pressure is less critical for hydrogen than for oxygen electrodes, and it is therefore possible to operate them with less gas break-through. This accounts for the higher current

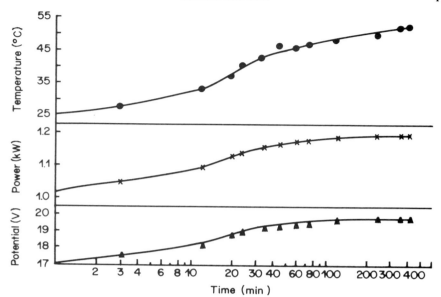

Figure IV.15. Terminal voltage, load and temperature of a 1 kW battery from cold start at 50 A (after Gillibrand and Gray[37c]).

efficiencies (75–98 %) of hydrogen than of oxygen electrodes (60–85 %). The hydrogen electrodes were initially almost 100 % efficient, but small gas leaks in the electrodes reduced the efficiency of both electrodes after 2000 hours of operation.

The power consumption of the auxiliary circuits was 192 W, of the electrolyte pump 48 W, and of the radiator fan for temperature control 84 W.

Table IV.3. Current densities and potentials (relative to a hydrogen electrode in the same solution) of silver-impregnated carbon/oxygen electrodes in 7N KOH at 25 and 60°C (after Barak, Gillibrand and Gray[37]).

Current density	Potential (mV)	
(mA cm^{-2})	at 25°C	at 60°C
100	770	820
200	730	760
400	630	670
600	544	600
800	—	530
1000	—	460

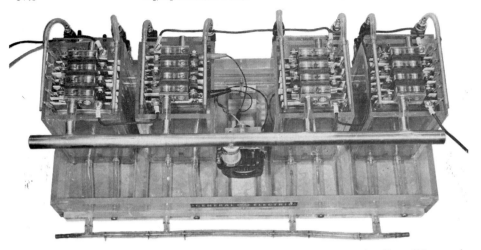

Figure IV.15a. Two hundred and twenty five watt hydrogen–air battery (after Kent, Nilson and Moran[38]).

IV.1.1.2.1   Hydrogen–air cell for alkaline electrolytes

Kent, Nilson and Moran[38] have constructed a battery of 225 watt rating, operating with hydrogen and air. The hydrogen is not circulated, joulean heat and water being removed by means of the air stream. Carbon dioxide is not removed from the air, so that carbonate is formed in the electrolyte. Electrodes, 0.3 mm thick and 100 cm^2 in area are of platinized carbon (anode) and platinized nickel (cathode). Hydrogen is supplied at 25 cm and air at 6 mm water pressure in excess of atmospheric. The battery (Figure IV.15a) consists of four groups of 8 cells in parallel, and is fed with 11 to 17 l min^{-1} of air. The electrolyte is 12N KOH.

In a test run of 225 hours at 40°C, the cells provided 100 mA cm^{-2} at 0.7 volt. After this period, carbonate in the electrolyte had built up to 1.7N, and the KOH concentration had fallen to 7.8N; simultaneously the output decreased from 75 to 64 mW cm^{-2}.

Kordesch[35b,36a] has recently made a detailed study of the effect of the CO$_2$ content of air (0.03 %) on the behaviour of hydrogen–air cells. The most important results are:

1. A carbonate content of the electrolyte (6N KOH) of, for example, 1M K$_2$CO$_3$, is completely harmless to hydrogen–oxygen cells. The electrolyte can therefore, if necessary, be used directly to absorb CO$_2$.
2. The operational life of air cathodes depends primarily on current density.
3. Metal-catalysed carbon electrodes show a better performance but are shorter-lived.
4. Cathodes operating in NaOH outlive those in KOH by a factor of three, but at cost of the potential level (Figure IV.15b).

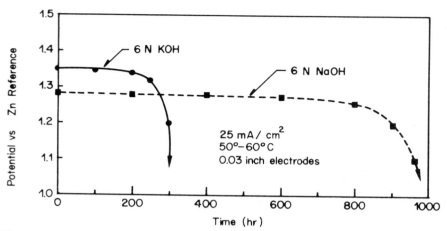

Figure IV.15b. Operational life and performance of hydrogen–air cells with CO_2-containing air, comparing Pt-catalysed cathodes in 6N KOH and in 6N NaOH (after Kordesch[36a]).

It seems that the extent of the liquid-film interface determines the rate of CO_2 pick-up from the air; NaOH is less wetting than KOH, and metal-catalysed electrodes are less repellent than carbon electrodes without metal addition.

Carbonate-plugged 6.3 mm carbon electrodes are usually permanently damaged, but the new composite electrodes (0.7 to 1.0 mm, see above) have frequently been washed free of carbonate and reused with success. Still newer electrodes of 0.5 mm thickness can be washed quickly *in situ* without suffering at all.

IV.1.1.3   *Porous Plastic Substrate Electrodes*

Thin, microporous electrodes based on a non-conducting plastic substrate have been developed by Williams and Gregory of Shell Research Ltd[38a]. 'Porvic M' (Porous Plastics, Ltd.) is a very suitable material; it is made from unplasticized polyvinyl chloride, 0.76 mm thick, and has a very uniform pore size of about 5μ. These properties allow the three-phase zone to be stabilized by a pressure difference of 0.15 to 0.25 kg cm^{-2}. The initial metallizing of the surface is carried out by vacuum evaporation of silver (for alkaline electrolytes) or of gold (for acid electrolytes), and the metal film so produced is thickened by electroplating. Finally the catalyst (noble metal) is applied by electrodeposition, or by incorporation in a binder. The electrodes are vacuum-impregnated with electrolyte before mounting in the battery.

An alkaline, 250 watt battery constructed of 21 cells is shown in Figure IV.15c: the electrode area is about 460 cm^2 (0.5 ft^2). Each cell has a working voltage of 0.6 V at a current density of 45 mA cm^{-2} [38b].

Figure IV.15c. Two hundred and fifty watt hydrogen–oxygen battery with 460 cm^2 porous plastic substrate electrodes (after Williams, Pearson and Gressler[38b]).

In extended experiments on these lines, a 5 kW hydrogen–air battery has been built (see chapter X, Figure X.4a). It is to be noted that a temperature of 65°C cannot be exceeded with the plastic material (Porvic) used. If polyethylene membranes of similar structure could be prepared, working temperatures up to 80°C would be practicable.

IV.1.1.4   Electrodes with Asbestos Membranes as Surface Layers ('supported' electrodes)

The preparation of sintered double-layer electrodes of large surface area is extremely difficult if the catalyst material is required to provide the electrode with its main structural support; this is particularly so when the electrodes must be as thin as possible. The Siemens Company (Erlangen, Germany)[38c] has therefore approached the construction of their fuel cell in another way, using the so-called 'supported electrodes'. In these, the catalyst is not in the form of a sintered layer, but is used as a porous layer of powder embedded between a metal screen and an asbestos membrane. The asbestos does not serve as an electrolyte carrier as in the Allis-Chalmers fuel cell (cf. section IV.1.2.2), but supports an otherwise mechanically unstable electrode and defines the boundary which it forms with the electrolyte. The asbestos hinders the entry of gas to the electrolyte, that is, it takes the place of the fine-pored surface layer of a double-layer electrode. This method of construction is suitable for making larger and, despite the small thickness of the active layer, more stable electrodes.

A 200 W battery made by Siemens consists of 15 cells (electrode area 300 cm^2), working at 30°C, with 1.2 to 1.3 atm (on the oxygen side), and a terminal voltage of 0.7 V per cell at 60 mA cm^{-2}. A current density of 15 mA cm^{-2} at 0.7 V is available at -20°C. The largest single cell with supported electrodes yet made has an electrode area of 0.25 m^2. The practical application of such batteries has been demonstrated by their use in motor boat propulsion; two alternative connections of 360 or 500 watts to a dc motor provided 0.5 hp at the propeller shaft.

IV.1.1.5   Gas and Electrolyte Circulation; Control of Temperatures and Water Content of Electrolyte[33]

The control of electrolyte concentration and temperature, for hydrogen–oxygen cells working at ambient temperature, is facilitated to some extent by a tendency to self-regulation. Dilution of the KOH by the water formed in the cell reaction is partially counteracted by an increase of the vapour pressure, and a rise in cell temperature facilitates the transfer of joulean heat.

There are three possible methods for the removal of water; they are as follows:

1. Operation of the cell under pressure above the boiling point (cf. the Bacon cell, IV.2.3).
2. Removal of the water by means of the gas-circulation system. Water evaporates into the hydrogen or oxygen delivered to the gas sides of the electrodes, and the gas streams become saturated with water vapour (at 20°C, 17 mg H_2O per litre of gas; 161 mg at 65°C). Subsequent passage of the gas through a condenser can be used for the continuous removal of water (Figure IV.16). Condensation at the gas sides of the electrodes must be avoided, or there will be obstruction of gas flow to the reaction zone.

For a battery working at 65°C, with a condenser at 20°C, one litre of gas can transport $161 - 17 = 144$ mg of water. Since 330 mg of water is produced per A hr, 2.3 l of gas is required to remove it. Hydrogen and oxygen consumed per A hr, however, amount only to 0.622 l. In this example, therefore, the quantity of gas required for water transport is four times that required for electrochemical reaction. This ratio is expressed by a factor F_{gas} (>4).

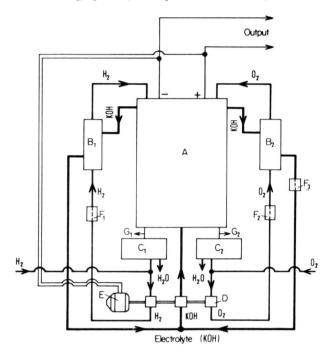

Figure IV.16. Diagram of a hydrogen–oxygen battery with gas and electrolyte circulation systems (after Kordesch[33]). A, hydrogen–oxygen battery; B$_1$, B$_2$, heat exchangers; C$_1$, C$_2$, condensers; D, pumps; E, motor; F$_1$, F$_2$, F$_3$, controls; G$_1$, G$_2$, outlets.

Kordesch and co-workers[34] have related limiting current densities, for a given gas flow, with the excess of gas (in terms of F_{gas}) required for water removal, for the case of a 6.3 mm thick carbon diffusion electrode operating in 9N KOH at various temperatures; the data are shown in Table IV.4.

Table IV.4. Limiting current densities (mA cm^{-2}) for 6.3 mm-thick carbon diffusion electrodes in relation to removal of water (after Kordesch[34]); 9N KOH, condenser (gas) temperature, 20°C, gas stream, 0.3 l hr^{-1} cm^{-2}.

Temperature °C	Current density (mA/cm^{-2})	Gas excess F_{gas}
35	10	50
50	30	17
65	60	8
80	100	5
100	180	3

3. Removal of the water by way of the *electrolyte*. At very high current densities, even the strongest gas stream could not alone remove all the water formed. The simplest solution to this problem is a separate distillation circuit for the electrolyte, distillation being facilitated by heat from the cell reaction, or by use of reduced pressure.

The Union Carbide Corporation has produced a 500 watt hydrogen–oxygen battery in which 7 % of the total energy produced is used for operating associated equipment (pumps, etc., cf. Figure IV.16). Kordesch[36] has successfully developed an energy-economizing process for gas and electrolyte circulation in hydrogen–oxygen batteries, without the use of pumps. The so-called *gas lift* is used in the electrolyte circuit. Nitrogen is used for the purpose on the laboratory scale, but in larger installations either the hydrogen or the oxygen is employed in a suitable unit included in the gas circuit. The gas used in this way as a carrier gas enters beneath a branched tube, which is connected on the one hand to the cell, and on the other to a vessel situated above the cell. The carrier gas carries part of the electrolyte with it into the upper vessel, and thus maintains a continuous flow of electrolyte back to the cell.

Recently Kordesch[35e] has used a venturi-jet at the gas entry to a module (Figure IV.16a) to effect gas circulation. Reaction water is removed in the hydrogen circulation circuit.

For purification of air from CO_2, Kordesch[35e] advocates asbestos paper impregnated with NaOH. Frysinger[38d] claims that soda-lime will reduce the

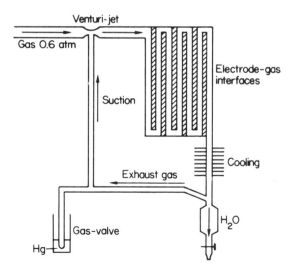

Figure IV.16a. Gas circulation by means of a venturi-jet (after Kordesch[35e]).

CO$_2$ content of air to less than 10 ppm, and will deal with 900–1000 lb of air per hour per pound of soda-lime, thus giving a scrubber weight lower than attainable with liquid absorbents.

IV.1.2　Cells with an Electrolyte Carrier

In these cells, a partition of porous material is situated between the electrodes— a diaphragm which will absorb an aqueous electrolyte and hold it, constrained in geometry and location, where required. This device eliminates the possibility of circulating the electrolyte, but offers the following advantages:

1. Substantial simplification in the construction of electrodes and cells.
2. The isolating effect of the diaphragm; this favours closer spacing of the electrodes, with a resulting reduction of resistance and of ohmic potential loss.
3. Particular suitability for cells of small volume and low weight.

Ion-exchange membranes and asbestos diaphragms have come into use for this purpose.

IV.1.2.1　Cells with Ion-exchange Membranes

The use of ion-exchange material as electrolyte, or electrolyte matrix, for fuel cells was first proposed by Grubb[39]. These materials have been mainly investigated for use with hydrogen–oxygen cells[40-46], but they have also been studied in relation to cells consuming other fuels, such as hydrocarbons and alcohols[47,48].

Since Niedrach and Grubb[49] have recently published an excellent, detailed review of this particular field, the present account will be limited to a summary of the more important results and reference to some recent developments.

The original concept of the use of ion-exchange membranes for fuel cells envisaged the transport of electrical charge *only* by the exchangeable ions (H$^+$ or OH$^-$). If the excess acid or alkali that has been used to convert the ion-exchange resin into its fully acidic or basic form is completely removed by washing, there remains a constant ionic content and hence a constant conductance of the resin. The ions are held by the coulombic forces exerted on them by the oppositely charged groups built into the three-dimensional polymeric network. The water produced by electrochemical reaction is not, in any case, formed at a location where it could wash away the residual ions. A further advantage depends on the fact that exchange membranes can take up only limited quantities of water, and thus reject the excess of water generated by the cell reaction; they can be regarded as *invariant* electrolytes. This property facilitates elimination of the water formed in the cell reaction.

In practice, however, it was soon found to be essential to use the membranes with addition of *unbound* electrolyte, since the maximum current densities otherwise attainable at technically useful terminal voltages were only of the order of 1 to 3 mA cm^{-2}. The membranes are therefore immersed in a suitable acid or alkaline electrolyte until the Donnan equilibrium is established. They are then removed and superficially dried (blotting paper) before being mounted in the cell; the membranes act as a gel-like matrix for the unbound electrolyte. This treatment substantially improves the conductance of the membrane and the operation of the electrodes. The cell contains no mobile electrolyte.

It is a disadvantage that the water produced in the cell reaction cannot be removed by evaporation; electrolyte removed by water flowing from the cell must be periodically replaced.

To overcome these difficulties, Juda, Tirrell and Lurie[50] have proposed the use of two ion-exchange membranes with *free* electrolyte between them, the outer surfaces being in contact with the electrodes. In a further modification of this scheme, they replaced the oxygen electrode with a bromine–bromide electrode. A current density of 70 mA cm^{-2} at 0.6 volt was attained at 25°C with 0.5N Br$_2$ solution in 31% sulphuric acid. It was, however, very difficult to obviate diffusion of bromine to the anode in prolonged operation.

The ion-exchange membrane ideal for use in fuel cells should have the following properties:

1. high ionic conductance,
2. zero electronic conductance,
3. very small permeability to gases,
4. resistance to deformation,
5. mechanical strength,
6. low water transport,
7. chemical stability,
8. insensitivity to partial drying-out.

Hitherto, two *types of membrane* have been predominantly used: heterogeneous membranes containing exchange material based on cross-linked polystyrene, and homogeneous membranes based on condensed phenol–formaldehyde resins. Since membranes of particularly good conductance are usually not of satisfactory mechanical stability, a non-conducting synthetic fabric is frequently used to strengthen them.

Commercially available membranes which have proved serviceable are, amongst others, Permaplex A20 (7 Ω cm^{-2} at thickness $d = 0.9$ mm) and Bayer A351 (6 Ω cm^{-2} at $d = 0.3$ mm) as alkaline types, and TNO C60 (4 Ω cm^{-2} at $d = 0.35$ mm) and Permaplex C10 (7 Ω cm^{-2} at $d = 0.7$ mm) as acid ion-exchange membranes. The membranes used by the General Electric Company have even better properties, and this is also particularly so for the membranes recently developed by the American Cyanamid Co., for hydrogen–oxygen cells with alkaline electrolyte matrices. The resistance per cm^2 of the membrane, furnished with unbounded electrolyte, is said to be about 0.5 ohm for a thickness of 0.3 mm.

IV.1.2.1.1   Construction and operation of membrane cells

The general principles of construction* of a hydrogen–oxygen cell with an ion-exchange membrane as electrolyte matrix are illustrated in Figure IV.17. Electrodes are arranged on both sides of the membrane (thickness, 0.3 to 0.9 mm). The greater part of the electrode surfaces is in direct contact with the membrane. The nature of this contact is a critical factor in relation to electrical properties, and this is one of the major problems in the development of technically useful cells. Plastic or metal plates, suitably recessed for the supply of gas to the electrodes, form the walls of the cell. When washed-out membranes are used, the vents of the cell are normally closed; they are used for running off reaction water and for blowing foreign gas out of the gas chambers. When the cell is used with addition of unbound electrolyte, water is removed by evaporation into a direct or circulating stream of gas. A typical laboratory-scale cell is shown in Figure IV.18. The effective geometrical electrode area is 10 cm².

Good contact between *electrode* and *membrane* is achieved by combining them into a single mechanical unit, with use as catalyst of a suitable metallic powder, (platinum and palladium for acid cells; skeleton-nickel and silver in addition to platinum metals for alkaline cells), by processes involving pressing, metal evaporation, or incorporation into the polymer structure[52]. The metal to be used can also be electro-deposited on a graphite substrate[53].

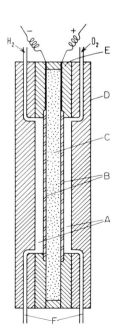

Figure IV.17. Principles of construction of a hydrogen–oxygen cell with ion-exchange membrane (after Niedrach and Grubb[49]). A, gas chambers; B, electrodes; C, membrane; D, front plate; E, gasket; F, vents.

---

* L. W. Niedrach, *Brit. Pat.* 894538.

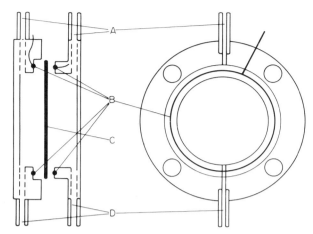

Figure IV.18. Constructional sketch of a typical laboratory cell
with acrylic resin housing (after Jahn, Vielstich and Wolf[51]). A, gas
entry; B, silver collar for current conduction; C, membrane-
electrode combination; D, gas exit.

Krumbein and Russell[53a] describe a method for the metallographic preparation of water-impregnated metal-polymer laminates. The technique involves the
replacement of the water component by epoxy-resin, by use of a series of water/
alcohol/resin baths and vacuum-impregnation.

The resistance of the current collectors is generally negligible. If, however, the
contact between the collector and the catalyst layer is purely mechanical, sufficient pressure must be provided to keep the contact resistance low. Collector
plates are necessary in cells with electrodes of large area (cf. Figure IV.24).

The major voltage losses are due to *overpotentials* at the electrodes and to
membrane resistance. The best short-term current–voltage curve obtained by

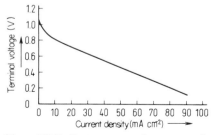

Figure IV.19. Best current–voltage curve of a
hydrogen–oxygen membrane cell, with H_2SO_4
as unbound electrolyte, at room temperature
(after Niedrach and Grubb[49]).

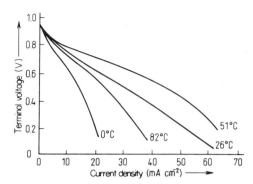

Figure IV.20. Temperature dependence of the current–voltage curve of an acid hydrogen–oxygen membrane cell with 8.8 mg cm^{-2} of palladium black (after Niedrach and Grubb[49]).

Niedrach and Grubb[49] with a membrane cell using a catalyst layer of platinum black (8.8 mg cm^{-2}), is reproduced in Figure IV.19. The influence of *temperature* on the current–voltage curve of a hydrogen–oxygen cell with 8.8 mg cm^{-2} of palladium black as catalyst is shown in Figure IV.20. At *lower* temperatures, loss of water by evaporation is relatively slow. The catalyst layer of the oxygen electrode, in particular, becomes covered with a layer of water which obstructs the supply of gas. The fall of current density with increase of temperature from 51 to 82° is attributed by Cairns, Douglas and Niedrach[46] to the considerably increased vapour pressure of water, and the consequent reduction of the partial pressure of oxygen.

According to Cairns, Douglas and Niedrach[46], the loss in current efficiency by diffusion of oxygen and hydrogen across the membrane is negligible. Their measurements for a reinforced, washed-out membrane based on phenol–

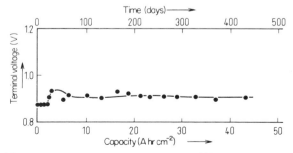

Figure IV.21. Prolonged operation of an acid hydrogen–oxygen membrane cell (22 mg cm^{-2} of platinum) at 3 mA cm^{-2} (after Niedrach and Grubb[49]).

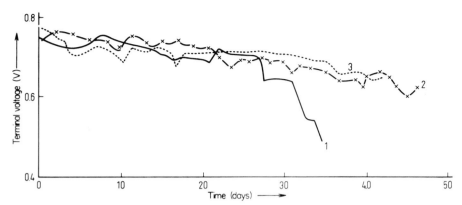

Figure IV.22. Prolonged operation of an acid hydrogen–oxygen membrane cell (22 mg cm^{-2} of platinum) at 10 mA cm^{-2}, at room temperature (after Jahn, Vielstich and Wolf[51]). Curve 1: home-made, phenol–formaldehyde-based membrane. Curves 2 and 3: TNO-C-60 membrane, American Machine and Foundry Co.

formaldehyde indicated a loss of gas which corresponded to a current density of only 25 μA cm^{-2}.

IV.1.2.1.2   Endurance tests of individual cells

Figures IV.21–23 show the behaviour of hydrogen–oxygen membrane cells with platinum as catalyst, in prolonged operation at room temperature, at current densities of 3, 10 and around 30 mA cm^{-2}. The working life, τ, is quite well estimated in terms of the charge provided in A hr cm^{-2}, i.e. the product $j\tau$, where j is current density, is approximately constant. In general, the limit lies at about 40 to 50 A hr cm^{-2}, but in individual cases a performance up to 150 A hr cm^{-2} was reached.

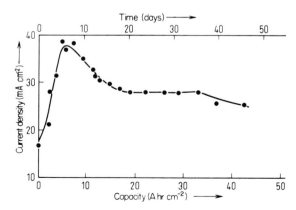

Figure IV.23. Current density–time curve at 0.5 volt terminal voltage emf for an acid, hydrogen–oxygen membrane cell, 22 mg cm^{-2} of platinum (after Niedrach and Grubb[49]).

An improvement in performance is seen to occur up to 10 A hr cm^{-2} (Figure IV.23). A slow change in membrane structure then causes a falling-off, until, eventually, the cell emf breaks down relatively suddenly and completely.

The working life is shorter when palladium is used as catalyst. A yellow colour developing in the electrolyte suggests that the palladium is being slowly corroded in the acid electrolyte, probably mainly on the oxygen side[49].

IV.1.2.1.3   Battery designs

Cairns, Douglas and Oster[42,46] paid particular attention to the design and construction of a battery of up to 37 cells (electrode area ca. 20 × 30 cm^2). For simplicity and ease of construction and use, rectangular cells were arranged in series on the filter-press principle (Figure IV.24). The collector plates are bipolar, and provide cooling surfaces for the removal of joulean heat. They are ribbed the better to make contact with the catalyst layers, the ribs on either side being at right angles to each other to provide mechanical stability.

Since the battery becomes warm in operation, it is no easy matter to ensure that the reaction gases remain saturated with water vapour. A gas stream of insufficient moisture content can dry out membranes locally, to the extent of causing perforations or cracks. This hazard is all the greater when air is used. The use of porous material to distribute moisture over the membrane surfaces has proved effective.

A portable 200 watt battery for military use is illustrated in Figure IV.25. Air is used as the source of oxygen, and also for cooling and removal of water. Hydrogen is supplied from an associated generator (on the left of the illustration), in which sodium borohydride is allowed to react with 35 % sulphuric acid. The battery consists of 37 cells and weighs 10 kg. Further development of this system culminated in the 1 kW batteries used in the space vehicle Gemini V.

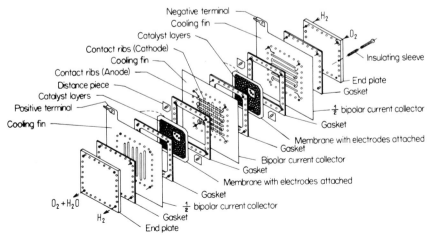

Figure IV.24. Structure of a hydrogen–oxygen battery with ion-exchange membranes (after Oster[42]).

Figure IV.25. Portable 200 watt battery of the General Electric Co[49].

In the last five years the Compagnie Française Thomson Houston[53b] have developed a hydrogen–oxygen battery by using ion-exchange membranes as electrolyte. The electrodes consist of ultra-fine, platinized nickel screens (3600 mesh cm^{-2}), hot-pressed to the thermo-plastic membranes, and provided with platinum black (or palladium black for the oxygen side) catalyst by electrodeposition. The cells are closed by gold-plated nickel plates, with gold-plated contact ribs pressing against the electrodes, since the cells are packed together on the filter-press principle. The end-plates of battery modules are similarly gold-plated, strengthened nickel plates. The ion-exchange membranes are of sulphonated polystyrene, made by polymerization of styrene-*m*-sulphonic acid. Recently, attempts have been made to improve the working life of membranes, already good, by perfluorination of the vinyl group before polymerization.

Figure IV.25a shows a 50 watt continuous rating prototype, with 20 cells (100 cm^2 electrodes) connected in series; it is 40 cm long, excluding the pressure vessel. The temperature of the module rises above 50°C on continuous load of 50 W. The terminal voltage is 16 V (0.8 V per cell) at 1 A, 11 V at 5 A (50 mA

Figure IV.25a. Hydrogen–oxygen (air) battery of the Compagnie Française Thomson Houston[53b], 50 W continuous rating. Battery weight, 9 kg.

cm^{-2}). The battery can be operated with air instead of oxygen by use of a blower consuming 15 W.

IV.1.2.1.4   Teflon-bonded flexible electrodes

The American Cyanamid Company has developed a membrane cell for hydrogen and oxygen[54] which differs from the usual construction. Electrodes rendered flexible by Teflon bonding are pressed mechanically against the electrolyte matrix.

An organic membrane, or an asbestos diaphragm of only 0.5 ohm cm^{-2}, is used for *alkaline* electrolytes (5N KOH). The electrodes consist of platinized nickel gauze embedded in Teflon; they are flexible and extraordinarily strong. The platinum content is 7 to 10 mg cm^{-2}.

Laboratory cells of 25 cm^2 effective surface area, with a strong gas stream through the gas chambers (cf. Figure IV.17)—no circulation—provided 100 mA cm^{-2} at 25°C. More than 1000 working hours with a terminal emf of 0.8 to 0.7 volt were achieved. Increase of temperature to 70°C raised the terminal voltage to 0.85 volt.

Fibreglass diaphragms are used in *acid* cells (5N H$_2$SO$_4$), and a tantalum gauze as the electrode support for the platinum catalyst. Under similar conditions to those mentioned above, a laboratory cell also worked for 1000 hr at 100 mA cm^2 and 25°C, with a terminal voltage of about 0.7 volt.

It is aimed to develop an electrode material containing less platinum. Electrode costs could be further reduced if the tantalum gauze could be replaced by active carbon (on a metal screen) as catalyst support. It is hoped to attain 80 mW cm^{-2} with 2 mg Pt cm^{-2}. This would correspond to a platinum investment of 25 g per kW.

IV.1.2.1.5 Cells with inorganic ion-exchange membranes

Polymeric inorganic substances such as $(ZrO)_n(H_2PO_4)_{2n}$ imbibe water like gels, and have ion-exchange properties. The conductance of inorganic ion-exchange materials is similar to that of organic ion-exchange resins, and they are more stable to increased temperature (up to 100°C), nuclear radiation and oxidative decomposition.

The work of Dravnieks, Boies and Bregman[55] has led to the development of a practicable membrane, consisting of a mixture of Teflon with $(ZrO)_n(H_2PO_4)_2$. Preliminary experiments with platinum black as catalyst have given 3 mA cm^{-2} at 0.6 V.

Later experiments by Berger and Stries[55a] have shown that membranes of the zinc phosphate type have, at 50 to 65°C, properties nearly as good as those of organic membranes; they obtained, for example, 30 mA cm^{-2} and 0.7 V at 65°C with hydrogen and oxygen. The membranes proved to be stable up to 150°C, but, of course, the properties of the hydrogen–oxygen cell deteriorate at temperatures above 60°C; the water content of the membrane falls, and the increasing partial pressure of water vapour decreases the rates of reaction of hydrogen and oxygen at the electrodes. At a current density of 30 mA cm^{-2}, terminal voltage was found to fall to 0.6 V at 100°C and 0.35 V at 150°C.

The stability of these membranes at temperatures over 100°C made it worthwhile to study the oxidation of hydrocarbons. The experiments of Berger and Stries showed, however, that only inadequate current densities could be obtained at technically practicable terminal voltages. Thus, at 120°C and 0.4 V, propylene gave 8 mA cm^{-2}, propane and butane 6 mA cm^{-2} and ethane 2 mA cm^{-2}, in each case after 5 minutes cell loading.

IV.1.2.2 Asbestos Diaphragms as Electrolyte Carriers

Recent work indicates that asbestos diaphragms have a number of advantages as electrolyte carriers over ion-exchange membranes. They are stable to 150°C in acid or alkali and are insensitive to drying out.

Because of their fibrous structure, asbestos membranes can take up, and hold like a sponge, considerable volumes of electrolyte. The fibres are cylindrical, of diameter about 200 Å. The *capillarity* of asbestos is a particularly important property; several atmospheres pressure are required to force gas through an asbestos membrane soaked with electrolyte. If such a membrane is brought into

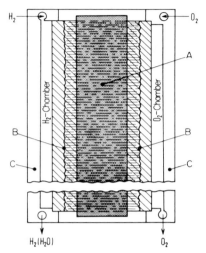

Figure IV.26. Diagram of a hydrogen–oxygen cell with asbestos diaphragm as electrolyte carrier (after Wynveen and Kirkland[56]). (A) Asbestos diaphragm; (B) porous nickel electrodes; (C) electrode support, used simultaneously for current collection and gas distribution.

contact with a porous electrode of lesser capillary activity, the electrolyte enters the pores of the electrode to but a slight extent, so that the electrode need not be water-repellent, nor of double-layer structure (cf. section IV.1.1). The pore distribution is therefore not critical (Figure IV.26).

The highly absorbtive properties of asbestos carry the further advantages that:

1. loss of water by evaporation is reduced, and
2. the mobility of the electrolyte is greatly reduced (the hydrostatic effect itself is negligible even for electrodes as large as $50 \times 50 \text{ cm}^2$).

IV.1.2.2.1 Construction and properties of hydrogen–oxygen cells with asbestos diaphragms

The properties of asbestos, outlined above, as an electrolyte carrier are exploited in the construction of the hydrogen–oxygen cell due to Wynveen and Kirkland (Allis-Chalmers Manufacturing Co.)[56], shown diagrammatically in Figure IV.26. The asbestos membrane, 0.75 mm thick and $15 \times 15 \text{ cm}^2$ in area, is soaked with 5N KOH. Porous nickel plates (porosity 85%), of about the same thickness, serve as electrodes; they consist of nickel gauze, for mechanical support, embedded in sintered carbonyl nickel.

The hydrogen electrode is platinized and the oxygen electrode palladized or silvered; the latter may be made of nickel gauze with an intermediate layer of sintered silver. *Homogeneous* dispersion of the catalyst through the thickness of the electrode plate is secured by impregnation of the sintered nickel by noble metal. This is essential to avoid localization of the reaction zone (see below and Figure IV.27).

Each of the porous nickel electrodes is welded into a nickel-plated, light metal mounting, which has the following functions:

1. to hold the electrodes in position so that they are lightly pressed against the asbestos diaphragm,
2. to distribute gas, by means of a suitable arrangement of grooves, uniformly over the reverse side of the electrode,
3. for current conduction and for holding the cell together mechanically,
4. for dissipation of heat.

When cells are assembled in series (Figure IV.30), the electrode mountings are bipolar.

The cell construction is such that the electrodes exert a certain pressure on the asbestos diaphragm and some electrolyte penetrates into the porous electrodes. An idealized representation
of the initial depth of penetration is given in Figure IV.26. The desired location of the *three-phase
zone* can be attained by adjustment of the quantity of electrolyte and the pressure of the electrodes on
the diaphragm. It is necessary to allow for this penetration by using electrolyte slightly in excess of
that corresponding with the pore volume of the asbestos membrane itself.

Reaction water, produced in continuous operation of the cell, is removed by evaporation in the gas circulation system. Since, in the case of alkaline electrolytes, the water is produced at the hydrogen electrode, it is adequate, for current densities of about 100 mA cm^{-2}, to circulate the moist hydrogen through a condenser. The oxygen is not circulated, but from time to time accumulated impurities must be expelled by a blast of gas.

Whereas the quantity of water formed is directly proportional to the quantity of electricity withdrawn from the cell, the rate of collection of water at the electrode depends on several factors, such as temperature, gas pressure, electrolyte composition, pore structure of the electrode, rate of gas stream, and difference of temperature between the cell and the condenser. It is therefore very difficult to keep the water content of the electrolyte at a constant value for prolonged periods, particularly under conditions of varying electrical load on the cell.

If less water is removed than is formed, the excess penetrates through the asbestos membrane
from the hydrogen to the oxygen electrode until a new equilibrium is established. The three-phase
zone then moves as shown, in an idealized way, in Figure IV.27a–c. The symmetry of the profiles

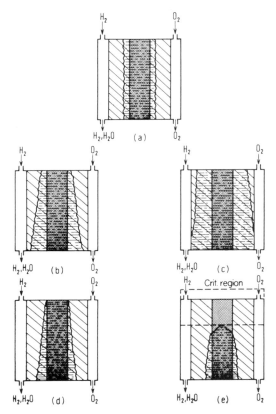

Figure IV.27. Displacement of the electrolyte boundary in a hydrogen–oxygen cell with asbestos diaphragm and porous metal electrodes (after Wynveen and Kirkland[56]). (a) Initial distribution; (b) and (c) too little removal of water; (d) and (e) excessive removal of water.

is retained if hydrogen and oxygen are under the same pressure, and if both electrodes have the same pore structure.

If, on the other hand, too much water is removed from the cell, then movement of electrolyte from the oxygen towards the hydrogen electrode takes place, and the equilibrium profile is displaced as indicated in Figure IV.27 d and e. If the electrolyte boundary reaches the asbestos diaphragm, the vapour pressure drops because the electrolyte is then confined to the finer-pore system. Only when the water contained in the asbestos begins to evaporate does any danger of mixing of the reaction gases enter.

It is therefore essential to allow the water content of the cell to vary only between certain limits. Suitable methods of regulation are described later. It is, of course, a prerequisite that the electrochemical properties of the cell shall not be too greatly dependent on change in electrolyte concentration and profile. Figure IV.28 provides a typical example showing that this is the case.

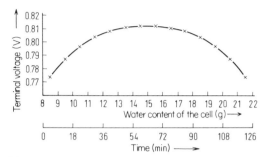

Figure IV.28. Terminal voltage as a function of water content or of operation time (after Wynveen and Kirkland[56]). Hydrogen–oxygen cell with asbestos membrane, with no gas circulation for removal of water; 77°C, load 20 A (ca. 100 mA cm^{-2}), gas pressure 0.35 atm above atmospheric.

With increase of water content from 8 to 22 g per cell, the terminal emf passes through a maximum, but the greatest voltage variation amounts only to about 5%.

A four-cell unit, with suitable regulation equipment, was operated for 700 hr at approximately 100 mA cm^{-2}. The average terminal emf was 0.8 volt per cell at a working temperature of 65 ± 3°C, and the total gas pressure was regulated between 1 and 1.35 atm. The current–voltage curves before and after the test are shown in Figure IV.29.

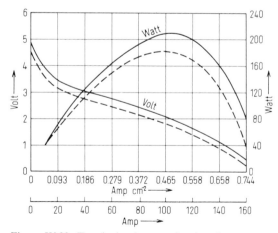

Figure IV.29. Terminal voltage emf and performance as functions of load after 1 hr (———) and after 700 hr (– – – –) operating time (after Wynveen and Kirkland[56]). Four-cell unit; 65°C; 1.35 atm (total); ohmic resistance, 1.4 × 10^{-3} ohm, electrode area 230 cm^2.

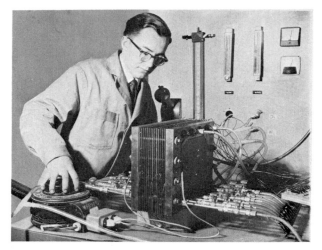

Figure IV.30. 500 watt hydrogen–oxygen battery of the Allis-Chalmers Manufacturing Company (after R. Jasinski and T. G. Kirkland[57]).

The type of cells just described, used in batteries of up to several kW capacity, are connected in series by the filter-press principle (Figure IV.30). In operation over several hundred hours, at 150 mA cm^{-2} and 80°C, a terminal emf of 0.8 volt is obtained. The current–voltage curve of a single cell with 0.25 mm thick asbestos diaphragm is shown in Figure IV.31.

Only a slight deterioration in performance was observed in more than 4500 hours of operation. For working with *air* instead of oxygen, the 1963 performance data[59] are 100 mA cm^{-2} and 0.78 volt, under otherwise similar conditions. Carbon dioxide was removed from the air by means of a NaOH trap.

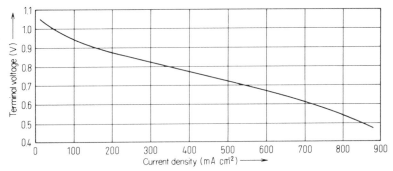

Figure IV.31. Typical current–voltage curve of a single cell (after Jasinski and Kirkland[58]); $p_{H_2} = p_{O_2} = 1$ atm total pressure; 80°C; 0.25 mm asbestos; active surface area, 15 × 26 cm^2.

IV.1.2.2.2   Control of water content in continuous operation

The Allis-Chalmers Manufacturing Company has proposed three methods for the control of the water content of hydrogen–oxygen cells with asbestos membranes as electrolyte carriers. The first method depends on the relationship between terminal voltage and electrolyte concentration (cf. Figure IV.28). The water content is continuously controlled by *sensing the terminal voltage*; corresponding adjustment of the temperature of the condenser in the hydrogen circuit then determines whether more or less water is removed.

The second process also works by adjustment of water removal by control of the condenser temperature, but instead of using emf measurement for the purpose, the *water vapour pressure* in the hydrogen and oxygen circuits is determined every three minutes by gas chromatography[60].

> The hydrogen circulated through the cell takes up water from the electrolyte to an extent which brings the water vapour pressure in the gas stream almost to the vapour pressure of the electrolyte. The gas is cooled in the condenser and part of the vapour condenses. The measurement in the gas chromatograph is independent of the carrier gas used, so that a given quantity of water provides the same signal, irrespective of whether hydrogen, oxygen, or air conveys it.
>
> In experiments to the present, a Beckmann GC-2 Chromatograph, with conductance cell and detector has been used, with helium (42 cm^3 min^{-1}) as carrier gas.
>
> Chromatographic measurements have also been used to demonstrate the existence of a concentration gradient of OH$^-$ ions across the asbestos membrane. At a current density of 47.5 mA cm^{-2}, the KOH concentration at the oxygen electrode becomes 5 to 6 % greater than on the opposite side of the membrane (cf. Figure IV.33).

Both of these methods have the advantage that a wide range of electrolyte concentrations provides an adequate terminal voltage (cf. Figure IV.28). Correct choice of the condenser temperature ensures that the 'dry limit' is not exceeded, independently of the rate of hydrogen circulation. On the other hand, a sufficiently rapid flow of hydrogen precludes transgression of the 'wet limit', even at very high current densities.

Experiment shows that the *optimal* KOH concentration lies between 32 and 38 wt %. The range between 27 and 45 % KOH is, however, acceptable, since it involves a variation in terminal voltage of less than 10 %.

In the third method, water is withdrawn from the cell with a second asbestos membrane (Figure IV.32). This *water transport membrane* is initially wetted with a KOH solution of somewhat higher concentration. It is open to the hydrogen chamber on one side and is in contact with a porous nickel partition on the other. Since the water-transport membrane is completely filled with liquid, loss of hydrogen through it by diffusion is negligible.

If the content of water vapour of the hydrogen chamber increases, the water transport membrane absorbs additional water, the volume of liquid increases and the space behind the porous nickel plate accumulates water vapour. If, now, the water vapour pressure in this space is reduced to the desired value, water is withdrawn from the cell. In this way the KOH concentration in both membranes and the rate of removal of water from the cell become, at a given temperature,

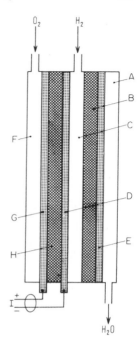

Figure IV.32. Diagram of a hydrogen–oxygen cell with two asbestos membranes for automatic control of water vapour pressure (after Platner and Hess[61]). A, H_2O chamber; B, transport membrane; C, hydrogen chamber; D, porous H_2 electrode; E, porous nickel electrode; F, oxygen chamber; G, porous O_2 electrode; H, membrane soaked with electrolyte; I, electrical output.

directly dependent on the pressure in the chamber behind the porous nickel partition. An example of the manner in which the KOH concentrations adjust themselves in a cell with two membranes is shown in Figure IV.33.

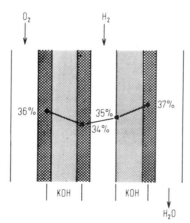

Figure IV.33. Example of the distribution of KOH concentrations in a cell with two asbestos membranes (after Platner and Hess[61]).

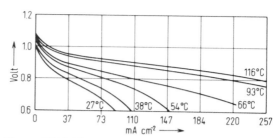

Figure IV.34. Temperature-dependence of the current–voltage curve of a hydrogen–oxygen cell with asbestos membrane (after Platner and Hess[61]); $p_{H_2} = 0.35$ atm atmospheric pressure.

This method of water removal is particularly suitable for application in *space vehicles*, water vapour being expelled directly to space, except for a control mechanism which would adjust the water vapour pressure to a value appropriate to the working temperature. Circulation of gases is unnecessary, but gas-conveyed impurities must occasionally be expelled.

Figure IV.34 shows the strong dependence of the current–voltage curve on the *working temperature*, at a total gas pressure of $p_{H_2} = 1.35$ atm.

Further data (e.g., energy/volume and weight) of developed prototypes are given in chapter X, where possible applications are also discussed.

IV.1.3   Sources of Hydrogen for Hydrogen–oxygen Cells

Two processes come under consideration for the cheap preparation of hydrogen in large quantities. The first is the so-called 'Steam hydrocarbon process'

$$C_nH_{2n+2} + nH_2O \rightarrow (2n + 1)H_2 + nCO$$
$$CO + H_2O \rightarrow CO_2 + H_2 \tag{IV.1}$$

and the second is the *decomposition of ammonia*. Installations for conducting the first of these reactions operate with a capacity of 750 to 60 000 m³ day⁻¹; for lesser quantities and discontinuous operation the ammonia decomposition is more economic.

For vehicle propulsion by fuel cells, size and weight of the generator are decisive factors. Kordesch and co-workers[34,35] have developed a small, laboratory scale reformer for the steam-hydrocarbon process, which produces 7 m³ of hydrocarbon per hour (Figure IV.35). This is sufficient to supply a battery of 10 kW rating. The column on the right-hand side of the figure serves to remove carbon dioxide by adsorption in ethylamine. The unit has already worked successfully for several months using methanol, propane, natural gas and petroleum.

Figure IV.35. Hydrocarbon reformer with ethylamine
column to the right in the photograph; output 7 m^{3}H$_2$
hr^{-1} (after Kordesch[34]).

Recently attempts have been made to secure a more compact construction
by using palladium–silver membranes to separate hydrogen and carbon dioxide.
At 400 to 500°C, hydrogen should diffuse through the metal at a rate corresponding to several A cm^{-2}* (cf. section III.1.7). For this purpose, the palladium
cylinder is mounted concentrically inside the stream-hydrocarbon reformer[151a].

Meek, Baker and Allen[151b] have proposed a *three-stage* steam-hydrocarbon
process. In the first stage, natural gas is steam-reformed at 800°C to produce a
mixture of hydrogen, carbon monoxide and carbon dioxide. This, after cooling,
passes to a 'carbon monoxide shift reactor,' working at about 270°C, which
reduces the carbon monoxide content from a typical 15 % (dry basis) to about
2000 ppm. This gas might be an acceptable fuel for certain acid cells working
above 100°C, but further reduction of the carbon monoxide content is desirable
for lower-temperature cells. It is reduced to about 20 ppm in the third stage,
which consists of a low temperature (190°C) 'selective methanation reactor'
in which carbon monoxide and hydrogen react to produce methane.

* Besides wall-strength and temperature, factors determining practical rates of hydrogen transfer
  are hydrogen partial pressure in the crude gas, pressure of the pure hydrogen produced and hydrogen content of the residual gas.

Vertes and Hartner[151c] have shown that it is advantageous to couple the steam reactor, via a palladium–silver membrane, directly with the hydrogen anode, so that joulean heat is available to maintain the reforming process at a working temperature of 250°C. Methane, hexane and kerosene were studied as fuels for this '*internal reforming fuel cell*'; at 250°C and 0.8 V terminal voltage, about 50 mA cm^{-2} was obtained, whereas methanol at 200°C provided more than 200 mA cm^{-2}. In contrast to direct hydrocarbon cells, this kind of cell needs no platinum, but only palladium–silver and nickel.

A compact *ammonia decomposer* for supplying a 200 watt battery has been developed by Engelhard Industries Inc.[151]; in this, hydrogen is separated by means of a palladium–silver alloy membrane. Operating for 12 hours, the generator provides 107 l hr^{-1} of hydrogen; its total volume is 57 l and its weight 6.8 kg (excluding fuel).

For small, *portable* fuel cells (50 to 500 watt), it is also suitable to use metal hydrides as a source of hydrogen (cf. section IV.1.2.1 and Figure IV.25).

Finally, for economical electrical propulsion of vehicles over a restricted area (factory premises), generation of hydrogen and oxygen by *pressure electrolysis* can be used. The recharging of hydrogen and oxygen tanks requires very much less time than the charging of normal accumulators.

IV.2   FUEL CELLS FOR MEDIUM AND HIGH TEMPERATURES

Increase of cell temperature serves the main purpose of making practicable the oxidation of cheap but unreactive *hydrocarbons*. Molten carbonate (section IV.2.2), concentrated acids (section IV.2.3) and solid oxides (section IV.2.5) can be used as electrolytes.

Alkaline 'equilibrium electrolytes' (section IV.2.4) play a special role in the reaction of carbon-containing fuels in aqueous, *invariant* electrolytes. The Bacon hydrogen–oxygen cell, operating with concentrated KOH at 200° to 250°C, is appropriately discussed in section IV.2.3.

Since electrode reactions at temperatures above 100°C were not treated in chapter III, the present section also contains some basic theoretical discussion with, in section IV.2.1, a short explanatory introduction.

IV.2.1   Introductory and Historical*

An intrinsic difficulty in devising fuel cells to work with aqueous electrolytes at atmospheric pressure—and hence at temperatures below the boiling points of such electrolytes—is to find catalysts that will accelerate electrode reactions sufficiently to keep overvoltage losses small. This difficulty disappears at high

---

* By Dr. R. P. Tischer, Scientific Research Staff, Ford Motor Company, Dearborn, Michigan, U.S.A.

temperatures; thus at temperatures over 750°C, catalytic properties of electrode materials no longer play any significant part. By the same token, catalyst poisoning, such as caused by the impurities of commercial hydrocarbons is less of a problem. Hydrogen peroxide, the formation of which at the oxygen electrode at room temperature leads to voltage loss, is no longer stable even at moderately increased temperatures. It follows that there are many fuels (e.g. carbon) which can be brought into electrochemical reaction only at higher temperatures. The electrolytes suitable for use under these conditions have no worse a conductance—frequently better—than that of aqueous solutions, within their range of working conditions.

From this it would seem to be the only rational choice to operate fuel cells at elevated temperatures. Unfortunately, the advantages so gained are offset by a whole series of *disadvantages*. The higher the temperature, the greater the difficulties associated with the properties of materials. The working life of cells becomes limited by the wearing out of electrodes, containers and other components. The increase in reaction rates with rise of temperature also applies to undesired corrosion processes. Thermal expansion can lead to damage because of differing expansivities of components, even as a result of raising the temperature of the cell only once to a constant operating value. The increasing conductance of insulating materials accentuates the difficulties of maintaining good insulation. At high temperatures, porous electrodes cannot be provided with lyophobic surface coatings to guard against penetration by electrolyte.

Different temperature ranges call for different methods of cell construction, and the changes needed will vary according to the nature of the electrolyte that can be used at a particular temperature. Although reaction rates are increased by rise of temperature, equilibria are also displaced, and undesirable *side-reactions* are also accelerated. For instance, carbon may be oxidized incompletely to a mixture of carbon monoxide and carbon dioxide, corresponding to the Boudouard equilibrium. Carbon monoxide, used as a fuel, disproportionates to carbon dioxide and to carbon, which is much more difficult to oxidize. Hydrocarbons undergo transformation to substances which enter into electrochemical reaction more easily, but this can be accompanied by a troublesome deposition of carbon.

The requirements that a fuel cell should satisfy (high efficiency, adequate working life, stability of electrolyte composition, etc.) remain the same as at room temperature. For the electrolyte, this means that it must have a high conductivity, must remain stable at anode and cathode potentials and must therefore have a sufficiently large decomposition voltage, and it must not interact with reactants or reaction products in such a way as to undergo irreversible change in composition. The choice of electrolytes satisfying these requirements —as in the case of aqueous electrolytes—is very limited.

Alkaline hydroxides combine the advantages of considerable stability, good conductivity and low melting point. They allow electrodes of relatively base metals to be used, but they react with carbon dioxide.

An excellent review of early attempts to construct fuel cells was given in 1933 by Baur and Tobler[62], and only one of these pioneer works will now be mentioned. There was a turning point in the development of fuel cells in the period around 1900, which marked the beginning of a real understanding of the electrochemical processes that occur in cells. It is astonishing how much of our present knowledge of fuel-cell processes is to be found in publications that appeared in the first third of the century.

Even before the turn of the century, soon after Wilhelm Ostwald[63] had indicated the decisive advantage of the electrochemical utilization of fuel, Jacques[64] built, in Boston, a *1.5 kW battery with carbon anodes* and an iron cathode, that worked at 400° to 500°C with an air-bubbled alkali hydroxide electrolyte (Figure IV.36). Jacques himself believed that the air must be delivered to the carbon electrode.

The processes taking place in the *Jacques* cell were first clarified by Haber and Bruner[65]. The anodic process is not a simple oxidation of carbon according to

$$C + 4OH^- \rightarrow CO_2 + 2H_2O + 4e^- \quad \text{or}$$
$$C + 6OH^- \rightarrow CO_3^{2-} + 3H_2O + 4e^- \tag{IV.2}$$

(a process which occurs at a significant rate only at yet higher temperatures),

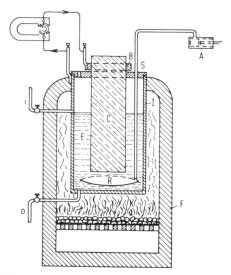

Figure IV.36. Fuel cell after Jacques[64]. C, carbon (anode); E, caustic soda melt; I, Norway iron pot (cathode); A, air pump; B, terminal, clamped to the carbon and resting on the insulating cover S; R, air-distributing rose; F, furnace; i and o, inlet and outlet for electrolyte.

but decomposition of the molten alkali,

$$C + H_2O + 2OH^- \rightarrow 2H_2 + CO_3^{2-} \qquad (IV.3)$$

followed by oxidation of the hydrogen produced:

$$2H_2 + 4OH^- \rightarrow 4H_2O + 4e^- \qquad (IV.4)$$

The oxygen electrode of the cell functions only in presence of manganate ions, but these were present in the impure materials used in amount sufficient for successful operation.

A current density of $100 \, mA \, cm^{-2}$ at 0.9 volt and an overall efficiency of 32% were adequate to demonstrate the technical and economic feasibility of the fuel cell, particularly at a time when power stations operated at an efficiency of 2.6%:

The cell was capable of intermittent operation for almost six months. This experiment, initially so successful, eventually failed because the oxidation of carbon led to the accumulation of carbonate in the electrolyte. This ultimately slowed the reaction to a standstill, because hydroxyl ions constitute an essential reactant (equation IV.3). The regeneration required rendered the cell uneconomic.

It is thus inexpedient to oxidize carbon, or any carbon-containing fuel, in an electrolyte that will suffer irreversible change by reaction with carbon dioxide. Molten carbonates, frequently chosen for use at moderate temperatures, are inapplicable at temperatures required for the direct oxidation of carbon, because they react with carbon by the Gay-Lussac reaction:

$$C + Na_2CO_3 \rightarrow 2Na + CO + CO_2 \qquad (IV.5)$$

This reaction occurs only at temperatures at which sodium has an appreciable vapour pressure; the entropy increment due to the formation of sodium vapour is alone sufficient to make ΔG negative and promote this undesired reaction.

Although there has been no lack of attempts at direct oxidation of carbon at high temperatures, the working life of the cells concerned has generally been very short. Carbon can be electrochemically oxidized even at temperatures below 100°C, but only at potentials that make this process impracticable for application in fuel cells[65a]. Neither direct electrochemical oxidation, nor indirect oxidation following reaction of carbon with the electrolyte—as in the Jacques cell, and in other medium-temperature cells to be described—have as yet been brought to success as practicable processes*.

In 1956, Justi[66] tried to oxidize coal directly, using a cell (Figure IV.37) similar to that used by Baur and co-workers[67]. Solid sodium carbonate served as electrolyte. This experiment also failed. The cell voltage (1.04 volt) collapsed on current load because carbon penetrated the sodium carbonate layer and led to short-circuit. The removal of ash and clinker would present one of the most serious problems in cells of this kind.

* The term 'carbon fuel cell' is meant to include cells in which carbon is brought into direct electrochemical oxidation, and also those in which carbon undergoes indirect reaction by establishment of the Boudouard equilibrium, $C + CO_2 \rightleftharpoons 2CO$, carbon monoxide reacting at the electrode, or undergoing a preliminary reaction with the electrolyte.

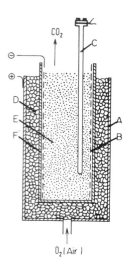

Figure IV.37. Cell with solid electrolyte (after Justi, Bischoff and Spengler[66]). A, chrome steel container; B, iron wire gauze; C, thermometer; D, CuO/Cu_2O or Fe_3O_4; E, granulated coal; F, Na_2CO_3 hollow cylinder.

Studies of high-temperature cells have since been mainly restricted to those concerned with electrochemical oxidation of simple *hydrocarbons* and of *carbon monoxide*. It has become evident that direct electrochemical oxidation of hydrocarbons first becomes possible at temperatures at which the hydrocarbon is no longer stable, or at least dissociates in presence of a catalyst, so that the real electrode reaction proceeds by way of decomposition products; this complicates the situation. It is for this reason that electrode potentials for hydrocarbons, calculated on the basis of thermodynamic data for the overall reaction, cannot be realized experimentally[67a]. Successful experiments involving oxidations at moderately high temperatures are obviously to be explained in terms of co-operative reactions involving the electrolyte or oxygen chemisorbed on electrodes (see chapter III); these aspects will be further discussed. In no case have intermediate products, such as alcohols, aldehydes or fatty acids, been observed.

Unfortunately, *decomposition reactions* catalysed by electrode materials lead to side products, incapable of electrochemical oxidation; these clog porous electrodes, or react in some undesirable way with the electrolyte. For instance, carbon may be deposited in the pores of an electrode. This may be countered by choice of suitable electrode material, or by addition of carbon dioxide and/or water vapour to the fuel gas. Unsaturated hydrocarbons contained in the fuel gas, or formed at the electrodes can react with acidic electrolytes, e.g.,

$$C_2H_4 + H_2SO_4 \rightarrow C_2H_5SO_4H \tag{IV.6}$$

or may be activated to polymerize at the catalyst surface, with consequent fouling of electrolyte, electrodes, or both.

Attempts can be made to avoid these undesirable side-effects by suitable choice of temperature, catalyst, electrodes and electrolytes. An alternative is to bring the hydrocarbon (if necessary, coal) first into reaction with *steam* and to apply the hydrogen formed to electrochemical use[68]. Such a process is not, in general, accompanied by more than minor loss of energy. If the reaction is taken as nearly as possible to completion, e.g.,

$$CH_4 + 2H_2O \rightleftharpoons CO_2 + 4H_2 \tag{IV.7}$$

which is possible with suitable catalysts at not too high temperatures (ca. 600°C)*[69], the hydrogen so obtained can be utilized in any hydrogen–oxygen fuel cell—if, however, an alkaline electrolyte is to be used, the carbon dioxide must be removed.

Another method of 'opening up' a fuel for electrooxidation has been successfully applied to, for example, JP-4—a jet-propulsion fuel, mainly $C_{14}H_{30}$[85]. This consists of partial combustion with a measured quantity of air (ca. 5%) before supply to the cell. Such a process is economic, especially for high-temperature cells, if the heat of combustion is used for heating the cells.

For the electrooxidation of fuel gases, containing carbon monoxide, carbonate electrolytes come into consideration first, since any alkaline electrolyte would be converted to carbonate. For high-temperature applications, *carbon dioxide* must be added to the carbon monoxide, since the latter otherwise disproportionates to CO_2 and undesirable carbon. For a similar reason, hydrocarbons should never be brought into reaction with steam at a temperature higher than that of the fuel cell, because the reaction, as mentioned above, does not proceed to complete conversion to CO_2. This applies only to fuel cells operating at a temperature at which the Boudouard reaction proceeds with noticeable rate.

Other troublesome side-reactions occur by the reversal of the methane reaction (equation IV.7) and by the occurrence of the water-gas reaction

$$H_2 + CO_2 \rightleftharpoons CO + H_2O \tag{IV.8}$$

at the anode in carbonate electrolytes. The equations do not indicate how these reactions proceed, but carbon monoxide and *methane* have been found in the anode gas of a cell supplied with pure hydrogen[70].

The more recent developments in the construction of high-temperature fuel cells stem from the work of Baur and his collaborators[67], as well as that of Davtyan[71]. Two different types of development are to be distinguished: cells with solid electrolytes, and those with liquid electrolytes, the latter including free liquid, or liquid sorbed in a porous matrix[72]—the name 'wick-stone cell' (Dochtsteinzelle) has been appropriately coined for the last-named device.

IV.2.2 High-temperature Cells with Liquid Electrolytes†

IV.2.2.1 *Principles of the Carbonate Cell*

A detailed investigation of the electrolyte used by Davtyan[71] initiated the successful research of Broers and Ketelaar[73] in this field. This electrolyte was not, as Davtyan thought, a solid at its working temperature; it consisted of a

---

* At lower temperatures the equilibrium is displaced to the left and reaction is incomplete.
† By Dr R. P. Tischer, Scientific Research Staff, Ford Motor Company, Dearborn, Michigan, U.S.A.

liquid fraction held in a solid matrix[74]. The constituents of the electrolyte (43% sodium carbonate, 27% monazite sand, 20% tungsten oxide and 10% water-glass) reacted with each other, and several of the products (e.g., sodium tungstate, sodium orthophosphate) formed eutectics melting between 600 and 700°C. The electrolyte proved to be unstable under reducing conditions at the cathode, oxidation and reduction phenomena occurred at the electrodes, and a completely new approach to the problem was indicated.

Of all salts melting at a suitable temperature, only *carbonates* have proved to be satisfactory for use under the conditions prevailing in fuel cells. The reasons are simple.

The cathodic reaction leads to enrichment of metallic oxide or of O^{2-} ions at the cathode:

$$2M^+ + \tfrac{1}{2}O_2 + 2e^- \to M_2O \to 2M^+ + O^{2-} \tag{IV.9}$$

The anodic reaction concentrates acidic oxide, with simultaneous formation of carbon dioxide and/or water vapour from the fuel gas, e.g.,

$$SiO_3^{2-} + CO \to SiO_2 + CO_2 + 2e^-$$
$$SiO_3^{2-} + H_2 \to SiO_2 + H_2O + 2e^- \tag{IV.10}$$

Many of these oxides (P_2O_5, SO_3, etc.)—like halogen hydrides—are volatile and escape with the products of the anodic reaction. Others (SiO_2, B_2O_3) remain in the melt, or are deposited as insoluble products. Some of these oxides are reduced by the fuel. In any case, a concentration gradient is established which cannot be dissipated fast enough by diffusion. In most cases the electrolyte suffers an *irreversible* change.

Concentration gradients also occur in carbonate melts, but are countered by simple means. It is sufficient to add *carbon dioxide* (from the gas evolved at the anode) to the oxygen or air[75], so that carbonate is continuously regenerated at the cathode, whilst carbon dioxide escapes from the anode with the oxidation products of the fuel. It is thus clear why carbonate melts have a fundamental advantage. Millet and Buvet[75a] have concluded from thermodynamic calculations that the addition of water vapour to the oxygen should have advantages over the use of carbon dioxide.

It has recently been proposed[76] to use an anode permeable to hydrogen (the palladium-diffusion electrode). The carbon dioxide produced, thus effectively separated from the hydrogen supplied to the anode, could if necessary be delivered to the cathode.

The numerous recent attempts to develop fuel cells with molten electrolytes (e.g.,[70,73,77–80]) have accordingly all been based on the use of carbonate melts. The *electrode processes* proceed according to the simple scheme:

$$\text{cathode} \quad 2e^- + \tfrac{1}{2}O_2 + CO_2 \to CO_3^{2-} \tag{IV.11}$$

$$\text{anode} \quad CO_3^{2-} + H_2 \to H_2O + CO_2 + 2e^- \tag{IV.12}$$

The emf of the corresponding cell is:

$$E = E^\circ + \frac{RT}{2F} \ln \frac{p'_{H_2} p_{O_2}^{\frac{1}{2}} p_{CO_2}}{p'_{H_2O} p'_{CO_2}} \qquad (IV.13)$$

where primed symbols relate to the anode*.

In this cell, the oxygen is transported from cathode to anode in the form of carbonate ions. *Solid electrolytes* are chosen so that the transference number of the oxygen ion is practically unity, and concentration changes at the electrodes are thereby avoided. In the case of liquid carbonate electrolytes, exclusive *transport of carbonate ions* arises from the working conditions of the fuel cell—cations do not react at the electrodes—and the electroneutrality requirement.

Gorin and Recht[80] have shown that restriction of the *contact area* between a porous electrode and the electrolyte can lead to a considerable increase in the effective electrolyte resistance.

Broers[81] gives the following example relating to a carbonate cell paste electrolyte (see below):
Electrolyte resistance, as function of conductance: 1.5Ω cm^2
Resistance measured by ac or interruption method[82]: 2.4Ω cm^2
dc resistance (including overvoltage) 3.8Ω cm^2.

It must be mentioned in this connection that Will's consideration[83] of current distribution in the region of the three-phase boundary cannot be applied to high-temperature systems without detailed investigation. For instance, the permeability of many metals to gases increases greatly with rising temperature, and may attain a value competitive with that of the molten salt; in this case mass transfer on the metal side of the interface may no longer be negligible (cf. section II.6). This applies, of course, only to H_2 and O_2—not to CO_2, nor to H_2O.

According to Broers[73] most of the voltage drop in a carbonate cell is caused by ohmic resistance. The kinetics of *electrode reactions* in carbonate melts has been studied in detail by Janz, Colom and Saegusa[83a], and also by Stepanov, Smirnov and their co-workers[83b]. Janz, Lorenz and Saegusa[83c] have also investigated the physical properties of carbonate melts. Tantram[77], Broers and their collaborators[73] have found a diffusion-controlled mechanism responsible for the electrode polarization in carbonate fuel cells. For kinetic studies it would be most useful to know the solubilities of all reacting species in the carbonate

---

* The following thermodynamically equivalent way of writing equation IV.13 has been suggested[77].

$$E = \frac{RT}{2F} \ln \frac{p_{O_2}^{\frac{1}{2}} p_{CO_2}}{p_{O_2}'^{\frac{1}{2}} p'_{CO_2}} \qquad (IV.14)$$

The cell is here being considered as an oxygen concentration cell. This is less acceptable from the kinetic standpoint, since under normal working conditions oxygen pressure at the anode would be of the order of 10^{-16} atm. Such conditions exclude any significant transfer process: there could be no physically reasonable concentration of oxygen atoms at the electrode surface, or alternatively, the mean life of an oxygen atom at the electrode would be less than 10^{-22} sec at significant current densities.

melt. Schenke, Broers and Ketelaar have measured the solubility of oxygen[83d]. The viscosity of carbonate melts has been determined by Karpachev and co-workers[83e].

IV.2.2.2 Cell Design

Three types of cell have been developed for use with carbonate melts; the matrix cell, cells with paste electrolytes, and cells with free electrolyte.

Figure IV.38 shows the construction of a matrix cell, after Broers[73]. Metal gauze and perforated metal plates press metallic-powder electrodes against a porous *magnesium oxide disc*, impregnated with the carbonate melt (lithium, sodium, and/or potassium carbonates). The pores of the sintered electrolyte matrix must be finer than those of the electrodes. The electrode materials so far used are silver for the air or oxygen side, platinum, nickel or iron (platinized if required) for the fuel side. Similar cells have been described by Sandler[70] and by Eisenberg and Baker[83f].

In extended time tests difficulties have been found with gaskets, as well as with maintenance of good contact between the matrix and the electrodes. The formation of fine cracks in the matrix also leads to failure.

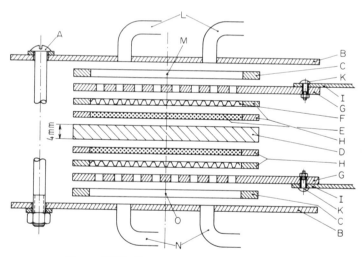

Figure IV.38. Components of a carbonate cell (after Broers[73]). A, screws with mica insulating washers; B, steel end-plates; C, asbestos gaskets; D, carbonate-impregnated magnesium oxide disc; E, layers of metal powder on either side of the electrolyte disc; F, wire gauze (silver at the cathode, iron, nickel or copper at the anode); G, perforated stainless steel plates; H, mica spacers; I, silver terminal wires; K, terminal screws; L, N, gas circulation pipes; M, O, gas chambers.

Figure IV.39. Laboratory-scale molten carbonate cell (after Trachtenberg and Truitt[85]).

A simplified construction for a matrix cell has been described by Trachtenberg and Truitt[85]*. A molten mixture of sodium and lithium carbonates is used, absorbed in a magnesium oxide matrix. Before impregnation with the electrolyte, a silver layer is sprayed on to the matrix, and the powdered-electrode metals are pressed and sintered on to it. Nickel is used on the fuel side, and silver or copper oxide on the oxygen side. The inferior performance of carbonate cells, as compared with low temperature aqueous cells, has been attributed to the instability of active surfaces due to sintering of finely-divided metal at the operating temperature[85a]. Figure IV.39 shows a small, laboratory-scale cell. The magnesium oxide disc is fitted into a steel housing. The two small electrodes serve only for measurements of overvoltage. The gas tubes and electrical connexions are attached to the end-plates. Working with hydrogen and air, this cell has attained an operational life of up to 4000 hr.

It is the electrodes which present the main problem in cells with *free* electrolyte. To avoid flooding of the electrodes, they can be made of double porosity—coarse pores on the gas side, fine pores towards the electrolyte. The gases must then be supplied under slight pressure, but this adds a complication because of the pressure-regulating equipment required. When free electrolytes are used, silver shows a very strong tendency to recrystallize, even if only minute quantities dissolve in the melt. Crystallites detached from the electrodes collect in the bottom of the cell, and finally lead to a short-circuit. Silver will deposit where fuel, diffusing from the anode, attains an appreciable concentration†. Recrystal-

---

* Cf. also the design of Kronenberg[84].

† We have, however, observed recrystallization of silver in molten salts even under conditions of the most thorough exclusion of reducing substances (local cell action).

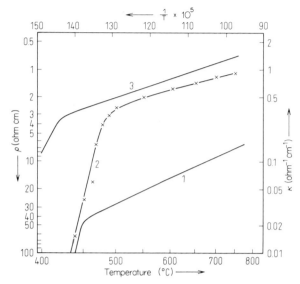

Figure IV.40. Specific resistance (and conductance) of liquid paste, and sintered matrix electrolytes (after Broers[73]). Curve 1: 70% MgO, sintered; 30% Na_2CO_3–Li_2CO_3–K_2CO_3 (1:1:1); Curve 2: 50% MgO + 50% $LiNaCO_3$ (unsintered mixture); Curve 3: Li_2CO_3–Na_2CO_3–K_2CO_3 (1:1:1, free liquid).

lization less frequently occurs in matrix-contained electrolytes, in which convection is restricted; in consequence, silver electrodes can only be used in cells with supported electrolyte, which do not, however, completely exclude the possibility of growth of dendrites.

Carbonate melts can also be prevented from invading porous electrodes by use of *paste electrolytes*[73,79]. Magnesium oxide powder is mixed with the carbonate, so that, above the melting point of the latter, a viscous mass is formed, which has shown little tendency to crack or alter in size and shape over long periods. This stability improves with increasingly fine particle size of the magnesium oxide powder. The conductance is considerably better than that of the same melt held in a magnesium oxide matrix, approaching that of the free electrolyte (Figure IV.40). The paste electrolyte is mostly preformed by pressing, whereas for the slurried electrolyte, described by Trachtenberg[85], the magnesium oxide powder in the cell is soaked by capillary action with carbonate melt. Electrolytes of this kind provide less support to growing silver dendrites, and correspondingly the cells are less prone to failure from dendrite growth. The idea of using partially-solid carbonate melts of non-eutectic composition has been abandoned because their consistency depends strongly upon temperature.

IV.2.2.3   Experimental Results

The curves in Figures IV.41, 42 and 43 show the characteristics of cells with *paste electrolytes* under various working conditions. Figure IV.44 illustrates the important point that tests of cells with porous electrodes must extend over longer periods if results are to be reproducible and assessments of performance significant. Recent results of Broers and Schenke[73] indicate that, after an initial formation period for the nickel anode, the decrease in performance is due only to loss of carbonate by evaporation. Trachtenberg[85] reports on cells running for 2000 hr without failure. In the cells described by Trachtenberg, the electrolyte can easily be replenished by filling carbonate into the reservoir from which the slurry is supplied. There is, however, still some decrease in performance with time in most cases.

For use in cells with paste electrolytes, preliminary steam reforming of hydrocarbon fuels over a separate catalyst has given good results. Figure IV.45a shows current-voltage curves obtained from a cell operating in this way; hydrogen is consumed at the electrode, but, obtained from the methane reaction, is diluted.

As the fuel is progressively used up, there is increasing dilution of the fuel gas, and the efficiency falls away rapidly (Figures IV.41, 42). If, then, oxidation of the fuel gas in a single cell is allowed to proceed as far as possible to completion, the efficiency that can be attained is only that characteristic of highly diluted fuel gas. It is therefore of considerable advantage to oxidize the fuel gas in a *stepwise* manner, passing it through a succession of cells. This can be seen clearly by inspection of Figure IV.46. This shows, for the case of hydrogen as fuel, the emf of the cell (neglecting overvoltage) for various stages of fuel consumption. The experimental points correspond well with the curve calculated from the equation

$$E = E^\circ - \frac{RT}{4F} \ln \frac{p_{H_2O}^2}{p_{H_2}^2 p_{O_2}} \tag{IV.15}$$

If oxidation in a single cell is taken to 97%, it can be seen that the energy obtained corresponds with the area of the rectangle ABCD. If, on the other hand, oxidation is carried out in two steps, first to 50%, then to 97%, the energy is given by the areas A'B'C'D + EBCC', and is greater by an amount A'B'EA. Division of the processes into more steps would lead to further improvement. This simple demonstration shows the considerable advantage of *successive oxidations*. In the case of the carbonate cell, it must be remembered that the difference in carbon dioxide pressure at the two electrodes has an influence on the emf (equation IV.13) and, further, that for each hydrogen molecule oxidized, a molecule of carbon dioxide is formed, as well as a molecule of water. If 50% of the hydrogen delivered to a cell is oxidized, the resulting gas mixture contains only about 33% of hydrogen; the hydrogen is therefore already considerably diluted. The same will apply, *mutatis mutandis*, to all fuel gases. *Solid-oxide electrolytes* (see section IV.2.5) have the advantage of absence of such additional dilution. Gaseous reaction products must be removed with a minimal loss of fuel and energy; this is facilitated by stepwise oxidation, which allows fuel to be utilized down to lower concentration limits. For a more refined analysis of successive oxidation, one has to take into account that progressive fuel depletion and product accumulation tend to increase the (mass transfer controlled) polarization[73]. It is the current density that limits utilization. For more complete utilization, the current density in the cells running with highly depleted fuel has to be reduced. Analogous considerations hold for utilization of oxidant; but in this case a high utilization is desirable mainly for heat economy and becomes less important in larger batteries.

Carbon dioxide production at the anode is a serious disadvantage of carbonate cells, as compared with aqueous electrolyte cells, for operation with hydrogen. Since, in the latter, water can easily be removed by condensation, the only impurity left in the fuel gas is nitrogen which may have diffused across from the air electrode. The quantity of impurity to be removed is smaller by orders of magnitude, than that represented by the carbon dioxide formed at the anode of the carbonate cell.

A most successful development in the field of carbonate cells has been described by Trachtenberg and Truitt[85], who have produced a *magnesium oxide*

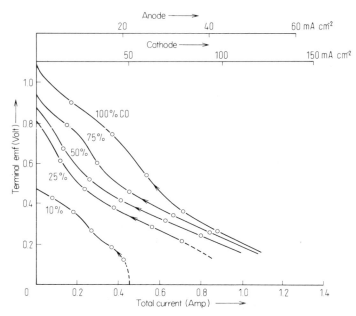

Figure IV.41. Experimental results with CO–CO_2 mixtures and paste electrolyte at 730°C (after Broers[73]). With the 10 % mixture complete conversion is attained in one pass. Tube cell, 4th week, temperature = 730°C, CO + CO_2 mixtures at Fe–Ni–Cu anode, 71 % air + 29 % CO_2 at a silver cathode.

matrix (see above) that will sustain, without damage, repeated heating and cooling cycles between room temperature and 600°C. The characteristics of the cell remain unchanged by the severe stresses thus imposed on it. More recently, this group has used the slurry electrolyte mentioned in section IV.2.2.2.

Various hydrocarbon mixtures, from natural gas to kerosene, have been tested as fuels—after partial combustion (with addition of steam) at 1100 to 1200°C. A chromatographic analysis of the dried fraction of the gas mixture so obtained from JP–4 jet-fuel gave:

38 % by vol nitrogen	3 % by vol methane
30 % by vol hydrogen	1 % by vol ethylene
23 % by vol carbon monoxide	5 % by vol carbon dioxide

It is, of course, desirable that the reaction with steam to produce hydrogen and carbon dioxide should be as complete as possible. Residual steam reacts in the cell or in the hot plumbing with carbon monoxide in the desired direction in the cell at 600°C.

Figure IV.47 shows a current–voltage curve for operation with hydrogen and air; Figure IV.48, the overpotentials of the individual electrodes. The high

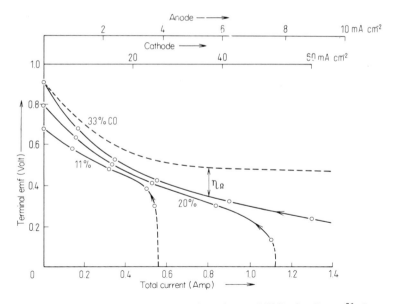

Figure IV.42. CO–CO_2 mixtures, past electrolyte at 660°C, after Broers[73]. Complete conversion is reached with 11 % and 20 % mixtures. Effect of ohmic drop is shown in upper curve. Tube cell, fifth week, temperature = 660°C, $CO + CO_2$ mixtures at Fe–Ni–Cu anode, 71 % air + 29 % CO_2 at a silver cathode; ↕ ohmic voltage drop.

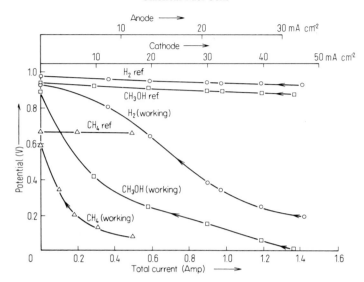

Figure IV.43. Characteristics and electrode polarizations for various fuels in cells with paste electrolyte (after Broers[73]). Potentials of the working electrode and the zero-current reference electrode are measured relative to the cathode (air). Anode 'flaked nickel'; cathode, silver. The cathode does not polarize at 4% final oxygen content. Tube cell, second week, temperature = 700°C.

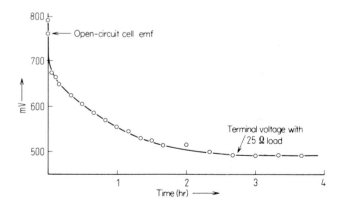

Figure IV.44. Slowness of the dc polarization process (after Broers[73]). Cell: temperature = 710°C, 27th day; internal resistance, 0.1 ohm at frequency of 400 sec^{-1}; anode, CH_4 and CO_2 (each at 1 l hr^{-1} measured at 20°C and 1 atm), 25% Ag + 75% ZnO; cathode, air (25 l hr^{-1}) and CO_2 (1 l hr^{-1}), 100% Ag.

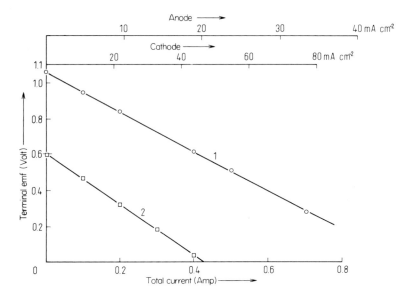

Figure IV.45a. Influence of a nickel catalyst on the methane–steam equilibrium and the resulting performance of a cell with paste electrolyte (after Broers[73]), catalyst in the anode chamber. Tube cell, sixth week, temperature = 730°C; 50% CH_4 + 50% H_2O at a Fe–Ni–Cu anode; 71% air + 29% CO_2 at a silver cathode; feed rate of CH_4, 1 l hr^{-1}, corresponding with 9.0 amp; feed rate of O_2, 0.5 l hr^{-1}, equivalent to 2.24 amp; curve 1, with catalyst; curve 2, without catalyst.

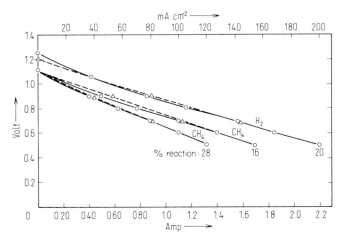

Figure IV.45b. Current–voltage curves with hydrogen and natural gas (steam reformed in a separate reactor) at two different rates of gas flow, for a cell with matrix electrolyte (after Sandler[70]). ○, Increasing current; △, decreasing current. Measurements taken at two-minute intervals (cf. Figure IV.44).

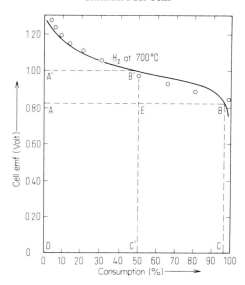

Figure IV.46. Cell voltage as a function of gas consumption (after Chambers and Tantram[77]). ○, Observed voltage including ohmic voltage drop; ———— theoretical curve.

rest-voltage of more than 1.4 volt depends on the low water content of the electrolyte (equation IV.13). Further reduction of the water content is, however, impossible, as long as the fuel gas contains water vapour.

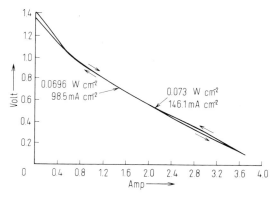

Figure IV.47. Current–voltage curve of a carbonate fuel cell with matrix electrolyte (after Trachtenberg and Truitt[85]); 5 days in operation; temperature, 600°C; feed rate of hydrogen, 480 cm^3 min^{-1} ($\equiv$ 65 amp for 100% reaction; at 1 amp and 0.6 volt, $\eta = 1.8\%$).

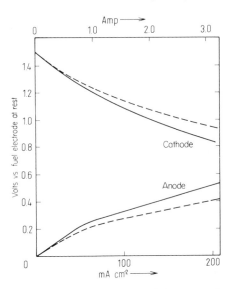

Figure IV.48. Current–potential curves for separate electrodes in a carbonate fuel cell, after Trachtenberg and Truitt[85]; cf. Figure IV.47. The broken lines indicate overpotentials corrected for leads resistance. Cell in uninterrupted operation for 15 days; temperature, 600°C, hydrogen–air.

Batteries rated at 100 watts have been built (Figure IV.49), and larger ones are planned. It can be estimated that a battery with a rating of about 2 kW would generate sufficient heat, with reasonable insulation, to be thermally self-

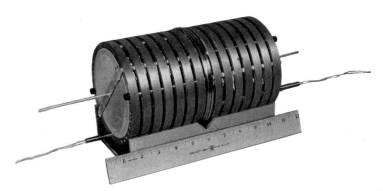

Figure IV.49. 25 watt battery with matrix electrolyte (after Trachtenberg and Truitt[85]); electrodes 60 cm²; partially oxidized Diesel oil as fuel gas; current density 40 mA cm⁻² at 600°C.

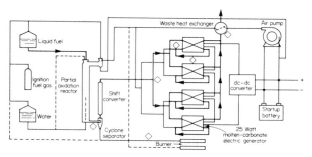

Figure IV.49a. Block diagram of 100 watt molten carbonate fuel cell system (after Frysinger and Truitt[85b]).

supporting; this would be necessary for competitive efficiency. Heat transfer characteristics have been calculated by Eisenberg and Baker[83f] and by Baker, Marianowski, Zimmer and Price[85a]. The maximal thermal efficiency of these cells, in relation to the calorific value of the fuel, taking all losses into account, is 22%—without allowance for heating.

Figure IV.49a shows a block diagram of a 100 W demonstration unit which is quite self-explanatory. Table IV.4a records its performance. The addition of water vapour to the fuel, for the prevention of carbon deposition, will eventually be replaced by recycling anode flue gas to the anode as well as to the cathode; the carbon dioxide in it has the same effect as water in preventing carbon

Table IV.4a. Demonstration Unit

Fuel	JP-4 jet fuel	
Oxidant	Air	
Start-up time	2 hr 40 min	
Partial-ox. start-up time	17 min	
Partial-ox. efficiency (%)	57	
Fuel utilization in cells (%)	25–38	
Power (watts)	*design*	*to date*
Generator output	100	73
DC–DC converter (73% efficient)	73	53
Parasitic blower power	45	45
Net output	28	8
Generator output efficiency		
(% net heat supplied)	4	3
Cell output (watts/ft^2)	18.7	13.7
Weight of four modules (lb)	71	
Weight of partial-ox. system (lb)	20	
Operating cycles—total system	10	
Operating time—total system	62 hr 25 min	
Operating temperature	600°C	

deposition. The low efficiencies are partially due to oversized auxiliary equipment drawing an excess of parasitic power. Such losses will rapidly decrease with larger units. This unit has operated equally well with a variety of commercial hydrocarbon fuels.

IV.2.3   Medium Temperature Cells (150° to 250°C)*

Operation at high temperatures involves numerous inconveniences and is impracticable for many applications. Attempts have therefore been made to find whether a moderate increase of temperature ($<300°C$) can lead to significant improvements in the efficiency of cells. For cells to work with hydrogen alone, an alkali melt or solution commends itself as electrolyte, because of the improved efficiency of the oxygen electrode in such a medium. Outstanding development work in this direction has been carried out by Bacon[86]. Carbon-containing fuels, however, require the use of neutral or acid electrolytes. Very promising cells with acid electrolytes have recently been developed, based on the fundamental work of Taitelbaum[87].

IV.2.3.1   Cells with Alkaline Electrolytes

There are two possibilities in the application of alkaline electrolytes under this heading; either the use of alkali melt of low water content ($<20\%$) under atmospheric pressure, or the use of an aqueous alkali solution under increased pressure. In either case the fuel must be free from carbon compounds.

Bacon[86] has devoted many years to the high-pressure cell. It is desirable at this stage to give a brief account of Bacon's theoretical considerations, since this will clarify the advantages and disadvantages of different working conditions.

The cell is affected by temperature in a number of ways. If it is assumed that the electrolyte remains under its own equilibrium vapour pressure at constant gas partial pressure, then the emf of the cell decreases with rising temperature. This effect can be compensated by a simultaneous increase in gas pressure. The loss-current, due to diffusion of gas through the electrolyte to the counter-electrode, may be increased by rise of temperature because of enhancement of diffusion coefficients, but there is, of course, a simultaneous decrease in gas solubilities. Reduction of surface tension leads to lesser penetration of the electrolyte into the pores of the electrode, with consequent decrease in the length of the diffusion path. The considerable reduction of overvoltage with increased temperature minimizes the effect of loss-currents (such as shunt currents through the ducts connecting the cells for electrolyte circulation) upon output, cell voltage and current efficiency. Increase of electrolyte concentration enhances the emf as a result of reduced vapour pressure, but cannot be taken too far because of difficulties associated with solidification of the electrolyte on cooling. Dilution before cooling is one way to deal with this problem; another will be mentioned below.

Electrode overpotentials and resistance overvoltage fall with rise of temperature, but so does the solubility of gas in the electrolyte, and this has an opposite effect. Increase of gas pressure enhances the exchange currents at both electrodes; increase of electrolyte concentration increases the cathodic exchange current in virtue of the higher OH^- ion activity (cf. section III.6.2.). Simultaneous reduction of the very large exchange current at the anode (cf. section III.1.2) is not significant, so long as the charge-transfer step at the anode is not rate-limiting.

* By Dr R. P. Tischer, Scientific Research Staff, Ford Motor Company, Dearborn, Michigan, U.S.A.

After all these influences have been taken into account, the construction of these cells still presents a whole series of technological problems.

IV.2.3.1.1   *The Bacon cell*[86]

As a result of experimental and theoretical studies, Bacon chose the following *working conditions*. The potassium hydroxide electrolyte had a concentration between 37 and 50%, the temperature was about 200°C, and pressure lay between 20 and 45 atm (usually about 28 atm).

Nickel was chosen for *electrode* material. Since no water-repellent agents could be used under the working conditions, double-layer electrodes were used, with fine pores facing the electrolyte, and coarser pores, of 30 μ or more diameter, on the gas side. The electrodes were 1.6 mm thick. The electrolyte was prevented from entering the coarser-pored layer by suitable regulation of gas pressure. To prevent corrosion, the oxygen electrode was preoxidized; the nickel oxide so formed, however, had much too low an intrinsic conductance (p-type), but this was increased by doping with lithium oxide.

For the construction of *batteries*, Bacon used bipolar electrodes with apertures and recesses for gas delivery, distribution and circulation. These were assembled in series on the filter-press principle (Figure IV.50). Particular difficulty was encountered in finding suitable gaskets which would withstand the joint attack of oxygen under pressure and hot alkali. Resistance heating elements were provided for bringing the cell to its working temperature, and for maintaining the temperature under conditions of small load.

Figure IV.50. 40-cell hydrogen–oxygen battery of filter-press type (after Bacon[86]). Rating about 5 kw, 240 amp, 0.5 A cm^{-2}.

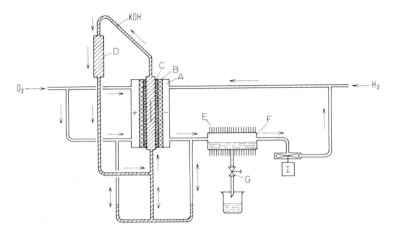

Figure IV.51. Schematic diagram of a hydrogen–oxygen battery
with gas- and electrolyte-circulation systems (after Bacon[86]). A, cell;
B, porous nickel (pores ca.30μ); C, porous nickel (pores ca.16 μ);
D, electrolyte circulator; E, cooling fins; F, condenser; G, condensate release valve; I, hydrogen circulator.

Figure IV.51 is a schematic representation of Bacon's experimental system. A propeller is used
for hydrogen circulation. A differential pressure transducer between the electrolyte and hydrogen-
circulation system regulates the speed of the propeller so that just as much water is condensed as is
produced at the anode. Other pressure transducers regulate the gas feeds. Any gas entering the
electrolyte circuit (by leaks at the electrodes, or by electrolysis in the connecting parts between
the cells) is automatically expelled. Any cell which fails must be automatically taken out of operation.
All these control mechanisms have operated satisfactorily, but, despite this, the necessity for such
complicated ancillary devices is a considerable hindrance to practical application of batteries of
this kind.

The major losses in this cell occur at the oxygen electrode. The distribution of
overvoltage losses is illustrated in Figure IV.52. Figure IV.53 shows the improvements that were effected in the performance of the Bacon cell in the year 1960.

With *air*, the cell output is only slightly lower than with oxygen. The reduction
current is proportional to $\sqrt{p_{O_2}}$, and, on transfer from oxygen to air, the current
falls to about a half. Since, however, the current–potential curve of the oxygen
electrode is very flat, the cell voltage at a *constant current density* decreases by
only 25 to 50 mV.

There is, of course, an energy loss involved in the compression of the nitrogen in the air, but the
residual energy in the exhaust gas can be utilized for the initial compression of the air. Care must be
taken to ensure good circulation to prevent accumulation of nitrogen; carbon dioxide must be
carefully removed from the air supplied to the cell.

A decisive disadvantage of the hydrogen–oxygen cell developed by Bacon
is its limited *working life*. At 250°C, with 45% KOH under relatively high pressure, corrosion makes it impossible to operate the cell for more than a few
hundred hours.

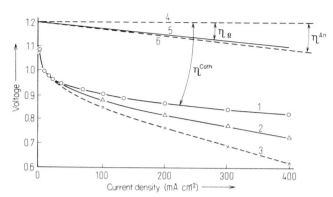

Figure IV.52. Allocation of overvoltage losses in the Bacon cell, 1954 data. Curve 1: overpotential of the oxygen electrode. Curve 2: cell characteristic curve with highly activated hydrogen electrode. Curve 3: cell characteristic curve with poisoned hydrogen electrode. Curve 4: theoretical rest voltage, E° (40% KOH, 200°C, 40 atm). Curve 5: ohmic voltage drop due to electrolyte resistance. Curve 6: overpotential of the poisoned hydrogen electrode (the overpotential of the highly activated electrode is negligible).

The cell further developed by the firm of Pratt and Whitney* in the last few years has a working life of about 1000 hours. Use of 85% KOH at a working temperature of 200 to 250°C allows the cell to be operated without increased pressure.

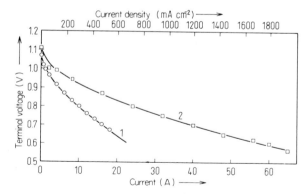

Figure IV.53. Improvement of the Bacon cell effected in 1960[86]. 7 cm electrodes. Curve 1: earlier cell (17.3.1960); Curve 2: later cell (29.11.1961).

---

* Division of United Aircraft Corpn., East Hartford, Connecticut, U.S.A.

In comparison with the Bacon high-pressure cell, however, the Pratt and Whitney cell has the disadvantage that 85 % KOH is solid at ambient temperature, and a special technique is required to fill the cell with electrolyte. Before the cell will operate it must be brought to a working temperature of 200°C, at which the electrolyte is molten.

A restriction of anodic overpotential is necessary when either 45 % or 85 % KOH is used as electrolyte because, at $\eta \geqslant 230$ mV, corresponding to a current density > 200 mA cm^{-2}, corrosion of the nickel anode usually occurs. The terminal voltage for continuous operation at this current density is about 0.9 V.

The relatively high temperature at which these cells work facilitates heat abstraction, which presents a critical problem, particularly for batteries of higher rating/volume ratio. The upper limit of temperature is set by water vapour pressure; it is at 260°C for the Pratt and Whitney cell.

A 500 watt battery made by Pratt and Whitney is shown in Figure IV.54. It consists of double cells, of the 'folded can' type[93] (Figure IV.55), connected in series. Each cell is made in the form of a nickel canister, with two hydrogen electrodes mounted on opposite walls. Inside the canister, a unit is mounted consisting of two oxygen electrodes (lower section of Figure IV.55), arranged

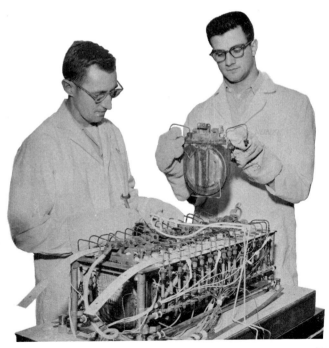

Figure IV.54. Five hundred watt hydrogen–oxygen battery by Pratt and Whitney[93]; a double cell being dismantled.

Figure IV.55. Hydrogen–oxygen cell on the 'folded can'
principle by Pratt and Whitney[93].

back-to-back and welded together peripherally. The space between these
electrodes is supplied with oxygen. The whole unit thus consists of two cells
with a common electrolyte, connected in parallel; this arrangement has the
advantage that an insulating gasket between the electrodes is necessary only
at the top of the canister. A further advantage is that each double cell is an
independent unit which can be readily replaced.

IV.2.3.2   Cells with Acid Electrolytes

IV.2.3.2.1   Hydrogen–oxygen cells

Experiments have shown that, for hydrogen–oxygen cells operating at atmospheric pressure, use of acid electrolytes such as concentrated sulphuric or
phosphoric acids at temperatures up to 200°C, offers no advantage over the
use of aqueous electrolytes at temperatures below 100°C[87,88].

Comparative experiments have been made[89] to study the behaviour of one
and the same platinized graphite electrodes in 4N sulphuric acid at room
temperature, and in phosphoric acid at 150°C*. The curves in Figure IV.56
are derived from individual electrode overpotential curves, by using a measured
cell voltage. The curve measured by Elmore and Tanner[88] is included for
comparison. It is clear that, in the case of acid electrolytes, increase of temperature offers no advantage when hydrogen is used as fuel. Figure IV.57 shows
current–potential curves for the oxygen electrode alone. There is a barely
significant improvement over the performance at room temperature, more

_____
* For purposes of this comparison, no special attempt was made to use electrodes of high activity.
  The graphite electrodes are not of unlimited stability under the experimental conditions, as
  already shown by Taitelbaum's experiments[87].

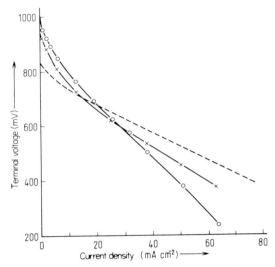

Figure IV.56. Comparison of hydrogen–oxygen cells using phosphoric acid at 150°C, with a hydrogen–oxygen cell with 4N sulphuric acid as electrolyte at room temperature, the same electrodes being used in each case (after Tischer[89]). ○—○, Phosphoric acid, 150°C; ×—×, 4N sulphuric acid, 25°C; — — —, phosphoric acid, 150°C (Elmore and Tanner[88]).

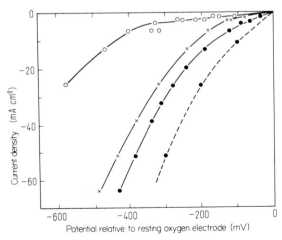

Figure IV.57. Current–potential curves for the same oxygen electrode in phosphoric acid at 150 and 200°C, and in 4v sulphuric acid at room temperature (after Tischer[89]). ●—●, Phosphoric acid, 150°C; ●— —●, phosphoric acid, 200°C; ×—×, 4N sulphuric acid, 25°C; ○—○, molten $(NH_4)HSO_4$, 200°C.

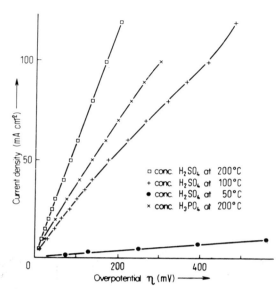

Figure IV.58. Anodic current–potential curves for the same hydrogen electrode in phosphoric acid and in concentrated sulphuric acid at various temperatures (after Tischer[89]).

than compensated by the greater loss at the hydrogen electrode apparent from Figure IV.56. If it can be assumed that transport of dissolved gas molecules in the electrolyte is rate-limiting, then, for a given electrolyte, increase of temperature should lead to improved electrode performance, provided that there is not too great a decrease in gas solubility. On the other hand, if operation at atmospheric pressure is to be conducted at a higher temperature, higher acid concentrations must be used, and the considerably higher viscosity then offers greater hindrance to diffusion. This is illustrated for the case of the hydrogen electrode in Figure IV.58. At 50°C only trivial current densities can be attained with concentrated acid. Even at 150°C the values are less than those obtainable for 4N sulphuric acid at room temperature. The situation is somewhat different for the oxygen electrode because the charge-transfer step is strongly hindered at low temperatures, and rise of temperature leads to net improvement in the electrode characteristics because of the reduction in this hindrance.

IV.2.3.2.2 Hydrocarbons as fuels

The improvement in electrode characteristics with rise of temperature applies in much greater measure to the oxidation of hydrocarbons. Their low reactivity as fuel gases renders direct oxidation at room temperature extremely difficult, if not impossible (cf. section III.4). A rise in temperature of 100°C brings considerable advantages. As early as 1910 Taitelbaum effected complete oxidation of carbon, hydrogen and various organic substances in concentrated sulphuric acid at temperatures up to 250°C. He realized that the sulphuric acid played an essential part in these reactions. It is reduced by the fuel to sulphur dioxide and the anode works as a sulphur dioxide electrode; since this electrode has a potential 0.37 volt more positive than that of the hydrogen electrode[90], this mode of reaction involves a loss of energy.

Analogously, redox systems can be coupled with the oxygen electrode. Haber and Bruner[65] had shown that the cathode of the Jacques cell worked reversibly only in the presence of manganate ion. Correspondingly, Taitelbaum discovered that cathodic overpotentials in concentrated sulphuric acid systems are decreased by the presence of redox couples such as VO_2^+, VO^{2+} or Tl^{3+}, Tl^+. This does not apply, however, to porous gas electrodes, at which overpotentials are not diminished by addition of redox systems to the electrolyte. The effect, which has recently been reexploited in relation to the use of nitric acid as an oxidant[145], depends on the oxidation of the redox couple by oxygen bubbling through a large volume of solution, and its subsequent reduction at the electrode. It is useful only when the oxygen is not confined to a narrow zone of contact with the cathode.

No potential loss occurs, apart from that due to electrode overpotential, because the concentration ratio of the redox couple can adjust itself to the oxygen potential. In the case of the SO_2–SO_3 equilibrium at the anode, the SO_2 would have to be at more than atmospheric pressure if the hydrogen potential in concentrated sulphuric acid were to be maintained.

The energy loss in hot concentrated sulphuric acid increases with rising temperature since the reducing action of sulphur dioxide decreases (the SO_2 potential is displaced in a positive direction). Moreover, sulphuric acid reacts with, for example, methane or propane to give side products such as H_2S, which acts as a catalyst poison. Side reactions also rule out perchloric acid as an electrolyte. In contrast, no direct reduction of phosphoric acid by hydrocarbons in homogeneous solution has yet been observed. The present situation therefore seems to be that phosphoric acid alone is suitable for use as an electrolyte at temperatures above 100°C* (cf. IV.2.4).

IV.2.3.2.3   Cells with concentrated phosphoric acid as electrolyte

Hydrocarbons, including the aliphatic constituents of common engine and Diesel fuels, have been successfully oxidized to more than 90% in a cell with phosphoric acid as electrolyte. A 'yield' of $98 \pm 4\%$ has been reported for propane[91]. Aromatic and olefinic compounds must, however, be removed for continuous operation[59]. Just as special fuels have been developed for internal combustion engines, specially suitable mixtures are to be sought for use in fuel cells.

The cell developed by Grubb and Niedrach[91] was of conventional construction. The electrodes consist of finely-divided platinum (about 10 mg cm^{-2}) on a porous polytetrafluoroethylene substrate, with a built-in metal gauze

---

* Direct reaction of hydrocarbons has been successfully achieved with sulphuric acid as electrolyte at concentrations $\leqslant 3N$ in the temperature range 70 to 100°C, but adequate current densities have so far required the use of Raney-platinum as catalyst[154]. A propane–oxygen cell at 72°C, with electrodes of 4 cm diameter (continuous gas supply at 1.5 atm total pressure), has provided 370 mA at 330 mV and 600 mA at 200 mV.

(tantalum or platinum) for rigidity and electrical contact[91a]. Gold as an electrode support material would have the advantage of much higher conductivity[91b]. The gas sides of the electrode are made hydrophobic by means of an additional layer of polytetrafluoroethylene. The electrolyte is circulated*.

Depending on the water content the boiling point of aqueous orthophosphoric acid lies between 100 and 261°C. Loss of water occurs by evaporation and also by consumption at the anode; since lack of water leads to carbon deposition, water vapour must be added to the fuel gas. Reliable control of the water balance in continuous operation requires a temperature not exceeding 150°C. The appropriate phosphoric acid concentration is 85% (14.6M).

Figure IV.59 shows current–voltage curves for a *propane–oxygen cell* operating with various concentrations of phosphoric acid at several temperatures. The broken line shows the small voltage drop due to internal resistance. Since the overpotential at the propane electrode at 200°C and 200 mA cm^{-2} amounts only to 50 mV, the main loss of cell voltage with increasing current density is attributable to the oxygen electrode. At 150°C, 50 mA cm^{-2} is obtainable at 0.4 volt. The limiting current density of a single electrode at 150°C reaches more than 500 mA cm^{-2}. As already mentioned, propane is oxidized completely to CO_2:

$$C_3H_8 + 6H_2O \rightarrow 3CO_2 + 2OH^+ + 20e^- \qquad (IV.16)$$

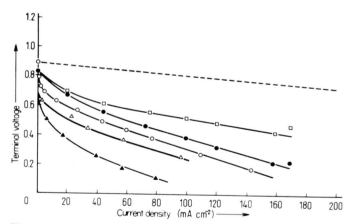

Figure IV.59. Current–voltage curves for a propane–oxygen cell with phosphoric acid as electrolyte (after Grubb and Neidrach[91]). ▲, 75% H_3PO_4, 120°C; △, 85% H_3PO_4, 150°C; ○, 91% H_3PO_4, 175°C; ●, 95% H_3PO_4, 200°C, ohmic voltage drop eliminated.

* A variation of this cell using a Teflon–zirconium phosphate supported electrolyte, was described by Hamlen and Szymalak[91c].

Attempts to detect intermediate products have so far been unsuccessful. They are probably strongly adsorbed and have a short lifetime.

Jasinski, Huff, Tomter and Swette[92] have studied the oxidation of methane, n-butane, isobutane and propylene, as well as propane, in 85% H_3PO_4 as electrolyte at 150°C. The 7.5 × 7.5 cm² electrodes consisted of a platinum-activated tantalum gauze (a mixture of platinum black and a Teflon dispersion was applied to the tantalum); the cell walls consisted of Teflon supported by stainless steel end-plates. Between the electrodes was an asbestos diaphragm as an electrolyte carrier (cf. section IV.1.2.2.). Chemically stable diaphragms were developed for use with 3N H_2SO_4 at 60°C, as well as for 85% H_3PO_4 at 150°C. Tantalum frames served as electrode supports and current leads.

Figures IV.60 and IV.61 show the current–voltage curves obtained with the various gases. Surprisingly, *methane* could be brought into reaction remarkably well. No significant fall of terminal voltage occurred during 2 hr of operation at 20 mA cm^{-2}. Control experiments with helium as anode gas showed that corrosion currents were negligible.

Measurements by Grubb and Michalske[153] confirmed these results with methane; nearly identical current–voltage curves were obtained under the same conditions. For current densities below 20 mA cm^{-2}, however, the curve lay somewhat above that for propane. According to the yield of CO_2, the anodic reaction corresponded with complete oxidation of methane:

$$CH_4 + 2H_2O \rightarrow CO_2 + 8H^+ + 8e^-$$

The potential of the methane anode amounted to +500 mV, relative to the reversible hydrogen potential in the same electrolyte, at a current density of 40 mA cm^{-2}.

With propane as gaseous fuel, a terminal voltage of 0.51 volt is obtained at 20 mA cm^{-2}, corresponding to an efficiency of 47% (cf. Table II.3). The maximal output amounts to 15 mW cm^{-2}. The propane–oxygen cell has already been operated for periods up to a week, but without gas circulation. Oxygen, as well

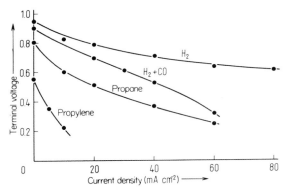

Figure IV.60. Current–voltage curves for various gaseous fuels with phosphoric acid as electrolyte at 150°C (after Jasinski, Huff, Tomter and Swette[92]).

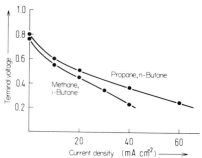

Figure IV.61. Comparison of normal and branched-chain hydrocarbons in phosphoric acid at 150°C (after Jasinski, Huff, Tomter and Swette[92]).

as carbon dioxide and propane, has been found in the exhaust gas from the anode, indicating that oxygen supplied to the cathode had penetrated the electrolyte-carrying asbestos diaphragm. Appropriate modifications can reduce the oxygen permeability of these diaphragms[59].

Propylene as fuel gives a very much inferior cell voltage which falls away rapidly; the authors suggest that the catalyst becomes poisoned by polymerization products. This poisoning effect by propylene is significant in so far as technical propane contains about 15% of propylene.

The current–voltage curve for *water gas* lies between those for propane and hydrogen. The direct reaction of water gas is of technical significance, since the

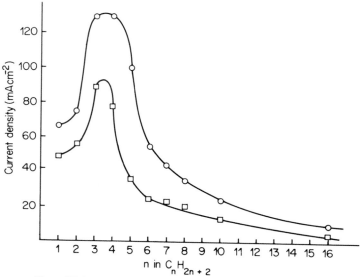

Figure IV.61a. Current densities at cell voltage E = 0.3 V .... ○ and at E = 0.5 V .... □ for various normal paraffin fuels (after Grubb and Michalske[92a]).

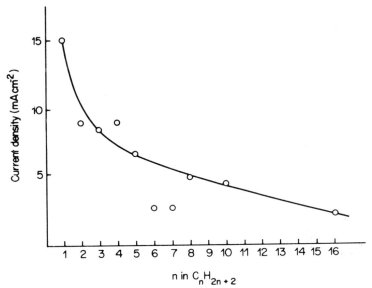

Figure IV.61b. Current densities at cell voltage E = 0.6 V for various normal paraffin fuels (after Grubb and Michalske[92a]).

results indicate that reformed hydrocarbons can be utilized without further purification.

The current–voltage curves of Figure IV.61 show that the normal hydrocarbons (propane, n-butane) are preferable as fuels to the branched-chain hydrocarbons (e.g. isobutane) and to methane. Grubb and Michalske[92a] compared the reactivities of various hydrocarbons in the phosphoric acid cell. Figures IV.61a and IV.61b show current densities for cell voltages of 0.3, 0.5, and 0.6 volts (ohmic voltage drop eliminated) as a function of the number of carbon atoms in the hydrocarbon molecule. Only the data for normal hydrocarbons are plotted, values for branched and unsaturated hydrocarbons were found to be lower. At low overpotential, methane yields the highest current, perhaps because it contains a higher percentage of hydrogen than any other hydrocarbon, whereas at higher overpotential, the reactivity exhibits a strong maximum at $n = 3$. Contrary to this electrochemical behaviour, the reactivity of normal hydrocarbons in the gas phase increases monotonously with chain length.

IV.2.4   Special Type Electrolytes

IV.2.4.1   Alkali-hydrolysed 'Equilibrium Electrolytes'

A large number of carbon-containing fuels can be electrochemically oxidized to carbon dioxide and there is no difficulty in removing the carbon dioxide from acid electrolytes. In the case of alkaline electrolytes, however, OH^- ions are consumed, with formation of carbonate. A process has been suggested by Grüneberg, Jung and Spengler[94] leading to the formation of gaseous CO_2 in alkaline media.

In electrolytes containing carbonate, OH^- ions are formed by the hydrolytic equilibrium

$$CO_3^{2-} + H_2O \rightleftharpoons HCO_3^- + OH^- \tag{IV.17}$$

which is displaced to the right by rising temperature; at 100°C a pH value between 10 and 11 is attained. The bicarbonate simultaneously takes part in the decomposition equilibrium

$$2HCO_3^- \rightleftharpoons CO_3^{2-} + CO_2\uparrow + H_2O \tag{IV.18}$$

which is also shifted to the right by rise of temperature.

If, then, carbonate is formed electrochemically under the above conditions in an appropriate 'equilibrium electrolyte', it undergoes the resultant reaction

$$CO_3^{2-} + H_2O \rightarrow 2OH^- + CO_2\uparrow \tag{IV.19}$$

The electrode reactions proceeding in equilibrium electrolytes, and the dehydrogenation reactions which precede them, are facilitated by increase in OH^- concentration. For the attainment of maximal current densities the most favourable temperature is therefore that which leads to the maximum pH value. According to Grüneberg's measurements this is approximately at the normal boiling point (Figure IV.62). The equilibria IV.17 and IV.18 are linked in such a way that rise of temperature causes carbonate ion concentration and pH value to increase rapidly, while the stationary bicarbonate ion concentration tends to zero.

The general reaction scheme incorporating an alkali-catalysed dehydrogenation can be illustrated by the formate reaction:

$$
\begin{array}{lclcl}
 & & & & 2HCOO^- \\
 & & & & + \\
2CO_3^{2-} + 2H_2O \rightarrow & & 2HCO_3^- & + & 2OH^- \\
 & & \downarrow & & \downarrow \\
 & & CO_3^{2-} & & 2CO_3^{2-} \\
 & & + & & + \\
 & & H_2O & & 4H_{ad} \\
 & & + & & \\
 & & CO_2 & &
\end{array}
\tag{IV.20}
$$

The essential electrochemical reaction

$$H_{ad} + OH^- \rightarrow H_2O + e^-$$

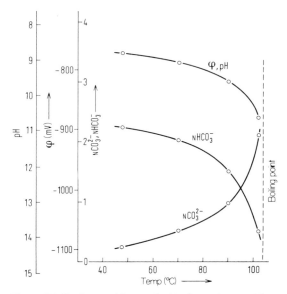

Figure IV.62. Composition and pH of a carbonate–bicarbonate equilibrium electrolyte as functions of temperature
(after Grüneberg[95]); K^+ ion concentration, 2.5N; OH^- ion
concentration measured potentiometrically (H_2 electrode
relative to SCE).

is thus preceded by reactions equivalent to

$$HCOO^- + H_2O \rightarrow OH^- + CO_2 \uparrow + 2H_{ad} \qquad (IV.21)$$

The disadvantage of this scheme is that the current densities attainable at the
relatively low pH values are an order of magnitude less than those available
with strongly alkaline solutions. For instance, Grüneberg, by using a formate–
oxygen cell with K_2CO_3–$KHCO_3$ electrolyte, obtained about 10 mA cm^{-2} at
0.7 volt, at a temperature of 105°C (Raney-nickel as anode catalyst, silver for
the cathode). This laboratory cell worked successfully for more than a month
under these conditions with continuous addition of formic acid to maintain
constant composition of the electrolyte. It is probable that increase of the
working temperature to 150 or 200°C in a pressure cell would lead to as much
as a ten-fold increase in current density.

Cairns and Macdonald[96] pointed out that caesium and rubidium carbonates
and bicarbonates have greater solubilities, by factors of 1.5 to 3, than those of
the potassium salts; increase of salt concentration is advantageous in improving
the conductance of the electrolyte.

A selection of results with use of Cs_2CO_3–$CsHCO_3$ as equilibrium electrolyte and C_2H_4, C_2H_6, CH_3OH and CO as fuels is presented in Figures IV.63

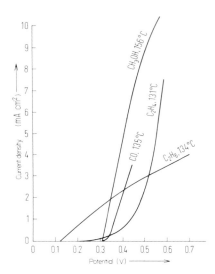

Figure IV.63. Anodic current–potential curves for various fuels in Cs_2CO_3–$CsHCO_3$ equilibrium electrolyte at platinum (after Cairns and Macdonald[96]). Potentials relative to the hydrogen electrode in the same electrolyte ($p_{H_2} = 1$ atm).

to IV.65. Anode and cathode were porous platinum-black electrodes (General Electric). The highest current density was obtained with ethylene (Figure IV.63). Ethane and methane were formed at the anode in addition to CO_2 (cf. section III.4). In prolonged operation (Figure IV.64), water vapour was added to the ethylene to obviate deposition of carbon at the anode (arising from CH_4 as intermediate product).

With ethane as fuel the current densities decreased with rising temperature from 134 to 172°C because methane formation increased two-fold over this temperature range. Figure IV.65 shows current–voltage curves for the ethylene cell in dependence on temperature.

IV.2.4.2 Fluoride Electrolyte

Evidently, unsubstituted hydrocarbon electrodes do not perform as well in carbonate solutions as they do in phosphoric acid. More promising are the results of Cairns[96a] with fluoride electrolytes. His choice of electrolyte was guided by three considerations: hydrofluoric acid enhances the solubility of saturated hydrocarbons in aqueous solutions[96b]; fluoride ion is probably not adsorbed at the electrode surface[96c]; alkali fluorides form stable complexes with hydrofluoric acid and thus depress the vapour pressure of the acid[96d].

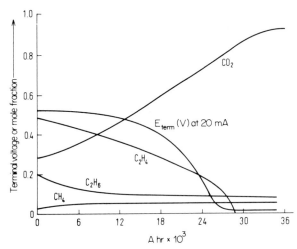

Figure IV.64. Analysis of circulated anode gas produced from
ethylene as fuel (after Cairns and Macdonald[96]).

Only rubidium and caesium fluorides are soluble enough, and reduce the vapour
pressure sufficiently, for their solutions to be used as electrolytes with high
boiling point.

To test the influence of fluoride on cell performance, hydrofluoric acid (48 %)
was added to the electrolyte of a caesium carbonate fuel cell. Additions of 25 ml
were made to the 200 ml of electrolyte in the system. Thereafter the electrolyte
was concentrated, until its boiling point reached 160°C. Figure IV.65a shows

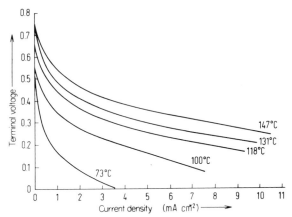

Figure IV.65. Current–voltage curves of an ethylene–oxygen
cell with Cs_2CO_3–$CsHCO_3$ equilibrium electrolyte at various
temperatures (after Cairns and Macdonald[96]).

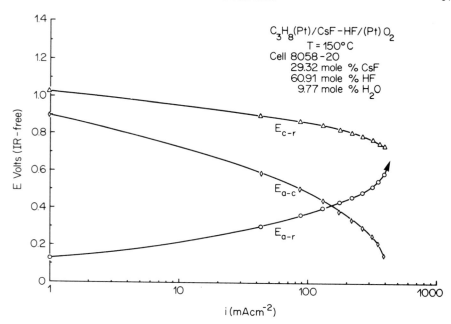

Figure IV.65a. Propane performance at 150°C using CsF/2.1 HF + 15 mole per cent water as electrolyte (after Cairns[96a]).

propane performances with the electrolytes so obtained at 150°C. In Figure IV.65b the logarithm of anode current density at a potential of 0.5 V (relative to a hydrogen electrode in the same solution) is plotted against the ratio $[F^-]/[Cs^+]$. The sudden sharp increase in performance of more than an order of magnitude does not start at a concentration ratio of one, but slightly above this point, when the pH reaches 7. It is the acidity that is important for efficient hydrocarbon oxidation. A cell with hydrofluoric acid (36 mole %) without caesium ion, operating at 105°C, was quite inferior in performance as compared with the optimal Cs–HF cell at 150°C, but considerably better than the caesium carbonate cell.

These investigations are of great theoretical interest but practical application of such cells will be hindered by the limited availability of caesium, and, for the hydrofluoric acid cells, by the limited availability of electrode materials compatible with a fluoride electrolyte.

IV.2.5   High-temperature Cells with Solid Electrolytes ($\sim 1000°C$)*†

Reference has been made in preceding sections of this chapter to the fact that it is desirable under some circumstances to operate fuel cells at high tempera-

---

* By Dr. Helmut Schmidt, Institut für Physikalische Chemie der Universität Bonn.
† A good introduction to this field has been given by Liebhafsky and Cairns[97].

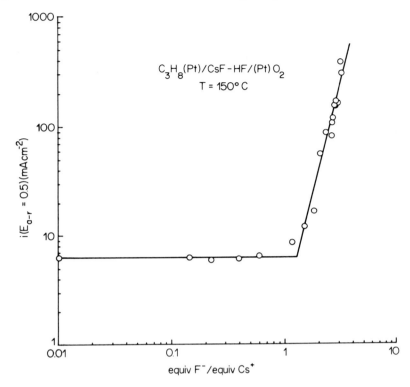

Figure IV.65b. Effect of F^-/Cs^+ ratio on propane current density at $E_{a-r} = 0.5$ V; 150°C (after Cairns[96a]).

tures, using, for example, molten salts as electrolytes. This is particularly the case if an electrode reaction involves a high charge-transfer overpotential. The combustion of carbon provides an example; although thermodynamically possible at low temperatures, it proceeds so slowly that considerable increase of temperature is indispensable on reaction-kinetic grounds.

From the industrial standpoint, high-temperature cells lend themselves to the utilization of heat energy, via heat exchangers, in the cracking or conversion of hydrocarbons, which thus become available for indirect use as electrochemical fuels.

The advantages of high-temperature cells using molten electrolytes are offset by a whole series of disadvantages. Undue rise of temperature may cause decomposition of the melt. The solubility of gases in melts is considerably smaller than in solutions. Above all, there are serious corrosion problems that, for prolonged operation of cells on the technical scale, may be extremely difficult, or even impossible, to solve.

For these reasons, thought was given at a relatively early stage to the use of *solid* electrolytes in high-temperature cells—substances with the highest possible electrolytic conductance at the temperatures concerned. They would offer the additional advantage of effectively separating the reacting gases from each other.

The *requirements* which a solid electrolyte must satisfy are exacting. It must be mechanically and chemically stable at high temperatures; it must not attack the electrodes; it must have the highest possible anionic conductance and the lowest possible electronic conductance. Overvoltages must be small and due to ohmic voltage drop alone. To minimize this voltage drop the thickness of the electrolyte and the resistance of its contacts with the electrodes must be kept low; thin plates of electrolyte, or, preferably, membranes, must be used, and means must be found for making electrolyte–electrode contacts secure and durable. Finally, cationic conductance is undesirable since it may lead to the formation of poorly conducting or insulating surface films, or even to complete destruction of the electrolyte. In other words, the transport number of the oxygen ions, which are generated at the cathode and migrate to the anode, must be as close to unity as possible. Consideration of economic factors, involved in working at temperatures providing adequate conductance and in costs of materials for electrolyte and electrode, indicates that the choice of materials suitable for use in this way is likely to be restricted.

IV.2.5.1 Choice of Electrolytes

In 1937 Baur and Preis[67] reported the investigation of a considerable number of ceramic materials; they were nearly all shown to be unsuitable. Only certain mixed oxides, mainly of zirconium, yttrium and calcium—sometimes in admixture with other similar oxides or salts—proved to be usable. Schottky[98] had already proposed substances of this kind for use as electrolytes in fuel cells; their high electrolytic conductance had been discovered by Nernst in 1900[99] ('Nernst-Masse' = 85% ZrO_2 + 15% Y_2O_3). Quantitative measurements have been published by Weininger and Zemany[100] indicating that the contribution of ionic conductance to the total current is strongly dependent on the pretreatment of the material, on the current density and, particularly, on the temperature. The mechanism of migration of oxygen ions in such oxides has been studied by numerous authors[101–108a]. Wagner[109] well understood that mixed oxides crystallizing with a fluorite-type lattice contain oxide ion vacancies on which the transport of oxide ions depends. This has been confirmed by Hund[110], by X-ray diffraction and pyknometric studies. Kingery and co-workers[111] have used ^{18}O to measure oxygen diffusion coefficients in these oxides and have also calculated oxide ion mobilities by use of the Nernst–Einstein relation. The agreement between their results and the measured conductances leads to the conclusion that practically pure ionic conductance is involved.

No further discussion of these and various similar investigations[112–116b] will be given here since they are concerned essentially with the fundamental problems of such solids and only to a minor degree with their application to fuel cells. The main conclusion is that, of the numerous substances that have been examined, practically only the mixed oxides of a few metals (Zr, Y, Ca, Ce, etc.), having a sufficient number of vacancies in their anion lattices, are practicable for use as solid electrolytes. The conductance depends on the number and distribution of vacancies, as well as on temperature. The objective of further research must be to discover other similar systems and, by variation in composition, to achieve the maximum possible conductance. The best results were obtained with $ZrO_2.CaO$ and $ZrO_2.Y_2O_3$, but, more recently, the mixed oxides $(HfO_2)_{0.90}(Y_2O_3)_{0.10}$[116c], $ZrO_2.Yb_2O_3$[116d] and $CeO_2.La_2O_3$[116e] have been shown to have equally good or even better properties. For this reason the following description of fuel cells* that have been subjected to practical tests is confined to those with the solid electrolytes mentioned above. Other systems have not proved themselves[118,119]; thus, to mention but one example, the cell of Bischoff and Justi[117], consisting of an oxygen electrode of granular CuO, Cu_2O in an iron vessel, with a thin cylinder of Na_2CO_3 as electrolyte (cf. Figure IV.37), worked at 750°C with carbon as fuel. Although it gave the theoretical emf, its terminal voltage collapsed in a very short time under the slightest load.

IV.2.5.2 *Solid Electrolyte Cells with Direct Supply of Fuel Gas to the Anode*

From the investigations of zirconia as a solid electrolyte, the work of Weissbart and Ruka[120–122] is the first to be discussed. Figure IV.66 illustrates the cell they used, operated with continuously flowing gases—oxygen to the cathode chamber, and a hydrogen–water vapour mixture of varied composition to the anode space, the two being separated by the solid electrolyte $(ZrO_2)_{0.85}(CaO)_{0.15}$. Current was passed by two platinum electrodes, 0.0025 cm in thickness. The electrolyte was 0.15 cm thick and about 2.5 cm^2 in area. Contact resistance was 1 to 2 ohms. With hydrogen the cell may be regarded as operating as an oxygen concentration cell:

$$H_2, H_2O, Pt|(ZrO_2)_{0.85}(CaO)_{0.15}|Pt, O_2$$

The emf is then a function of the different partial pressures of oxygen at the electrodes; at the cathode the pressure is that of the oxygen admitted to it (usually 1 atm), at the anode it is determined by the equilibrium $2H_2 + O_2 \rightleftharpoons 2H_2O$. Accordingly, the emf is:

$$E = E^\circ + \frac{RT}{nF} \ln p_{O_2(cath)} + \frac{RT}{nF} \ln \frac{p_{H_2}}{p_{H_2O(anode)}} \qquad (IV.22)$$

---

* High-temperature cells with solid electrolyte have been applied to measurements of partial pressures of oxygen (W. M. Hicksam, *Fuel Cell Progress* **2**(8), 1 (1964); J. Langer, *BB-Nachrichten*, **46**(9) (1964)). Devices using this technique are commercially available.

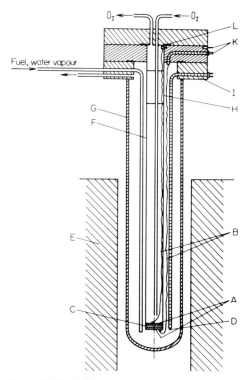

Figure IV.66. Solid electrolyte cell (after Weissbart and Ruka[120]). A, Platinum electrodes; B, platinum electrode connections; C, electrolyte $(ZrO_2)_{0.85}$ $(CaO)_{0.15}$; D, Pt/Pt–10% Rh thermoelement; E, furnace; F, tube of $(ZrO_2)_{0.85}(CaO)_{0.15}$; G, tube of Al_2O_3; H, Kovar to glass to ceramic seal; I, water-cooled metal flange; K, connections for potentiometer and milliammeter; L, .002″ Teflon gasket seal.

The values calculated from this equation for various H_2/H_2O ratios (with constant oxygen pressure at the cathode), agreed with the measured emf within experimental error (Figure IV.67). The hydrogen and water concentrations were obtained by occasional analysis of the gas passing out of the anode compartment. The current density–voltage curves for various temperatures (800 to 1100°C) were straight lines of slopes agreeing to within 5% with those to be expected from ac measurement (frequency 1000 sec^{-1}) of cell resistance. Change of the H_2/H_2O ratio at a given temperature caused only a parallel displacement of these lines (Figure IV.68).

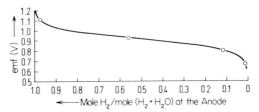

Figure IV.67. Comparison of measured terminal voltages (0) with theoretical emf values for the cell H_2, H_2O, Pt|$(ZrO_2)_{0.85}(CaO)_{0.15}$|Pt, O_2 at 1015°C. Oxygen pressure at the cathode, 731.2 Torr (after Weissbart and Ruka[120]).

These results show that:

1. the cell works as an oxygen concentration cell, and
2. the overpotential is almost exclusively due to ohmic voltage drop, which is, of course, the smaller the higher the temperature.

The operation of this cell with hydrogen has hardly any practical significance; from the economic point of view it is senseless to operate a hydrogen–oxygen cell at a high temperature. It has been mentioned that temperature increase is desirable only if an electrode reaction has a high charge-transfer overpotential and if the cell can be coupled with a heat exchanger. Neither of these conditions applies to hydrogen, which can readily be brought into direct electrochemical reaction at room temperature.

The results obtained with *methane* in the same cell are therefore of greater importance. In these experiments, gas mixtures of low methane content were used, passing through the cathode compartment at a rate high enough to ensure

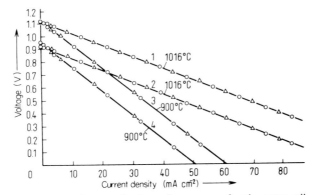

Figure IV.68. Current density–voltage curves for the same cell (Figure IV.67) for various temperatures and H_2/H_2O ratios. $H_2/(H_2 + H_2O)$ for curves 1 and 3 ~ 0.97; for curves 2 and 4 ~ 0.54. ○ = Increasing, △ = decreasing current (after Weissbart and Ruka[120]).

not more than 20% consumption of the methane. Gas issuing from the cell was analysed for H_2, H_2O, CO, and CH_4, and the results were used for calculation of theoretical emf values. For a mixture of 3.8% CH_4, 2.1% H_2O and 94.1% N_2, the rest voltage was within 5 mV of that calculated for the simple H_2–O_2 reaction, and about 50 mV less than that for the CO–O_2 reaction (Figure IV.69). The overpotential was again shown to be purely resistive. Higher rest voltages were obtained with slower passage of gas because a greater proportion of methane took part in the reaction

$$CH_4 + H_2O \rightleftharpoons CO + 3H_2$$

The analyses of the gas mixtures showed in each case that it was not CH_4, but H_2 and CO that were electrochemically oxidized. No carbon was deposited in the cell—this applied, however, only under the conditions described. At lower H_2O/CH_4 ratios, and lower rates of gas flow, carbon deposition did occur —a very unwelcome result to which further reference will be made.

Since these experiments were directed to fundamental rather than technical aspects no studies were made in relation to optimum efficiency and working life. Nevertheless, a hydrogen–oxygen cell of this type worked uninterruptedly at 1000–1200°C for two months, at a current density of 10 mA cm^{-2}, with only a trivial loss of platinum.

Extensive work on solid electrolytes based on zirconia has been carried out in the last few years in the Westinghouse Research Laboratories[122a,b,c]. As electrolyte, segmented tubes of tapered and 'bell and spigot' design are used, lending themselves to assembly into larger units, as illustrated in Figure IV.69a. Porous platinum layers, both inside and outside each segment, are used as electrodes. The main difficulty in building larger batteries in this way is the

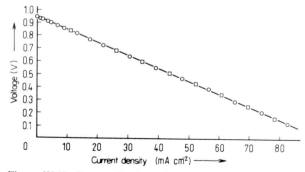

Figure IV.69. Current density–voltage curve for the cell CH_4, H_2O, Pt|$(ZrO_2)_{0.85}(CaO)_{0.15}$|Pt, O_2 at 1015°C. Fuel gas: 3.8% CH_4, 2.1% H_2O, 94.1% N_2. Oxygen pressure at the cathode = 731 mm. $\bigcirc$ = Increasing, $\square$ = decreasing current (after Weissbart and Ruka[120]).

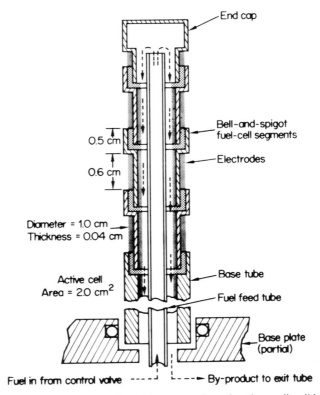

Figure IV.69a. Schematic axial cross-section of a three-cell solid electrolyte battery showing feed tube (after Archer, Elikan and Zahradnik[122c]).

maintenance of gas-tightness and good electrical contact between the units. The path taken by the fuel is shown in Figure 69a ; the oxygen, or air, is supplied to the outside of the cell.

Hydrogen, or hydrogen–carbon monoxide mixtures, can be used as fuel, the open-circuit emf being within 3% of the thermodynamic value. The somewhat higher overvoltage encountered when CO–CO_2 mixtures are used can be greatly reduced by adding water vapour to the fuel gas, and by incorporating a catalyst layer of Cr_2O_3 on the (inner) anodic surfaces of the electrodes (cf. Figure IV.69b).

To construct a 100 W unit, 20 batteries, each of 20 individual cells, were combined in two groups (Figure IV.69c), to give an open-circuit emf of 200 V. The maximal load (with hydrogen as fuel and air as oxidant) was 102 W at 1.2 A—about 20% less than expected from measurements on individual cells. This deficiency is attributed to non-uniform distribution of air within the battery, and to a fall of temperature due to heat loss.

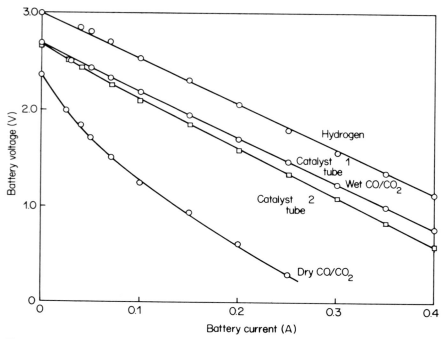

Figure IV.69b. Performance of three-cell battery using a Cr_2O_3 catalyst sintered to the fuel feed tube. Specifications: three-cell battery, temperature 1000°C; CO/CO_2 ratio 9:1; fuel flow 5.7 ml sec^{-1} (after Archer, Elikan and Zahradnik[122c]).

Carter and coworkers[123,124] have given an account of an interesting variation of the previously described cell. They found that 14.2 to 15.0% of CaO in the solid electrolyte gave a maximum conductance at 1000 to 1100°C (recently, additions of other oxides than CaO have about doubled the conductance).

The construction of the cell was as follows:

A nickel screen was fitted inside a zirconia cylinder about 1 cm in diameter and 5 cm long. The outer surface of the cylinder was provided with a metal catalyst and exposed to the air. Gaseous hydrocarbon supplied to the interior of the cylinder decomposed on the nickel screen, depositing carbon and forming hydrogen. Oxygen from the air ionized at the outer surface, migrated inwards in the form of O^{2-} ions, and reacted with the deposited carbon inside the cylinder:

$$C + O^{2-} \rightleftharpoons CO + 2e^-$$

The gas leaving the cylinder consisted of a mixture of the original hydrocarbon, CO and hydrogen; it was burned and used to maintain the working temperature. The fuel itself was used in this way for the initial heating.

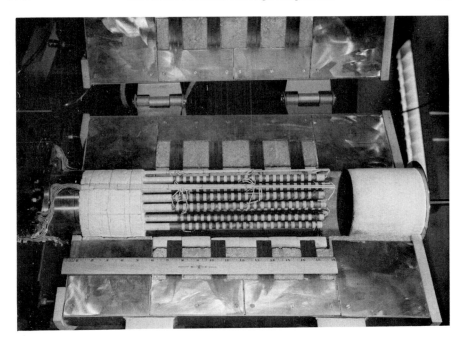

Figure IV.69c. Solid-electrolyte, 100 watt fuel cell power system with
furnace door open. (Westinghouse Res. Lab.)

With silver as the oxygen catalyst a rest voltage of 1.07 volt was obtained
with a terminal voltage of 0.6 to 0.7 volt at 25 to 50 mA cm^{-2}. Unfortunately,
silver is not serviceable as a catalyst in prolonged use; platinum is stable and
very suitable in its catalytic properties, but is too expensive. At present a metal
oxide less catalytically active than silver is used for endurance tests. With
methane, natural gas or propane as fuel, no cell deterioration has been found in
300 hours of operation at low current densities.

The most recent work of the General Electric Company[124a,b] has led to the
development of a solid electrolyte cell with a molten silver cathode and an
anode of continuously regenerated, pyrolytic carbon. The cathode initially used
was in the form of a crucible containing molten silver, through which oxygen
was bubbled, but this has been developed into the arrangement illustrated in
Figure IV.69d. The outer wall of the electrolyte tube (ZrO_2–CaO, 0.62 mm
thick) carries a layer of porous ZrO_2 flooded with molten silver which, saturated
with oxygen, is retained in the porous layer by capillary action. The fuel, mainly
methane, is passed along the electrolyte tube; air for the oxidation passes between
the outer wall of the electrolyte and the inner wall of a ceramic heating jacket.
Current is taken from the cell by means of a nickel wire screen, or a nickel

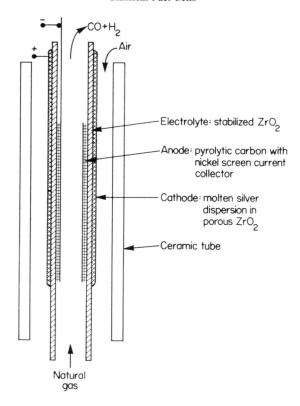

Figure IV.69d. Pyrolytic carbon-anode fuel cell with zirconia electrolyte tube (after White[124b]).

dispersion in zirconia. The silver electrodes have worked satisfactorily for
more than two months, loss of silver by evaporation being kept within tolerable
limits. The main difficulty with the pyrolytic carbon anode is to keep the
deposition and consumption of carbon in equilibrium, and, in this, not only
the gross quantity of carbon but also its distribution and structure are significant.
Addition of H_2O, H_2 or air to the fuel gas, regulation of its flow rate corresponding to the cell loading, are measures taken to minimize this difficulty.

The open-circuit emf of the cell, 0.95–1.05 V (with air at 965–1100°C) is
about 15–20% less than the thermodynamically calculated value. Current–
voltage curves are linear except for minor deviations at low current densities.
Load densities of up to 0.125 W cm^{-2} have been reached and could be increased
by reduction of electrolyte thickness, since electrolyte resistance has proved to
be the dominating source of cell loss. The thermal efficiency, i.e., the ratio of electrical work to the heat of combustion of methane, which should be attainable

with batteries of this kind, is estimated as 20%, or 45–75% if the calorific value of the exhaust gases (H_2 and CO) is taken into favourable account.

IV.2.5.3   *Solid-electrolyte Cell with Associated Reformer*

In the cells so far described the hydrocarbon has been supplied directly to the anode compartment where it has undergone reaction in presence of residual reaction products. How far the hydrocarbon itself, or the secondary reaction products, such as H_2, CO and C, are concerned in reaction with the oxide ions, no doubt depends on the temperature and the composition of the gas supplied, but no clear conclusions can be drawn about individual reaction steps and the intermediate products associated with them. Analysis of the exhaust gas from the cell can only provide information about the overall reaction. There is always the hazard that carbon, which is generally formed as an intermediate, will deposit uncontrollably and interfere with, or completely stop, the working of the cell. The work of Russian authors[71,126–128] may be quoted as an example; they developed various solid-electrolyte cells that worked quite satisfactorily, if at low current densities, with hydrogen as fuel. These cells, however, became inoperative in a very short time when used with hydrocarbon fuels because of carbon deposition. Considering again the work of Weissbart and Ruka, which was not adversely affected in this way, it is to be noted that they used only gas mixtures containing a small proportion of methane, which was only partially consumed.

Special interest attaches to experiments in which Binder and co-workers[125] combined a galvanic cell with a reformer. Their arrangement is schematically shown in Figure IV.70. $Zr_{0.85}Ca_{0.15}O_{1.85}$ was again used as electrolyte, in the form of a disc, 0.5 to 1.0 mm thick, separating two half-cells made of special steel. The electrodes were of porous platinum and the cell was sealed with two

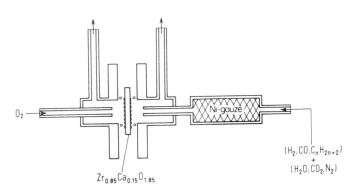

Figure IV.70. Solid electrolyte cell with reformer (after Binder and co-workers[125]).

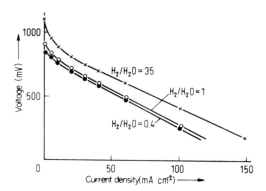

Figure IV.71. Current density–voltage curves for the cell represented in Figure IV.70, at 1000°C, as functions of H_2/H_2O ratio. Oxygen pressure at the cathode = 1 atm (after Binder and co-workers[125]).

rings of fine gold, which also served as electrical connections. Oxygen was supplied to the cathode compartment and the appropriate gas mixture to the anode, now, however, via a reformer which contained only a fine-meshed nickel gauze. The working temperature was between 800 and 1000°C.

Experiments with H_2, H_2O and CO, CO_2 mixtures as fuels gave results in conformity with the theoretical dependence of emf on gas composition. The current density–voltage plots were lines of slope corresponding with ohmic resistance (Figure IV.71). Small negative deviations of the rest voltage up to 10 mV were attributed to the minor incidence of electronic conductance and initially

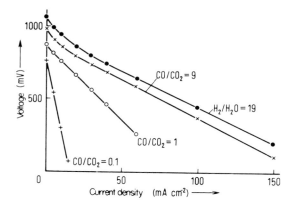

Figure IV.72. Current density–voltage curves for the cell of Figure IV.70, at 1000°C, as functions of CO/CO_2 ratio. Oxygen pressure at the cathode = 1 atm. A curve for hydrogen is included for comparison (after Binder and co-workers[125]).

larger overvoltages were identified as concentration overpotentials arising from accumulation of water vapour at the electrode–electrolyte interface. The corresponding curves for carbon monoxide (Figure IV.72) differ, in so far as an additional overvoltage is observed with increasing current density and decreasing concentration of CO. The probable explanation of this effect is that it is a concentration overpotential arising from disparity in diffusion rates. CO and CO_2 must certainly diffuse in opposite directions through the pores of the electrode as the reaction proceeds at the three-phase boundary: electrode, electrolyte, gas. In contrast to hydrogen, carbon monoxide is insoluble in platinum.

Again, experiments with *hydrocarbons* are of greater practical importance. As in previously quoted work it is impossible to calculate the emf of a hydrocarbon–oxygen cell because, at the prevailing high temperature of the cell, the hydrocarbon molecule is cracked, and it is essentially the secondary reaction products that are potential-determining. If, however, the hydrocarbon is subjected to previous reforming with H_2O or CO_2, the emf becomes a calculable function of gas composition and temperature: always provided that the gas mixture is in thermodynamic equilibrium at the working temperature of the cell. Binder and co-workers[125] have carried out calculations of this kind, starting with the equation for the reforming reaction

$$C_nH_{2n+2} + nH_2O \rightleftharpoons nCO + (2n + 1)H_2$$

and that for the water gas reaction

$$CO + H_2O \rightleftharpoons CO_2 + H_2$$

Taking these equilibria as a basis, the ratio p_{H_2}/p_{H_2O} can be calculated as a function of the degree of conversion, the equilibrium constant of the water gas reaction, the chain length, n, of the alkane and the relative proportions of the conversion reagents with respect to each other and to the alkane.

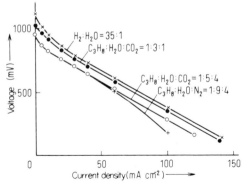

Figure IV.73. Current density–voltage curves for water vapour and/or carbon dioxide-reformed propane, with use of the cell of Figure IV.70, in dependence on the proportions $C_3H_8/H_2O/CO_2/N_2$ in the original gas mixture. A curve for hydrogen is included for comparison. Oxygen pressure at the cathode = 1 atm (after Binder and co-workers[125]).

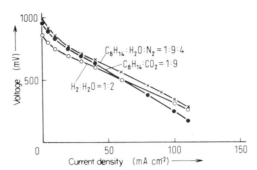

Figure IV.74. Similar curves to those in Figure IV.73,
but with hexane in place of propane (after Binder and
co-workers[125]).

The value of p_{H_2}/p_{H_2O} so obtained is used in equation IV.22 to calculate the emf of the cell. At 811°C the equilibrium constant of the water-gas reaction is unity, so that at this temperature it is immaterial whether H_2O or CO_2 is used in the conversion—the same emf is obtained in either case. This has been confirmed by experiment.

The calculations were experimentally tested by means of reformed propane and hexane (Figures IV.73 and IV.74). The observed rest voltages were at most 5 mV (i.e., 1%) less than those calculated, and the current density–voltage plots were parallel throughout with those found for H_2 and CO, indicating that again—with the reservations mentioned above—only resistance overvoltage was present.

These experiments served only the purpose of clarifying the basic problems of such high-temperature cells—investigations relating to working life, optimum efficiency and technical development were not carried out.

IV.2.5.4 Summary of the Present Position in the Development of High-temperature Cells with Solid Electrolytes

Solid electrolytes based on zirconia attain a specific resistance of about 50 ohm cm at 1000°C and they are therefore suitable for use in fuel cells.

Overvoltage is almost exclusively resistive. The specific resistance can thus be determined from current density–voltage plots, and the change of terminal voltage with temperature is calculable if the change in conductance with temperature is known.

In the electrochemical reactions of hydrogen and carbon monoxide the measured rest voltages agree with thermodynamically calculated emf values to within 1%. The same is true for cells operating with hydrocarbons, if the hydrocarbons are first brought into reaction with H_2O or CO_2 in a reformer. This indicates that the corresponding equilibria are also established at the electrodes.

The transition from experimental cells to technically developed batteries is thus a matter of technical rather than fundamental difficulty. The chief problems are those associated with the electrolyte itself—its durability, chemical stability and its conductance, which must attain an adequate value at as low a temperature as possible, and must be exclusively anionic. In addition, attempts must be made considerably to reduce the thickness of the electrolyte; then, if the only source of overvoltage is ohmic resistance, acceptable working voltages can be obtained in virtue of low electrolyte resistance. Closely related are problems of the electrodes; on the one hand, the difficulty of making secure and stable contacts with the electrolyte to ensure low contact resistance—on the other hand the question of stability of electrode materials; these must suffer no appreciable loss by evaporation at 1000°C, even when operating in flowing gas.

A series of recent studies[125a,b,c] (see also refs. 124a,b) has been specially concerned with electrodes for use with solid electrolytes. Of the metals, only palladium, platinum and (despite its comparatively high vapour pressure) silver come into question. Nickel, cobalt and chromium are usable as cathodes only for short periods because they rapidly oxidize, but nickel and cobalt can be used as anode materials. Metal oxides are mostly reduced when used as anodes, only titanium and uranium oxides showing some degree of stability. Solid solutions of other oxides (e.g. of Ce, Pr, U, Ni) in solid electrolytes are ideal in relation to electrical contact with the electrolytes and provide an electronically conducting surface layer. Unfortunately the conductance of such layers is lower by orders of magnitude than that of metals, so that irremediably high voltage losses must be faced. The use of carbon as an electrode material has already been discussed.

It is unrewarding to make prophecies about these cells. In 1935, viewing technical aspects, Schottky[98] was pessimistic. Recommending fundamental investigation he nevertheless saw more future for molten salt electrolytes. On the other hand, in 1937 Baur and Preis[67,119,129], having themselves worked with melts and observed numerous failures in this field, expressed the view that the future of high-temperature cells lay exclusively with solid electrolytes. Yet in 1962 Justi[8] stated that in his opinion solid electrolytes were fundamentally inadequate for the construction of cells of useful rating. The best position that can be adopted at the present time is one of guarded optimism.

IV.3   CELLS FOR LIQUID FUELS

The use of liquid fuels eliminates the necessity for a three-phase zone and therefore simplifies electrodes. A further advantage is that the fuel can be brought to the catalyst electrode in a highly concentrated form, so that, provided the charge-transfer reaction is fast enough, high current densities are attainable —up to 1 A cm^{-2} at room temperature.

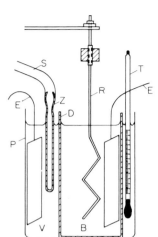

Figure IV.75. Cell for reaction of a liquid fuel (after Taitelbaum[87], 1910). B, fuel-electrolyte mixture; R, stirrer; E, platinum electrodes; Z, Beckmann bubbler (oxygen); V, catholyte; D, earthenware diaphragm; S, pressure tubing; P, porcelain beaker.

Two general procedures are to be distinguished for the oxidation of liquid fuels in galvanic cells.

Taitelbaum[87], in 1910, was the first to use a liquid fuel in a galvanic cell (Figure IV.75). The fuel was completely mixed with the electrolyte (H_2SO_4) in the anode vessel B, which contained a stirrer. Oxygen was passed through a fine-pored bubbler into the catholyte, V, where oxidation of vanadium (IV) (VO^{2+}) to vanadium (v) (VO_2^+) proceeded. Anolyte and catholyte were separated by a porous earthenware partition, D. If the redox system is eliminated and

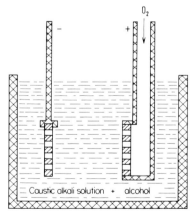

Figure IV.76. Cell with liquid fuel dissolved in the electrolyte (after Justi and Winsel[130]); a porous fuel electrode (negative terminal) and porous oxygen-diffusion electrode (positive terminal) immersed in the same fuel-electrolyte mixture.

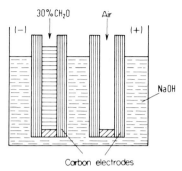

Figure IV.77. Formaldehyde–air cell (after Kordesch and Marko[131]). The porous electrodes dip into a fuel–electrolyte mixture, the fuel being fed to the pores of the anode from the opposite surface to that in contact with the electrolyte.

a cathode material is used which is inert to the fuel (cf. section III.6.5 and Figure III.63), the partition becomes superfluous (Janes*, Justi and Winsel[130], Figure IV.76).

A second procedure was proposed by Kordesch and Marko[131]. Besides an oxygen diffusion electrode, a porous fuel electrode dips into an acid or alkaline electrolyte. Liquid fuel or a fuel–electrolyte mixture penetrates, or is compressed into, the pores of the fuel electrode *from the outside* (Figures IV.77 and IV.86). In this way the active two-phase interface is enlarged. Further, with appropriately adjusted working conditions, the fuel concentration in the vicinity of the oxygen electrode can be kept relatively low without the use of a diaphragm (cf. section IV.3.2).

A modification of this procedure has, however, proved to be preferable; anode and cathode are separated by a diaphragm which acts as electrolyte carrier, the reverse side of the porous fuel electrode supporting the active catalyst mixture (cf. Figure IV.88). Circulation of the electrolyte allows the fuel concentration to be kept very low, and it falls still lower across the diaphragm, towards the oxygen electrode.

Of the large number of liquid fuels, besides glycol, glycerol, isopropyl alcohol, formaldehyde and alkali borohydride solutions, hydrazine, methanol and formate have proved to be the most serviceable on technical and economic grounds. The fuel costs in the case of hydrazine are certainly relatively high at the present time but there are good prospects for the development of economically favourable production methods.

* M. Janes, *Trans. Electrochem. Soc.*, **77**, 411 (1940).

IV.3.1   Alcohol Cells

Of the series of electrochemically active alcohols (cf. sections III.2 and III.3), methanol and glycol have been thoroughly investigated in cells with oxygen or air as oxidants. In acid electrolytes the potentials of fuel electrodes are more positive, and therefore less favourable, than in alkaline electrolytes (cf. chapter III). Since similar considerations apply to the oxygen electrode (section III.6), alkaline cells have hitherto been preferred, although these involve consumption of the electrolyte* :

$$CH_3OH + 2OH^- \rightarrow CO_3^{2-} + 6H_{ad} \qquad (IV.23)$$

$$CH_2OH.CH_2OH + 2OH^- \rightarrow C_2O_4^{2-} + 8H_{ad} \qquad (IV.24)$$

If acid electrolytes are used the whole of the carbon dioxide formed in the reaction escapes, but for cells of considerable rating this means that circulation of the anolyte is indispensable, otherwise there is danger of blocking the reaction zone with gas bubbles. This also applies to the alkaline hydrazine batteries to be mentioned later, but for these the exhaust gas is, of course, nitrogen.

Complete reaction to carbonate or oxalate provides, respectively, about 160 A hr per mole of methanol (32 g) and 215 A hr per mole of ethylene glycol. The electrical output per unit weight is thus particularly high in the case of methanol. About 5000 A hr are stored in a litre of methanol and 1000 A hr can be obtained from one litre of a 6.2M methanol solution.

IV.3.1.1   Cells with Circulation of Fuel—Electrolyte Mixtures

At room temperature the rates of oxidation of alcohols are relatively small compared with that of hydrogen. If current densities of 50 mA cm^{-2} and more are required, temperatures of 50 to 80°C are necessary†.

A comparison of current–potential curves[132] for some freshly-prepared fuel-electrolyte mixtures (14M KOH), obtained with platinized carbon electrodes, is shown in Figure IV.78. In protracted use current densities fall to about 10 to 15 mA cm^{-2}. When isopropyl alcohol is used as fuel the working life of the cell is only a few weeks; reference has already been made in section III.3 to the poisoning action which the reaction product, acetone, has on the fuel electrode. Operation at 88°C effectively removes the acetone (b.pt. 56.3°C) by volatilization but the electrode potential still deteriorates because the formation of polymerization products (the electrolyte goes brown), which block the electrode, cannot be avoided under these conditions. In any case this working temperature is already close to the boiling point (82°C for pure isopropyl alcohol).

Ethylene glycol (b.pt 197.2°C) is a suitable fuel for use at higher temperatures. Spengler and Grüneberg[134] have reported current densities of more than 500 mA cm^{-2} at 90°C (6N KOH + 2M glycol), with use as fuel electrodes of layers of coarsely granular Raney-nickel between metal sieves. By using silver

---

* Isopropyl alcohol provides an exception (equation III.21), but other special difficulties are associated with the use of this fuel—see below.
† An alcohol cell working under these conditions was studied for the first time by Justi and Winsel[130] ($C_2H_5OH–O_2$).

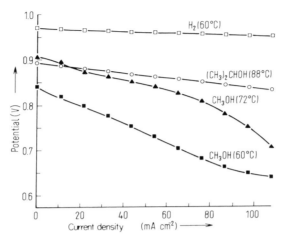

Figure IV.78. Current–potential curves for alcohol electrodes at raised temperatures (after Yeager[132]); ohmic voltage drop eliminated, potentials relative to HgO electrode.

electrodes* on the oxygen side, the following performance at 80°C was recorded:

$$200 \text{ mA cm}^{-2} \times 0.75 \text{ volt} = 150 \text{ mW cm}^{-2}$$

The electrical yield amounted to 150 A hr per mole.

In using *methanol* (b.pt 64.7°C) as fuel, it is expedient not to exceed 50°C as a working temperature. Murray and Grimes[133] carried out an important series of measurements in relation to the construction of a methanol–oxygen battery of 0.5 kW rating.

In the laboratory cell the fuel electrodes consisted of porous nickel plates (7×7 cm², thickness 0.75 mm) carrying a mixed Pt, Pd catalyst. The oxygen electrodes were of similar construction, but were silvered, and rendered water-repellent with polytetrafluoroethylene.

The fuel-electrolyte mixture was circulated between the electrodes throughout and the methanol concentration was kept constant by appropriate additions. The distance between the electrodes was 1 to 2 mm, giving at 30 mA cm⁻² and 50°C a voltage drop of 30 to 50 mV.

The overpotential of the oxygen electrode was but little affected by the composition of the electrolyte (Figure IV.79). The characteristic curve of the methanol electrode showed two noteworthy features. At about 8 mA cm⁻² (50°C), the overpotential increased considerably, and at low current densities the presence of carbonate in the electrolyte had an obviously beneficial effect.

In continued operation, the terminal voltage and performance at current densities of about 30 mA cm⁻² at 30°C, or 60 mA cm⁻² at 50°C decreased considerably (Figure IV.80). Change of electrolyte did not quite restore the original performance.

* Catalyst sieve electrode, cf. Podvyaskin and Shlygin, ref. 19, chapter III.

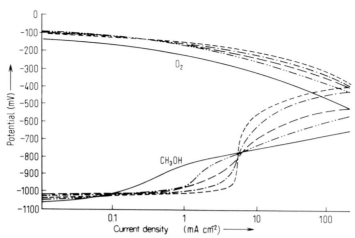

Figure IV.79. Influence of electrolyte composition on the characteristic curves of anode and cathode (after Murray and Grimes[133]); 50°C, ohmic voltage drop eliminated, potentials relative to SCE ——, 6M CH$_3$ + 6N KOH; — ·· —, 6M CH$_3$OH + 5.1N KOH + 0.5M K$_2$CO$_3$; ——, 6M CH$_3$OH + 3.9N KOH + 1.05M K$_2$CO$_3$; — · —, 6M CH$_3$OH + 2.8N KOH + 1.35M K$_2$CO$_3$; - - -, 5M CH$_3$OH + 2.3N KOH + 1.85M K$_2$CO$_3$.

Figure IV.81 shows that, in the first 250 hr of operation, the main effect is deterioration of the anode potential with increasing carbonate content of the electrolyte—Figure IV.79 indicates that this is to be expected at 65 mA cm^{-2}. The voltage jumps correspond with changes of electrolyte; the last of these coincided with regeneration of the anode.

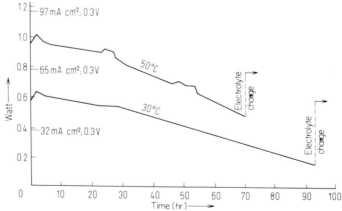

Figure IV.80. Typical performance–time curves of a methanol–oxygen cell at 50°C and 30°C (after Murray and Grimes[133]). Initial KOH concentration, 6N; methanol content of the electrolyte kept constant.

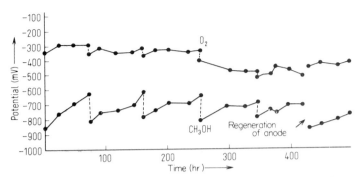

Figure IV.81. Potentials (relative to SCE) of individual electrodes during an endurance test over 500 hours (after Murray and Grimes[133]). Methanol–oxygen, 50°C, 65 mA cm^{-2}, ohmic voltage drop eliminated, electrolyte changes at the potential jumps.

At ACHEMA 1964, the Battelle Institut, Frankfurt, presented a methanol–oxygen battery (2 cells of diameter 110 mm and length 70 mm) that, at 32°C (attained in operation), could be continuously loaded at 1.9 A at a terminal voltage of 0.75 V (1.2 A, 0.95 V; 2.7 A ($\sim$ 50 mA cm^{-2}), 0.5 V)[152,154]. The electrolyte was 10N KOH + 10% CH$_3$OH and the 3 mm thick nickel electrodes (diameter 8 cm) were activated with palladium or silver. At 70°C, the rating was 2.5 watts, corresponding to a current density of 50 mA cm^{-2} and a terminal voltage of 500 mV per cell.

IV.3.1.1.1   Batteries for alkaline electrolytes

Bipolar electrodes have been used to construct a methanol–oxygen battery of the smallest possible volume (Figure IV.81a)[133]. Anode and cathode are attached to a nickel cell-partition; between this partition and the cathode is left a very narrow pocket into which oxygen is supplied from a cylinder, the excess escaping freely. The distributor for circulation of the fuel-electrolyte mixture is mounted

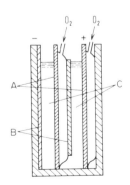

Figure IV.81a. Methanol–oxygen battery with bipolar 'pocket electrodes' and circulated fuel-electrolyte mixture (after Murray and Grimes[133]). A, porous oxygen electrode; B, fuel electrode; C, fuel + KOH.

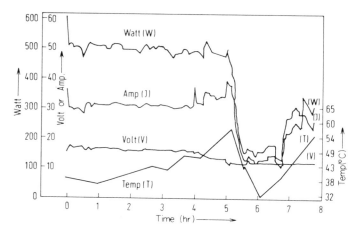

Figure IV.81b. Variation with time of the performance data of a 500 watt methanol–oxygen (after Murray and Grimes[133]).

in the base; the circulation system for the 40-cell battery includes a heat exchanger. The performance data for the battery are collected in Table IV.5.

Figure IV.81b shows the variation of the electrical characteristics of the 500-watt battery with time and temperature over several hours of operation. It is seen that the fall in output arising from increasing carbonate concentration could be compensated by an appropriate increase in temperature—achieved by reduced cooling. After five hours the KOH concentration had fallen to 2M; formate concentration remained constant at less than 0.1 molar.

If a battery of this kind is to be brought into practical use, and rejection of the electrolyte is excluded, it must be coupled with a regeneration unit. The use of lime suggests itself. About 2.5 kg of $CaCO_3$ would have to be precipitated for every kW hr at 0.3 V cell voltage[133].

Table IV.5. Performance data of a methanol–oxygen battery (after Murray and Grimes[133]).

	Working range		Overload
Power (watt)	440	570	750
Terminal voltage (V)	16	14	10
Current (A)	27.5	41	73
Cell voltage (V)	0.4	0.35	0.25
Current density (mA cm^{-2})	57	85	151
Power/volume (kW m^{-3})	13	16	21.5
Power/weight (W kg^{-1})	6.5	8.5	11
Oxygen pressure (atm)	0 to 0.35	0 to 0.35	0 to 0.35
Temperature (°C)	50	50	50

The Varta Company[133a] has studied alkaline methanol–oxygen batteries for several years; the most recent type uses 150 cm² Janus cathodes (cf. section IV. 1.1.1.2) supplied with oxygen at an excess pressure of 1 atm. The sintered-nickel anodes carry 1.5 mg cm⁻² of Pd–Pt mixed catalyst. A 20-watt demonstration battery[133b] consists of 5 modules, each containing 7 oxygen (Janus) electrodes and 6 methanol electrodes, cast in epoxy resin; it provides 4 V (0.8 V per cell) on open circuit, falling to 1.8 V at 10 A at room temperature. The battery operates with a pump-circulated[133a] mixture of 8M KOH and 4M CH_3OH.

IV.3.1.1.2   *Reaction of methanol in alkali-hydrolysed equilibrium electrolyte*

Cairns and Bartosik[96,141] have investigated the oxidation of methanol at platinum black electrodes with Teflon binder in an invariant alkaline caesium carbonate–bicarbonate equilibrium electrolyte (see section IV.2.4) at 115 to 130°C. At a terminal voltage of 0.55 V, the current density amounted to 20 mA cm⁻². The maximum output was 40 to 50 mW cm⁻². A successful endurance test of 500 hours was reported.

IV.3.1.1.3   *Methanol and formic acid cells with acid electrolytes*

Jasinski and co-workers[92] used the cell developed for the propane–phosphoric acid system at 150°C (cf. section IV.2.3) to study the reactions of methanol and formic acid in acid electrolytes (30% H_2SO_4 and 30% H_3PO_4) at 60°C. The experimental method was the same as that illustrated in Figure IV.88.

The curves shown in Figure IV.82 confirm the results discussed in section III.2. In acid electrolytes formic acid is a considerably better fuel than methanol but it affords only two electrons per molecule (53.6 A hr per mole).

Shell Research Ltd[133c] have developed a methanol–air battery with acid electrolyte using polyvinyl chloride-based plastic electrodes of the kind already

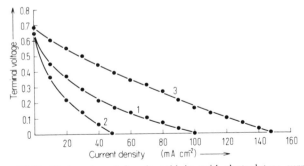

Figure IV.82. Methanol and formic acid in acid electrolytes; current–voltage curves from a laboratory cell (electrodes 7.5 × 7.5 cm²) at 60°C with oxygen in excess; cf. Figure IV.88 (after Jasinski, Huff, Tomter and Swette[92]). Curve 1, 30% H_2SO_4 + 10% CH_3OH; curve 2, 30% H_3PO_4 + 10% CH_3OH; curve 3, 30% H_2SO_4 + 10% HCOOH.

Figure IV.82a. Three hundred watt, 40-cell methanol–air battery (after Williams, Andrew and Jones[133d]).

described (cf. section IV.1.1.3), with Pt and Pt–Ru for cathode and anode catalysts, respectively. The battery[133d] shown in Figure IV.82a consists of 40 cells; it provides 300 W at 12 V working at 60°C.

A portable methanol–air battery intended for military use has been built by the Esso Research and Engineering Company[133e]. The series-connected module of 20 cells (of polypropylene construction) works with Teflon–platinum oxygen electrodes and tantalum–platinum metal anodes in 3.7M sulphuric acid at 60–70°C. The whole assembly, as necessary for its intended application, is fully automatic. Both air and fuel-electrolyte mixture are circulated within a system incorporating a heat exchanger and specially developed control and regulating devices which hold the methanol content (0.75M) and acid concentration at their optimum values. Temperature gradients within the module are avoided by appropriate distribution of electrolyte flow. Voltage is also regulated to remain at 6.0 ± 0.1 V up to currents of about 15 A. The usable output amounts to 82 W, leaving 15 W required for self-operation; only during the 3 minute

starting-up period are ancillary dry cells needed. The overall efficiency is 40%
at half-load and 23% at peak load.

IV.3.1.2   *Alcohol–air Cells with Periodically Renewed Electrolyte—*
Fuel Mixture

Glycol and methanol as fuels have the special feature of high A hr capacity
per unit volume and per unit weight of fuel; as mentioned above, 5000 A hr
are available from one litre of methanol, or 1000 A hr from a litre of a 6.2M
solution of methanol. Since these liquid fuels represent so compact a form of
stored energy, there is a possibility of using them in primary cells as well as in
continuously operating fuel cells. For this purpose it is convenient to combine
a fuel electrode with a hydrophobic air electrode, working at ambient temperature and pressure. The A hr capacity of such a cell is determined by the volume
of the fuel-electrolyte mixture supplied to it and, in contrast to commercial
primary cells, renewal of the mixture reactivates the cell after it has been discharged. There is the additional advantage of a very constant operating voltage
during discharge. Although the best A hr capacities are obtained with methanol
or glycol, formate has special merit as fuel for use at high current density and
low temperature; for certain practical applications, fuel mixtures (e.g., methanol
+ formate) can be used to effect a satisfactory compromise between the desirable
properties.

IV.3.1.2.1   *Glycol as fuel*

A glycol cell working with air at atmospheric pressure at room temperature was
studied in 1957 by Vielstich[27], with reference to its behaviour in prolonged use.
The cell contained 3 l of electrolyte (3M glycol + 8N KOH) in which were
immersed a hydrophobic carbon air electrode (cf. section II.6.2.2) of area
150 cm^2, and the fuel electrode of the same area. This consisted of coarsely
granular nickel-DSK material* held between perforated steel plates. At a
constant current output of 300 mA, the terminal voltage of the cell fell from
0.8 to 0.6 V in 3000 hours. The electrical yield amounted to about 70% (relative
to 8 electrons per molecule).

In extension of these experiments, Grüneberg and Jung[134] have made a
hydrophobic air electrode of such mechanical strength that it can be used to
form the outer walls of the cell.

The electrode is a pressing made from active carbon and polyethylene and has a double-layer
construction. The layer on the electrolyte side is only slightly hydrophobic and contains Ag_2O as
catalyst:

> 45 to 50 wt% active carbon,
>
> about 30 wt% Ag_2O
>
> 25 to 20 wt% polyethylene.

* This electrode construction and the use of glycol as fuel has been suggested by the experimental
work of Spengler and Grüneberg[134], described above.

The gas-side layer is strongly hydrophobic and consists of 65 to 70 wt % of active carbon with 35 to 30 wt % of polyethylene. The powder mixture is pressed at about 500 kg cm^{-2} for 5 to 10 min at 160°C. The electrodes are 10–12 mm thick.

Figure IV.83 shows Spengler and Grüneberg's six-cell battery[134]. Cell casings are made of Perspex with side-walls formed from circular air-diffusion electrodes. A fuel electrode of the same size is situated between each pair of air electrodes; it consists of coarse-grained Raney-nickel held between nickel sieves. The electrolyte is 6N KOH + 2M ethylene glycol. The rest voltage of such a cell is about 1.1 volt, and at room temperature it will provide 3 mA cm^{-2} at 0.8 V. Current densities up to 30 mA cm^{-2} can be withdrawn for short periods.

The working life of the cell is limited by the air electrode but it attained nearly 8000 hours of operation ($\sim$ 15 A hr cm^{-2}) under the following conditions:

> Current density 1.7 mA cm^{-2}
> Temperature 20°C
> Electrolyte change every 450 hr (corresponding with 50% fuel consumption at 8e^{-} per molecule)
> Terminal voltage 0.8 V (flat characteristic)
> Potential of the glycol electrode $-$1060 mV (SCE).

IV.3.1.2.2 Methanol as fuel

Use of methanol as fuel provides, at the same current density and temperature, some 100 to 200 mV lower terminal voltage than is the case for glycol as fuel.

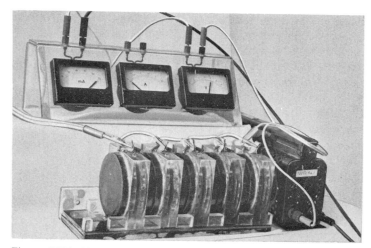

Figure IV.83. Six-cell glycol–air battery with hydrophobic air-diffusion electrodes of active carbon and silver with polyethylene as binder (after Spengler and Grüneberg[134]).

On the other hand the electrical yield per gram of fuel as well as per unit volume of fuel-electrolyte mixture is greater. For periods of operation exceeding 3000 hours without change of electrolyte, the fuel electrodes in methanol cells are less sensitive than those in glycol cells. Methanol is also much cheaper (see chapter X). It is on these grounds that the following methanol–air cell has been developed[135–136c] for the maintenance-free supply of electrical energy to navigation lights (buoys), weather stations, isolated observation posts and the like.

Construction of individual cells

The cell illustrated in Figure IV.84 is designed to contain the whole of the working materials for the desired period of operation. The rectangular plastic container of about 2 l capacity contains a hydrophobic carbon cathode and, concentric with it, a catalysed nickel-gauze anode. The electrolyte is 9M KOH + 4.5M CH_3OH, the concentration of KOH being chosen so that OH^- ion falls only to 1–2M after complete consumption of the fuel (cf. equation IV.23). Cells with NaOH as electrolyte show higher overvoltages, arising mainly at the oxygen electrode.

Platinum, or better, a palladium–platinum mixture, at 2–4 mg cm^{-2} serves as catalyst for the fuel electrode. The carbon air–diffusion electrode (geometrical

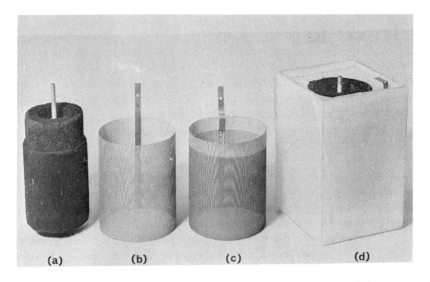

(a)   **(b)**   **(c)**   **(d)**

Figure IV.84. Electrodes and components of the methanol–air cell (after Plust[136a]). Volume of electrolyte 2 l, capacity 1400 A hr with 4.5M CH_3OH. (A), Carbon–air electrode; (B), nickel gauze electrode; (C), impregnated electrode; (D), complete cell.

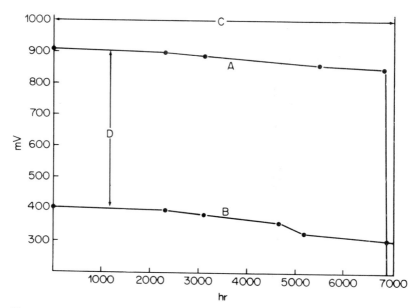

Figure IV.84a. Electrode potentials of a methanol–air cell up to complete consumption of methanol (after Plust[136a]). No metal addition to cathode, ambient temperature. (A), potential of the fuel electrode (rel. to SCE); (B), potential of air electrode; (C), theoretical operating time, 7020 hr; (D), terminal voltage.

area 200–250 cm², supplied by the CIPEL Company of Paris, is given additional hydrophobic treatment before use. With addition of silver catalyst continuous operation at 1 A at 15–20°C is possible. Without metal additions, potentials of −300 to −400 mV (relative to SCE) are established at 200 mA load. Intermittent use (as for light signals) permits 2-fold or 3-fold overload.

Each cell has a rest-voltage of 0.9 V and, on a single electrolyte-fuel filling, provides 7000 hr of operation at 0.2 A and 0.6 V terminal voltage (cf. Figure IV.84a); this corresponds to a capacity of 1400 A hr and stored energy of ca. 0.84 kW hr.

Prolonged experiments involving intermittent loading (2 sec at 0.5 A, 4 sec rest) have been conducted to explore the application to light signal systems. The periodic interruption of current not only facilitates the diffusion of atmospheric oxygen to the cathode, but also prolongs the activity and the working life of the fuel electrode. Figure IV.84b shows the terminal voltage of a cell loaded at 0.5 A for more than 12 000 hr; it is seen that there is but little change as the methanol is consumed (cf. also Figure IV.84a), even when the carbonate content of the electrolyte reaches a high value (cf. Figure IV.79).

Electrolyte analysis shows that the difference between theoretical and experimental current yields is mainly due to loss of methanol by evaporation

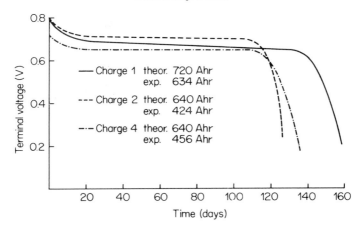

Figure IV.84b. Terminal voltage at periodic loads (two second load, four second rest) of methanol–air cell. Operating time 12,000 hours, with four electrolyte charges of 1 l; temperature 10–20°C; carbon electrode without silver (after Vielstich[136c]).

through the carbon; it is also shown that back-diffusion of CO_2 from the air by the same route is negligible in effect.

Batteries

Cells of this kind can be made in box-form and stacked like bricks to produce batteries of any desired rating, with series or parallel connection to suit any particular application. The main application of such batteries, however, is the provision over long periods of not very large supplies of electrical power (< 100 W), without servicing or attention, and some of the possibilities have been explored over the last few years (cf. chapter X). A difficulty arises in such cases in the seasonal variation of temperature, which can lead to undesirable variations in working voltage; means of obviating this are discussed in section IV.3.1.2.4.

IV.3.1.2.3   Formate as fuel

Formate has advantages over methanol as a fuel because it is less volatile and can be used at temperatures up to 100°C, and because—given suitable catalysts —it can provide higher terminal voltages (cf. Figure III.32a). It has the disadvantage of a substantially lower capacity because its oxidation yields 2 electrons per molecule, compared with 6 from methanol.

In studying formate cells Grimes and Spengler[136d] gave particular attention to the effectiveness of catalysts as a function of temperature (cf. section III.2.4); Pd was found to be best for the anode and Pd–Pt (1 : 1) for the cathode. Half-cell potentials for such electrodes in 4M KOH + 4M HCOOK are shown in Figure IV.84c.

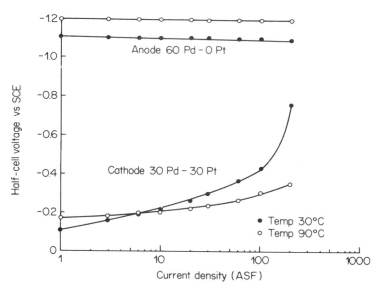

Figure IV.84c. Half-cell potentials of a suitable catalysed anode and cathode combined in a single cell containing 4M KOH + 4M HCOOK. Noble metal content of electrode, 60 mg in[-2] (after Grimes and Spengler[136d]).

Since oxidation of formate (or methanol) consumes hydroxyl and generates carbonate ions, continued cell operation is accompanied by fall in pH and rise in carbonate concentration. The effects of these changes were studied by examining the cell containing $4M\ K_2CO_5 + 4M\ HCOOK$[136d]; but slight increase in the overpotential of the formate anode was found, whereas the performance of the oxygen cathode was seriously impaired. At 90°C and 100–130 mA cm^{-2} a terminal voltage of only 0.5 V could be attained, compared with 0.8 V at 120–170 mA cm^{-2} with the equivalent cell using KOH in place of K_2CO_3.

The weakness clearly lay in the oxygen electrode and the development of a satisfactory cell of this kind waited upon the recent improvement, by an order of magnitude, in the performance of hydrophobic carbon diffusion electrodes. As a result of this improvement Vielstich and Vogel[136e] have developed a formate fuel cell of the same dimensions as a commercial D-size dry-cell; this is illustrated (in section) in Figure IV.85 and (in two alternative forms) in Figure IV.85a. A sintered nickel anode, catalysed by 1–5 mg cm^{-2} of Pd–Pt (4:1 to 9:1), is held by a perforated nickel screen against the inner surface of a steel or plastic cylindrical container, and is the negative pole of the cell. The carbon cathode, rendered hydrophobic with polyethylene, depends from the lid of the cell, and is provided with air through a silver-plated nickel grid which serves as the positive terminal of the cell. Two openings in the lid allow the introduction

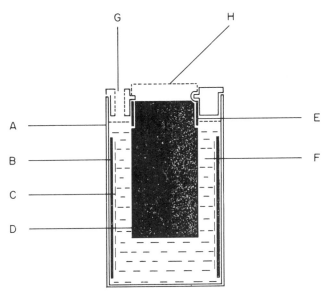

Figure IV.85. Section of liquid fuel–air cell. A, metal housing;
B, fuel electrode; C, perforated nickel screen; D, carbon–air electrode;
E, silver-plated nickel grid; F, electrolyte; G, opening; H, positive
terminal.

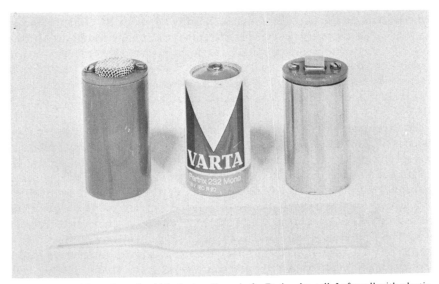

Figure IV.85a. View of two liquid fuel–air cells and of a D-size dry cell. Left: cell with plastic
housing. Right: cell with metal housing. Front: plastic spare tube containing 20 ml of fuel–
electrolyte mixture.

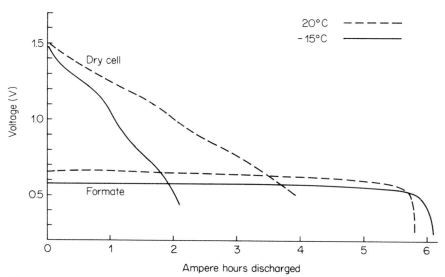

Figure IV.85b. Discharge voltages at 20 and − 15°C of a D-size formate–air cell compared with a commercial dry cell (after Vielstich and Vogel[136e]). Formate–air fuel cell: 7.5M KOH + 5M HCOOK, current 25 mA at − 15°C, 75 mA at 20°C. Dry cell: 1.5 V IEC R 20, current 100 mA at − 15°C, 150 mA at 20°C.

of the electrolyte mixture, 20–24 ml of 7.5M KOH + 5M HCOOK, equivalent to 5.5–6.5 A hr. Discharge curves for this cell at two temperatures are shown in Figure IV.85b, in comparison with those for a normal dry cell. Although the formate cell has a lower terminal voltage the voltage under load is very nearly constant for the whole period of discharge, and it has, for approximately the same weight and volume, a considerably greater capacity. Current yields are 90–97%.

The advantages of this new source of small power supplies over conventional dry cells of similar size are the extraordinarily constant discharge voltage, the high specific capacity, the unlimited shelf-life (in an unfilled state) and the easy 'rechargeability' by renewal of the electrolyte mixture. The low output voltage is a disadvantage, but this can be relatively simply met by means of transistorized voltage converters; experiments with these (very compact and of 60% efficiency) have succeeded in producing 10–30 mA at 6 V from a single primary cell of 0.6 V. The weight and volume of such an arrangement are considerably less than those of 10 primary cells in series. Reference should however be made to the fact that an equivalent cell with zinc as fuel gives a considerably higher voltage; this is discussed in detail in section V.3.

A considerably larger version of the formate cell has been built by Grüneberg and Barth[136f]; of total volume 1330 cm³, charged with 500 ml of 6M KOH +

Figure IV.85c. One hundred A hr formate–air fuel cell of Grüneberg and Barth[136f], operable in any orientation; 0.6 V, 1.0 A continuously at 23°C.

4M HCOOK, it has a capacity of 100 A hr and supplies 1 A continuously at a constant voltage of 0.6 V. Current of 7 A can be taken for short periods at 0.3 V. There is a novel arrangement of electrodes in this cell. The oxygen diffusion-electrode forms the outer wall of the cell, and is enclosed by a cast plastic housing bored with air-holes (see Figure IV.85c); the cell will provide heavier currents if the housing is removed, so as to allow free access of air from all sides. The cell is shock-proof, can be used in any orientation, and reaches a coulombic efficiency of 95–99 %.

IV.3.1.2.4   Formate–methanol mixtures[136a,136g]

As previously mentioned the methanol–air cell is suited for supply of energy to signal stations, buoys, etc., for long periods without attention because of its high A hr capacity. With the cycle of seasons, however, it is subjected to wide variations of ambient temperature, and because only low current densities are involved, the temperature of the battery must follow these external variations. It has been shown[136g] that in the significant temperature range below 10°C, the methanol electrode potential falls away very steeply for a given load, and it may fall low enough to be outside the range of regulation of any built-in constant voltage device (see Figure IV.85d, section a). A possible solution to this difficulty is to substitute formate for methanol, for the current–potential curves have a less unfavourable dependence on temperature (Figure 85d, section b). Against this, however, must be set the accompanying large decrease in capacity (2 electrons per mole for formate oxidation, 6 electrons per mole for methanol oxidation).

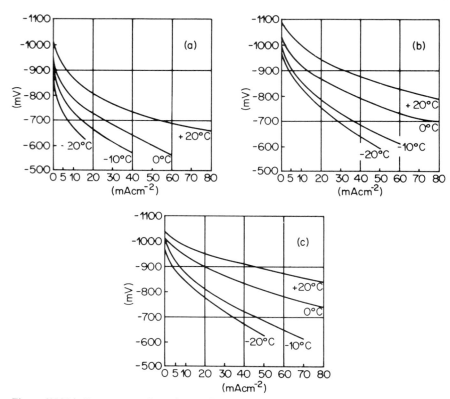

Figure IV.85d. Temperature dependence of potential–current curves of a nickel grid electrode with 2 mg cm^{-2} of Pd, Pt (4:1) anodically oxidizing (a), methanol in 6M KOH + 4M CH$_3$OH, (b) formate in 6M KOH + 4M HCOOK, (c) methanol + formate in 6M KOH + 2M CH$_3$OH + 2M HCOOK. Potentials relative to SCE (after Schmidt and Vielstich[136g]).

In practice, the following device is adopted. Figure IV.85d, section c, shows that for a formate–methanol mixture, the dependence of potential on temperature is hardly less favourable than that for a pure formate solution. If now, the filling and installation of the cell is undertaken during the coldest time of year, advantage can be taken of the higher voltage provided by the formate in the winter, whilst the higher capacity available from the methanol can be exploited in the summer. This involves predominant consumption initially of formate, of which the solution becomes gradually impoverished, leaving methanol as substantially the only fuel to power the cell.

The problem of suitable catalysts is important to the use of fuel mixtures; it is obvious that no one catalyst can be equally effective for formate and methanol (cf. Figure III.32a). A compromise must therefore be sought, and is found in the use of Pd–Pt mixed catalysts. Experiment shows that electrode potentials fall near the values to be expected from the ratio of Pd to Pt in the mixture. At present for practical operation, Pd–Pt 4:1 has been chosen and is used at 2–4 mg cm^{-2}.

IV.3.2   Hydrazine Cells

Hydrazine is a highly active liquid fuel, easy to handle in aqueous solution, about as poisonous as ammonia. The explosion hazard is less for a 60–80% aqueous solution than for petrol. Besides the relatively high price, the large gas evolution involved in its oxidation is a disadvantage; about 24 l of nitrogen are liberated per mole of N_2H_4 (32 g) at room temperature:

$$N_2H_4 + 4OH^- \rightarrow N_2 + 4H_2O + 4e^-$$

Complete electrochemical reaction thus gives about 107 A hr per 32 g of fuel.

A cell with hydrazine as fuel and nitric acid as oxidant was investigated for the first time by Lidorenko[155].

From the measurements that have been described in section III.5.1, it is clear that numerous metals, including silver and gold, can catalyse the electrochemical reaction of hydrazine. An alkaline electrolyte is to be preferred.

In order to suppress reaction of the hydrazine at the oxygen electrode, it is necessary to supply the fuel-electrolyte mixture (0.5 to 6% alkaline hydrazine solution) to the opposite side of the anode (Figures IV.77 and IV.86). Alternatively, an effective diaphragm is necessary to keep the N_2H_4 concentration low in the vicinity of the oxygen electrode.

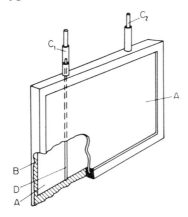

Figure IV.86. Double-walled, porous, hydrazine electrode (after Gillibrand and Lomax[137]). A, electrode plates; B, epoxy frame; C$_1$, filling tube for the fuel–electrolyte mixture, prolonged by tube D; C$_2$, nitrogen exit tube.

IV.3.2.1 Oxygen as Oxidant

The Gillibrand and Lomax hydrazine cell

Fuel electrodes of the Kordesch and Marko type (Figure IV.77) have been used by Gillibrand and Lomax[137] in a hydrazine–oxygen cell. The *fuel electrode* (Figure IV.86) consists of two porous nickel plates of dimensions $14 \times 11 \times 0.15$ cm, secured together in an epoxy frame with a gap between them.

The fuel–electrolyte mixture (7M KOH $+$ 0.5M N_2H_4) is introduced into the inside of the double-walled electrode, under pressure of nitrogen, through the opening C_1, and the tube D. In this way the fuel-electrolyte mixture is compressed into the pores of the electrode, where it undergoes reaction. The nitrogen formed issues from a second opening, C_2, at the top of the electrode. The electrodes are activated by treatment with a noble metal. Gas evolution at the resting electrode at 25°C is equivalent to a current density of 3 mA cm^{-2}.

The *oxygen electrode* is constructed similarly, the electrode plates being made of silver-impregnated, porous carbon (see Table IV.3). As for the hydrazine electrode, the attached tubes serve the purpose also of electrical connections, but since for the oxygen electrode only one opening is needed for gas entry, the other is closed off.

Figure IV.87 shows a battery of four hydrazine and five oxygen electrodes. It is charged with fuel–electrolyte mixture via the opening C_1. Electrolyte depleted of hydrazine is withdrawn through a suitable vent in the battery casing, and then, after addition of fresh hydrazine, is returned to the battery. To keep the hydrazine concentration around the oxygen electrode as low as possible the battery is supplied before use with only as much mixture as will

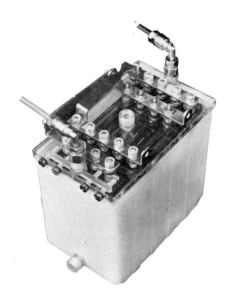

Figure IV.87. Hydrazine–oxygen battery (after Gillibrand and Lomax[137]). Four hydrazine and five oxygen electrodes connected in parallel.

lead to 80 to 90% fuel consumption during transport of the hydrazine through the anode.

The output of the battery is decisively limited by the oxygen electrode[138]; whilst in the absence of hydrazine in the electrolyte this electrode will provide 150 mA cm^{-2} at 60°C, in operation with hydrazine the current density falls to 50 mA cm^{-2}. The terminal voltage is 0.65 to 0.70 volt. The overpotential at the hydrazine electrode is negligible. The rating of the battery (Figure IV.87), nominally 40 watts, reaches a maximum of 60 watts at 60°C, but is only 24 watts at 20°C. Self-heating of the battery to 60°C takes about 45 minutes.

An undoubted advantage of this kind of cell is that no diaphragm is required but this exacts the penalty of considerable voltage loss at the oxygen electrode. The method of fuel supply can also be a disadvantage, and it would be difficult in larger units to arrange for all the anodes to be kept at optimum working conditions with a single hydrazine supply system.

The Tomter and Antony hydrazine cell, with circulation of fuel-electrolyte mixture

The first information on the properties of a hydrazine–oxygen cell was obtained by Tomter and Antony[139] with the experimental apparatus shown in Figure IV.88.

The fuel–electrolyte mixture [5.5M KOH (25%) + 0.9M N_2H_4 (3%)] was brought to 70°C in a storage vessel and circulated by means of a pump to the back of the anode (see below). The nitrogen formed in the electrochemical reaction, and by homogeneous decomposition, was carried along by the circulating electrolyte and released in the storage vessel. Oxygen was similarly circulated

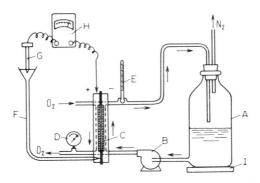

Figure IV.88. Laboratory cell with circulation of fuel–electrolyte mixture (after Tomter and Antony[139]). A, storage vessel for 25% KOH + 3% N_2H_4; B, pump; C, fuel cell; D, pressure release valve; E, thermometer; F, KOH bridge; G, calomel reference electrode; H, voltmeter; I, hotplate.

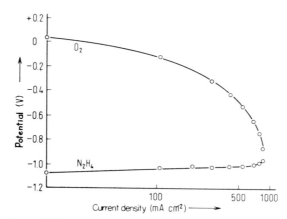

Figure IV.89. Current–potential curves for individual electrodes (after Tomter and Antony[139]); 25 % KOH + 3 %
N_2H_4, 70°C, potentials relative to normal hydrogen electrode.

over the backs of the oxygen electrodes. Single electrode potentials could be
measured with use of a calomel reference electrode.

Cell construction

The cell construction is identical with that of the hydrogen–oxygen cell described by Wynveen and
Kirkland[56] (section IV.1.2; Figure IV.28). An asbestos diaphragm with a plastics frame serving as
a gasket is used as an electrolyte carrier, and sintered nickel electrodes are pressed against it by
means of their mountings. Grooves in these mountings are arranged to provide even distribution
of the fuel-electrolyte mixture, or of oxygen, to the backs of the appropriate electrodes.

The nickel electrodes of the laboratory cell have an effective surface area of about 7×7 cm^2 and
are 0.75 mm thick; the porosity is 80 %. Palladium is used as catalyst for the fuel electrode—
platinum is too active in decomposing hydrazine at the rest potential. Recent experiments by
Jasinski[140] have shown that nickel boride, Ni_2B, provides a potential some 60 mV more favourable
than does palladized nickel (2 mg Pd cm^{-2}). The oxygen electrode is silvered and hydrophobic.

Experimental results

The current–potential curves for individual electrodes, obtained at the beginning of a series of experiments, are shown in Figure IV.89. The rest voltage of
the cell is between 1.1 and 1.2 volt but it can decrease to below 1 volt in prolonged operation. Small quantities of hydrazine reaching the oxygen electrode
can lead to the establishment there of a mixed potential (cf. the results of
Gillibrand and Lomax). Figure IV.90 illustrates the effects of *temperature* on
current density (at 0.7 V) and on terminal voltage (at 108 mA cm^{-2}). The main
effect of increase in temperature is to reduce the overpotential at the oxygen
electrode.

From Figure IV.91 it can be seen what *minimum concentrations* of hydrazine
are necessary for current densities of 50 to 300 mA cm^{-2}. The rate of circulation

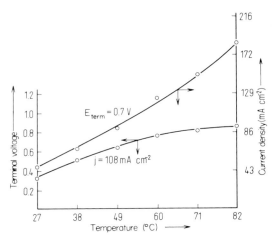

Figure IV.90. Influence of temperature on current density and terminal voltage, 25 % KOH + 3 % N_2H_4 (after Tomter and Antony[139]).

of the fuel–electrolyte mixture must of course be appropriate for each current density.

Figure IV.92 shows the behaviour of the laboratory cell over a period of 1800 hours at 70°C and a current density of 110 mA cm^{-2}. Electrolyte was renewed on several occasions when leaks developed in the system, or the electrolyte pump failed. Over 5 hr the current efficiency was 80% at 100 mA cm^{-2} and 77.5% at 200 mA cm^{-2}, calculated on a 4-electron reaction. The loss is to be attributed to spontaneous decomposition of hydrazine and to its oxidation at the silver cathode (cf. section III.5.1). The average terminal voltage at 100 mA cm^{-2} was 0.63 V; this gives an energy output of about 2 kW hr kg^{-1} of hydrazine.

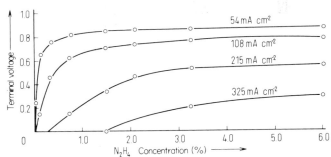

Figure IV.91. Terminal voltage as a function of hydrazine concentration at various current densities (after Tomter and Antony[139]).

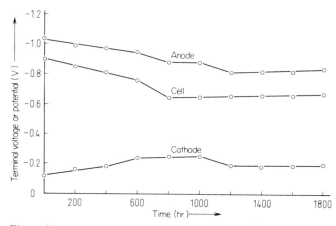

Figure IV.92. Terminal voltage and potentials (relative to standard hydrogen electrode) of individual electrodes during operation for 1800 hr at 110 mA cm^{-2}, 25% KOH + 3% N_2H_4, 70°C (after Tomter and Antony[139]).

Construction of a 3 kW unit

Since the planned 3 kW unit was intended for use in an experimental vehicle— golf cart (see chapter X)—it was decided to cater for a *working voltage* of 36 V at 100 A. The relatively high voltage carries a number of advantages:

1. Parallel connection of cells is eliminated. The series arrangement minimizes voltage loss at connections.
2. Bipolar electrode assembly can be used, minimizing weight and volume.
3. A larger voltage range is available for operation of associated equipment. The only disadvantage is the increased necessity for good insulation.

The number of cells to be connected in series to provide the required working voltage depended on the size of the electrodes, since current density and the terminal voltage of each cell were associated variables (Figure IV.39). The choice of electrode area of 0.5 ft^2 (28 × 16.5 cm^2) and of about 66 cells was determined by the requirement to adjust the joulean heat to maintain the desired working temperature of 70 to 80°C.

Other factors significant to electrode area are, of course, the mechanical structure of the electrodes and the necessity for uniform distribution of liquid and gas to them.

In the event, eight cells were combined in a *module* (wt 10 kg) for stability and better control. Each module had extra strong end-plates and common distributors for gas and liquid. The end-plates, and the electrode holders, were made of nickel-plated magnesium. The electrolyte was introduced at the bottom of the cells and issued from the top to facilitate filling and to allow the escape of

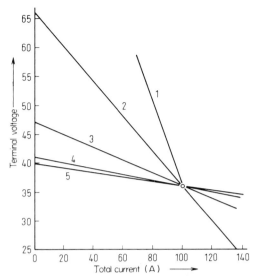

Figure IV.93. Current–voltage characteristics of a 3 kW battery as a function of the number and area of electrodes (after Tomter and Antony[139]). Curve 1, 106 cells, area 300 cm^2; curve 2, 66 cells, area 450 cm^2; curve 3, 47 cells, area 900 cm^2; curve 4, 41 cells, area 1800 cm^2; curve 5, 40 cells, area 4500 cm^2.

nitrogen. Oxygen was passed downward through each cell; this ensured the expulsion of traces of liquid from the gas distributor.

10 modules were combined in the 3 kW battery. Two steel frames held the assembly together (Figure IV.94) and the pressure they exerted served to maintain good sealing of the battery.

Operation of the battery and ancillary units

The interconnections of the battery and its ancillary operating equipment can be seen in Figure IV.95. A centrifugal pump draws the fuel-electrolyte mixture (25% KOH, 3–6% N_2H_4) from its container and delivers it to the battery at 19 to 23 l min^{-1}. The *electrolyte* then passes through two heat exchangers, where a fan provides air-cooling, and returns to the container; nitrogen carried with it issues from a gas valve. Because of the poisonous nature of the hydrazine vapour it carries, the nitrogen then passes through a wash-tank containing 30% H_2SO_4—a liquid path of 8 to 10 cm is adequate. Over the electrolyte container there is a storage vessel containing pure hydrazine hydrate (65% hydrazine); the connexion between the two contains a relay-operated valve.

Oxygen is provided from a steel cylinder fitted with a reducing valve controlling the gas pressure to 0.7 to 1 atm above atmospheric. On safety grounds

Figure IV.94. Three kW hydrazine–oxygen battery (after Tomter and Antony[139]).

excess oxygen leaves the system via another H_2SO_4 wash-tank. The gas is not circulated but the excess over that used electrochemically is controlled by a valve between the battery and the wash-tank.

The *water content* of the electrolyte is increased first by the electrochemical reaction (2 moles of water per mole of N_2H_4), and secondly by the water contained in the hydrazine hydrate (35%). The excess water is removed as far as possible by the nitrogen stream.

The load on the battery can be varied in four steps by relay-operated connection of alternative numbers of cell modules. Mechanical brakes also work on the electric motor supplied by the battery.

Electrically operated ancillaries—electrolyte pump, fan, relays—are supplied from the battery via a voltage regulator. The 1/8 hp pump takes about

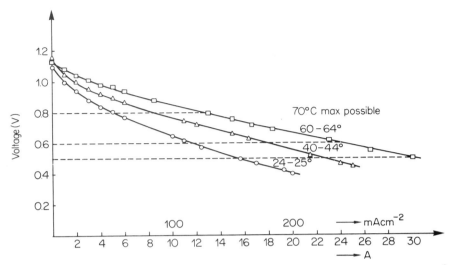

Figure IV.94a. Current–voltage characteristics of a hydrazine–oxygen cell. Electrodes of 100 cm^2 area, self-circulating anolyte (6M KOH + 1M N$_2$H$_4$). Total oxygen pressure 1.5 atm (after Grüneberg and Dünger[157a]).

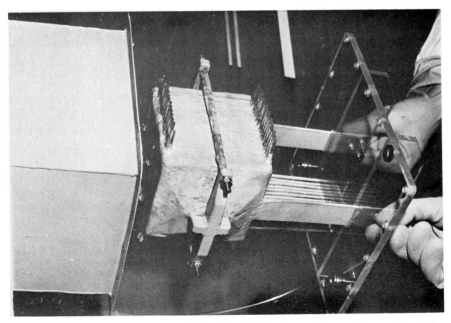

Figure IV.94b. Hydrazine–oxygen module of 15 cells lifted out of the anolyte container (after Grüneberg and Dünger[157a]).

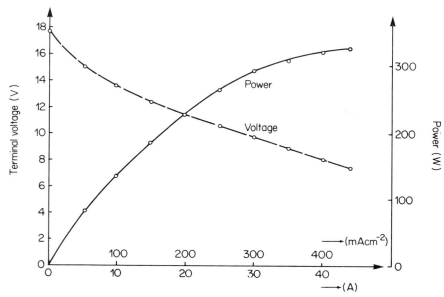

Figure IV.94c. Terminal voltage- and power-current density curves of a 16-cell hydrazine–oxygen module at 70°C (after Grüneberg and Dünger[157a]).

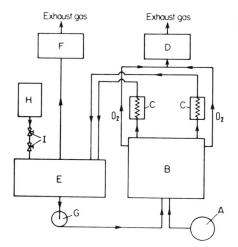

Figure IV.95. Lay-out of the battery and its ancillary equipment (after Tomter and Antony[139]). A, oxygen (air) store; B, battery; C, heat exchanger; D, oxygen washer; E, electrolyte container; F, N_2H_4 absorbing wash; G, electrolyte pump; H, hydrazine store; I, relay-operated valve.

3.5 % of the 3 kW output; for starting it is driven by a small nickel–cadmium battery.

The Grüneberg hydrazine battery

Grüneberg[157] has built a hydrazine–oxygen battery with a peak output of 320 W. The 1.5 l gross volume, 4.5 kg, unit consists of 16 cells with electrodes of 100 cm^2 area, catalysed with flame-sprayed Raney-nickel or with electro-deposited silver. Current–voltage curves at various temperatures for a single cell are shown in Figure IV.94a. The fuel–electrolyte mixture (6M KOH + 1M N$_2$H$_4$) is circulated by means of a 'gas-lift'; Kordesch had already adapted this device to hydrazine oxidation by using the nitrogen evolved in the reaction as the 'transport gas'. A new variation of this scheme can be seen in Figure IV.94b; the module is completely enclosed in a plastic housing with inlet and outlet for electrolyte and is completely immersed in the fuel-electrolyte mixture, in an outer container. Voltage– and power–current density curves are shown in Figure IV.94c.

The Eisenberg hydrazine battery

Eisenberg[158] has described a hydrazine–oxygen cell adapted for use in space vehicles. Working with 6.9M KOH + 0.05–0.2M hydrazine at 20–35°C, it provides 80–110 mA cm^{-2} at a terminal voltage of 0.6–0.7 V. A complete system for extended operation in orbiting satellites has been designed, including means for heat rejection and by-product removal under zero-gravity conditions. The weight of a 300 W unit is 55–60 lb and it is estimated that fuel and oxygen for 14 days operation would, with their containers, weigh 159 lb and occupy 2.9 cu ft; these figures compare with 177 lb and 6–7 cu ft for a hydrogen–oxygen battery of the same rating.

IV.3.2.2   Air as Oxidant

Saving in weight and volume of hydrazine-fueled cells can, as for methanol cells, be effected if they can be operated with air instead of oxygen. Because of their higher voltage and more favourable characteristics under load, hydrazine cells are already more compact than their methanol analogues and lend themselves better to the construction of larger batteries on the technical scale. The use of alkaline electrolyte also allows variation of catalysts from the platinum metals.

The Monsanto portable hydrazine–air battery

The Monsanto Company has developed a portable 60 W hydrazine–air battery with an alkaline electrolyte[159,160]. Palladium-catalysed porous nickel is used for the anode. The cathodes are of the thin, flexible type first developed by Haldemann and his co-workers (cf. section IV.1.2.1.4), made of Teflon and platinum-activated carbon. Silver gauze is used to provide good electrical

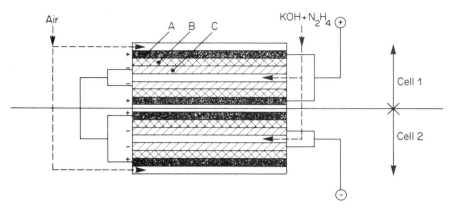

Figure IV.95a. Schematic diagram of hydrazine–air battery construction (after Terry[159,160]).
A, Porous air electrode; B, asbestos membrane; C, fuel electrode.

contact and to confer mechanical stability. The electrolyte is 5M KOH + 0.5–3M N_2H_4, and the electrodes are separated by an asbestos membrane 0.4 mm thick.

The cell construction, similar to that used by Tomter and Antony (see above) is indicated in Figure IV.95a. The whole 60 W battery consist of 36 cells, in 4 groups each of 9 cells in series, and has a working voltage of 7 V. Each cell, working at 75°C, contributes 0.78 V at 65 mA cm^{-2} to the total 60 W load. Continuous operation of a single cell for 1000 hr at 100 mA cm^{-2} is recorded as leading to a voltage drop—from 0.9 V to 0.77 V.

Figure IV.95b shows the scheme of operation of the battery and its ancillary equipment. Fuel-electrolyte mixture is pumped from a reservoir to the battery, and then, loaded with nitrogen, is returned to the reservoir via a liquid–gas separator. Although the hydrazine concentration is not very critical it is held approximately constant by an 'electrochemical pump', which works in the following way.

Two silver electrodes (1 in^2) are situated in a small sump at the bottom of a reservoir of fuel solution ($N_2H_4.H_2O$) to which 5 g l^{-1} of KOH has been added. They are supplied with a current proportional to the fuel cell load (ca. 25 mA A^{-1}) from a proportionating circuit or 'load senser', and generate hydrogen and nitrogen in amounts also proportional to the load. These gases displace fuel solution into the battery 'pick-up' tube.

The 'chemical air', i.e., that used as oxidant, is furnished by a fan, in quantity four times the stoichiometric requirement for 60 W output. Only sufficient pressure is needed (some tenths of an inch of water) to force air through the module and ensure even distribution over the cells. A second fan is used to cool the electrolyte tank.

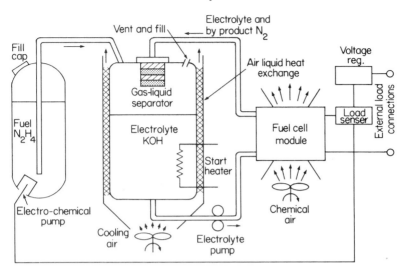

Figure IV.95b. Hydrazine–air battery with ancillary units (after Terry[159]).

Batteries up to 5 kW have been built on this principle[161], designed to deliver 180 A (100 mA cm^{-2}) at 28 V (0.8 V per cell, 2 × 35 cells in series). 8 such batteries have been used to drive a 3–4 ton truck, the complete 40 kW power plant weighing 450 kg without fuel. The auxiliary equipment used about 650 W.

The Union Carbide Corporation portable hydrazine–air battery

Although originally intended for military applications, the 300 W hydrazine–air battery developed by the Union Carbide Corporation (Figure IV.95c) has been built and offered for commercial use[161,162]. The cell construction is similar to that described above (Figure IV.95a). The 'metal-composite' nickel–carbon electrodes developed by Kordesch (see Figures IV.14b, IV.14e, section IV.1.1.2) are used for cathodes; these are separated from palladium-catalysed porous nickel anodes by 1.2 mm thick asbestos membranes. The cells operate with relatively low hydrazine concentrations of about 0.3% in order to minimize the effects on the oxygen electrodes[156]. The circulated fuel-electrolyte (6–9M KOH) mixture passes first through the space between two anodes, then through a condenser where reaction water is removed, and finally, via the circulating pump, back to the cells. Between the pump and the battery a small 'senser cell' measures the fuel concentration and sends an approximate control signal to an injector which draws concentrated hydrazine solution (65%) from a storage tank and injects it into the circulation system according to demands. In this way the hydrazine concentration is kept very nearly constant, independently of load, at 0.3% on entry to the module and 0.2% at the exit. This allows an

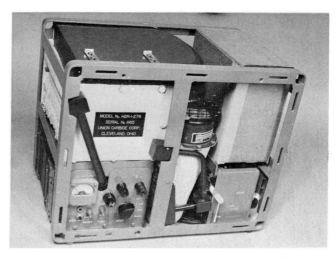

Figure IV.95c. Commercial version of the Union Carbide 300 W hydra-zine–air battery (after Gills[161]). A disposable hydrazine tank takes the place of the cardboard box.

efficiency (at 50 mA cm^{-2}) of 90 % to be attained. The weight of the whole assembly, including fuel for 12 hours' operation, is 15 kg; its volume is about 25 l, and it provides a regulated dc output voltage of 28 ± 1 V. An electrically driven motor cycle, powered by two such modules, each of 400 W, is reported in chapter X.

The Shell hydrazine–air battery

Shell Research Ltd has developed a hydrazine–air battery[163], analogous in construction to that used for operation with methanol (cf. section IV.3.1.1.3) and with similar electrodes. A unit of 25 cells in series provides, at 15°C, 20 A at 18 V; rather more at higher temperatures. Electrolyte circulation provides for heat abstraction, and reaction water is removed in the air stream. Ancillary accumulators are used to provide power for pump and fan only at the start of operation.

IV.3.3   Liquid–liquid Fuel Cells

IV.3.3.1   Hydrogen Peroxide as Oxidant

Liquid–liquid cells with, for example, hydrogen peroxide as oxidizing agent have not yet been so thoroughly investigated as the cells with liquid fuels discussed above.

A *methanol–*H_2O_2 *cell* of particularly simple construction has been described by Grimes, Fiedler and Adam[143]. A multi-cell battery (Figure IV.96) consists of a

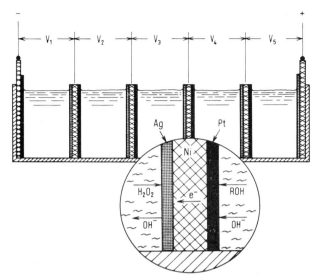

Figure IV.96. Section through a five-cell demonstration model of a methanol–KOH–H_2O_2 battery (after Grimes, Fiedler and Adam[143]).

series of compartments, the oppositely-facing walls of which are *bipolar* electrodes. Each of these is a nickel sheet with platinum deposited on one side and silver on the other. The compartments are filled with a mixture of electrolyte, methanol and hydrogen peroxide (e.g., $6N$ KOH + $1M$ CH_3OH + 0.2% H_2O_2). The silver electrode is inactive to methanol but is a good catalyst for the reaction of H_2O_2 (cf. section III.7). The platinum electrode is active to both fuel and H_2O_2 and assumes a mixed potential about 500 to 600 mV negative to the silver electrode*. Single electrode potentials and cell voltage are very dependent on the composition of the solution.

To test the performance of these cells, a 40-cell battery with 650 cm^2 bipolar electrodes was built. The electrolyte was circulated through the compartments and continuous additions of methanol and peroxide were made. The cell voltage was 0.3 V at 50 mA cm^{-2}. The rating of the battery was thus 400 watts (12 V × 32.5 A) and its power/volume ratio 14 kW m^{-3}.

Disadvantages of this cell are the rapid decomposition of H_2O_2, its direct reaction with methanol at the anode, and also the high water content of the oxidant. It has, however, found wide application as an ideal demonstration cell (Figure IV.96)—without electrolyte circulation[167].

* By appropriate pretreatment, a platinum electrode can be made selectively to catalyse only the reaction of H_2O_2 in an alkaline alcohol–peroxide mixture[144]. If an oxygen layer is generated on the platinum by an anodic pulse of current it becomes inactive to the alcohol, but still acts as a reversible H_2O_2 electrode. It is thus possible to set up a cell with two platinum electrodes in the solution specified above, having similar properties to a cell using silver as catalyst for the cathode.

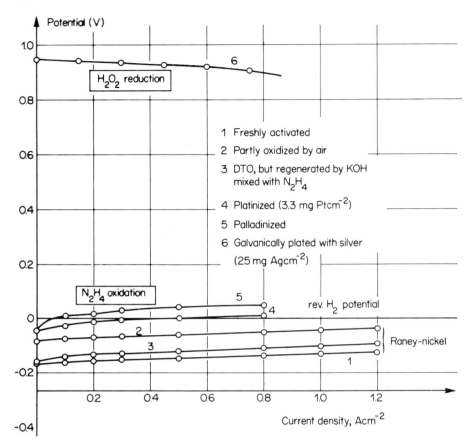

Figure IV.96a. Potential–current density curves from half-cell experiments (6N KOH + 0.5M N_2H_4 or + 0.5M H_2O_2) at 40°C (after Grüneberg and Weddeling[157]).

The most recent technical advance is a fuel-cell engine of 2 kW net power output using hydrazine as fuel and hydrogen peroxide as oxidant. Hydrazine as fuel has the advantage of providing high capacity per unit area of electrode and per unit volume of cell. The hydrazine–peroxide cell has the advantage over the hydrazine–oxygen cells described above of requiring no warming-up period. The full capacity is available at any time since the hydrogen peroxide electrode, even at room temperature, will accept a load of more than 100 mA cm^{-2} (see Figure IV.96a).

Dünger, Grüneberg and Weddeling[165] have developed a 2 kW prototype based on the 100 W laboratory unit with duplex circulation (hydrazine–KOH and H_2O_2–KOH) of Goebel, Nippe and Vielstich[164]. It operates at 40°C with

6M KOH + 0.5M N_2H_4 or +0.5M H_2O_2. The anodes are flame-sprayed Raney-nickel electrodes (ca. 50 mg cm^{-2} of Raney-Ni) and the cathodes are nickel screens carrying 25–30 mg cm^{-2} of electrodeposited silver. As seen from Figure IV.96a, current densities at the hydrogen peroxide electrode are attainable, comparable with the anodic current densities of more than 1 A cm^{-2}. The cells are assembled with cathode and anode held less than 1 mm apart by means of an asbestos membrane. Fuel and oxidant are supplied to the outer surfaces of this assembly. Suitable choice of electrode porosity and also of hydrazine and peroxide concentrations ensures that only very dilute electrolyte mixtures meet within the membrane, so that purely chemical reaction between fuel and oxidant is kept to a low limit. The exterior of each cell is formed by a pressed nickel sheet (Figure IV.96b), suitably grooved to provide the necessary spacing from the electrodes, and to distribute the flow of electrolyte over their surfaces; they also serve for current collectors. Each cell is 8 mm thick and, with other dimensions 18.5 × 25 cm^2, provides an effective electrode area of 350 cm^2.

A module consists of 12 such cells in series, and is sealed with fibre-glass reinforced epoxy-resin; no other support being necessary. Such a module is illustrated in Figure IV.96c. For electrolyte circulation, the cells are in parallel. The inlets are very narrow with ducts of non-conducting material, and since the

Figure IV.96b. Separated parts of a hydrazine–hydrogen peroxide cell (after Grüneberg and Weddeling[157]); left to right: pressed nickel cell wall, Raney-nickel anode, asbestos membrane, silvered cathode, nickel cell wall.

Figure IV.96c. One kW N_2H_4–H_2O_2 modules of 12 cells, each with electrodes of 350 cm² area (after Grüneberg and Weddeling[157]); older type on the left, recent model on the right.

gas carried by the considerably larger outlets largely prevents short-circuiting through the electrolyte, there is very little loss from leakage of current by this arrangement.

Two specially constructed pumps in the electrolyte tanks are used for circulation of anolyte and catholyte, and are driven by electronically controlled, 12 V dc motors. Two systems have been developed for this control. In the first a small quantity of electrolyte is diverted to a cell in which its limiting diffusion current is measured under precisely defined conditions, and is proportional to fuel concentration. This cell is illustrated in Figures IV.96d and e. The limiting current is compared electronically with a standard value, and the difference signal, via an amplifier, controls the pump motor. In the second system, all the anode and cathode potentials are sensed, as functions of the load on the module; an average is again compared with a standard, and the difference used to regulate the pump motors in a similar way.

The module provides 726 W on continuous, and 1040 W on peak load; its weight of 10 kg and volume of 6 l lead to the following specific data: 240 mW cm⁻², 10 kg kW⁻¹ and 6 l kW⁻¹. At 700 W, the efficiency relative to hydrazine is 70 %, that relative to hydrogen peroxide 50 %. These values fall away considerably with increasing load, particularly for the peroxide because the accompanying rise of temperature enhances its rate of decomposition. It would be advantageous to combine two modules in parallel to form a single unit, since taking into account the power demand of the ancillary equipment and loss of energy as joulean heat, such a unit would have a net capacity of 2.1 kW.

A remarkably compact hydrazine–peroxide battery recently produced by the Alsthom Company (Paris)[166] is shown in Figure IV.96f. A version consisting of 180 cells, 30 in series forming a module 2–3 cm thick, provides a rest voltage of 26 V, and a working voltage of 20 V at 50 A (1 kW) at 20°C. A duplex pump

Figure IV.96d. Cell for continuous electrochemical measurements of hydrazine or hydrogen peroxide concentrations in KOH electrolyte (after Grüneberg[157a]).

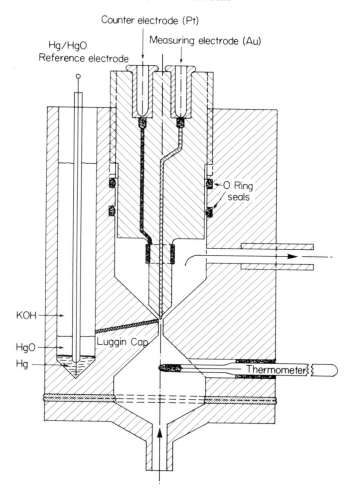

Figure IV.96e. Sectional diagram of the cell shown in Figure IV.96d.

(taking 5W) circulates both solutions at $0.09 \, l \, sec^{-1}$. The $100 \, cm^2$ electrodes are of silver foil, the hydrazine electrode being provided with a cobalt catalyst.

IV.3.3.2   Nitric Acid as Oxidant

Heath[145] has described laboratory experiments with a *methanol–nitric acid–air* cell. Anode and cathode compartments are separated by a porous membrane. The electrolyte is 30% H_2SO_4 and the electrodes are of platinum. The fuel-electrolyte mixture is circulated through the anode compartment. Nitric acid is reduced to NO at the cathode (cf. equations III.46–48). The catholyte, which

Figure IV.96f. The 180 cell hydrazine–peroxide battery of the Alsthom Company (after Warszawski[166]).

is also circulated, is regenerated with air (or oxygen) in a compartment external to the cell (cf. the experiments of Shropshire and Tarmy, section III.7.2). An efficiency of 96% was reached in these experiments. Loss of oxidant can be made good by appropriate addition of nitric acid.

The rest voltage of the cell at 82°C was 0.65 V; at a terminal voltage of 0.48 V, 16 mW cm^{-2} was attained.

A *hydrazine–nitric acid* cell, of 20 watt rating for 8 hours operating period, has been developed by the Monsanto Research Corporation[146]. Anolyte and catholyte are separated by an anion-exchange membrane (Ionics). The electrodes are of platinized platinum gauze (area 120 cm^2). The electrolyte solutions ($5M N_2H_4 + 10M NaOH$; $10M HNO_3$) are introduced from below through channels drilled in the plastic cell (at rates of about 0.6 cm^3 per minute per cell); initially the cell contains water.

A three-cell battery was successfully tested over a period of eight hours at room temperature. The emf amounted to 5.95 V (note the very large pH difference), and, delivering 5A, the battery had an average terminal voltage of 4V. The diffusion of hydrazine or nitric acid through the membrane was clearly negligible during the short working period. Studies with half-cells showed that impurities of $0.1M N_2H_4$ or $1.0M HNO_3$ affected the electrode potentials only slightly (< 100 mV).

IV.4 REDOX CELLS

The mode of operation of a redox cell, the oldest type of *regenerable* fuel cell[147], has already been explained in section I.3.4. The substances X and Y (see Figure I.9) concerned in the electrochemical process are continuously recovered *by chemical means* from the products of the cell reaction. Two cells studied in detail in recent years are described in the following sections.

IV.4.1   The Rideal and Posner Carbon–oxygen Cell

Rideal and Posner[148,149] studied the oxidation of powdered carbon with the aid of the redox systems Sn^{2+}, Sn^{4+} (HCl solution) and Br_2, Br^- (HBr solution). This involved the following partial reactions:

Anode $2Sn^{2+} \longrightarrow 2Sn^{4+} + 4e^-$

Reactor R(X) $2Sn^{4+} + 2H_2O + C \longrightarrow CO_2 + 2Sn^{2+} + 4H^+$

Cathode $2Br_2 + 4e^- \longrightarrow 4Br^-$

Reactor R(Y) $4Br^- + 4H^+ + O_2 \longrightarrow 2Br_2 + 2H_2O$

Total reaction $C + O_2 \longrightarrow CO_2$

Posner first constructed a cell without the reactors R(X) and R(Y) (Figure IV.97). Anode and cathode compartments are separated by a semi-permeable membrane; Teddol Dow-Corning Silicone fluid MS 1107 was chosen as a material conferring properties of effective separation and low ohmic resistance. The electrodes were of porous carbon through which the tin or bromine solutions percolated from the back so that concentration polarization effects were avoided. The distance between the electrodes was 1 mm.

As catholyte 1.2N Br_2 in 0.7N HBr + 3N HCl + 1.2N $SnCl_4$ was used; as anolyte, 1.2N $SnCl_2$ in 1.9N HBr + 3N HCl.

At a flow rate of 3 and 8 cm^3 min^{-1} respectively, the following results were obtained at room temperature. The emf of the cell was about 0.8 V. The current–voltage curve was determined by the low exchange current density of the anode reaction. The overpotential of the cathode was negligible (cf. Juda and co-workers[50]). The terminal voltage was 0.6 V at 15 mA cm^{-2} and 0.4 V at 30 mA cm^{-2}.

If, instead of the stannous–stannic redox system, the ferrous–ferric system* is used, the anodic overpotential becomes very much smaller (cf. chapter VIII), but the emf decreases to an unsatisfactorily low value.

The *regeneration processes* were investigated separately. For the reduction of the Sn(IV) solution with powdered carbon at adequate efficiency and rate, an autoclave at 160 to 180°C had to be used; this gave an efficiency of 30% in

---

* A biochemical cell operating with this redox system is discussed in Chapter V.

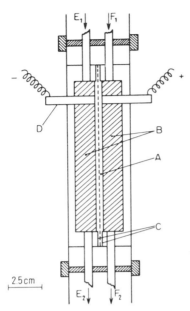

Figure IV.97. Redox cell with flowing anolyte and catholyte (after Posner[148]). A, membrane; B, porous carbon electrodes; C, packing; D, current connexions; E_1, E_2, anolyte inlet and outlet; F_1, F_2, catholyte inlet and outlet.

relation to the utilization of carbon. The oxidation of hydrobromic acid to bromine by air proceeds readily in presence of nitrogen oxides at room temperature. One disadvantage, amongst others, is the high vapour pressure of bromine; this leads to considerable loss in the circulation through the reactor.

IV.4.2   The Carson and Feldmann Hydrogen–air Cell with Sulphuric Acid Electrolyte

Carson and Feldman[149,150] developed a Ti^{3+}–Br_2 cell in which the electrochemically active components were regenerated by hydrogen and air:

Anode	$2Ti^{3+} + 2H_2O \longrightarrow 2TiO^2 + 4H^+ + 2e^-$
Reactor R(X)	$H_2 + 2TiO^{2+} + 2H^+ \longrightarrow 2Ti^{3+} + 2H_2O$
Cathode	$Br_2 + 2e^- \longrightarrow 2Br^-$
Reactor R(Y)	$2Br^- + 2H^+ + \frac{1}{2}O_2 \longrightarrow Br_2 + H_2O$

Total reaction $H_2 + \frac{1}{2}O_2 \longrightarrow H_2O$

The electrodes were of platinized tantalum (anode) and graphite (cathode). The emf was about 1 volt. At 80 to 85°C, the terminal voltage was 0.8 V at 40 mA cm^{-2}.

The regeneration of the anolyte was carried out with the help of a palladized alumina catalyst; oxidation of bromide ion was, as usual, by air, in presence of nitrogen oxides.

These investigations have not led to the development of a cell of technical interest[*]. As yet, no membrane of satisfactory selectivity and chemical resistance has been found for the purpose of separating catholyte and anolyte. Nevertheless, a 100 watt unit has been operated for more than a week.

IV.4.3   Advantages and Disadvantages of Redox Cells

The principle of the redox cell has the advantage of separating the electrochemical and the regeneration processes, so obviating all the difficulties associated with the direct reaction of solid or gaseous fuels at catalyst electrodes. Further, the avoidance of alkaline electrolytes and the formation of carbonate which they usually entail widens the possibilities of the use of air as oxidant.

On the other hand, it is disadvantageous that the continuous circulation of the components calls for ancillary equipment. A further difficulty lies in the stringent requirements in properties of diaphragms. Finally, for each cell, no fewer than four reactions must be synchronized with each other.

REFERENCES

1. W. R. Grove, *Philos. Mag.*, **14**, 139 (1839).
2. F. T. Bacon, *Ind. Eng. Chem.*, **52**, 301 (1960).
3. A. Schmid, *Die Diffusions Elektrode*, Enke, Stuttgart, 1923.
4. A. Travers and J. Aubry, *Atti del X° Congresso Internazionali di Chimica*, Rome, 1938.
5. M. Raney, *Ind. Eng. Chem.*, **32**, 1190 (1940); *US Pat.* 1563587 (1925).
6. E. Justi, W. Scheibe and A. Winsel, *Ger. Pat.* 1019361 (1954).
6a. G. Wurzbacher, *Chem. Ing. Tech.*, **37**, 532 (1965).
7. E. Justi, M. Pilkuhn, W. Scheibe and A. Winsel, *Hochbelastbare Wasserstoff-Diffusions elektroden für Betrieb bei Umgebungstemperatur und Niederdruck*, *Abh. Mainzer Akad. Nr. 8.* (1959), Steiner, Wiesbaden.
8. E. Justi and A. Winsel, *Fuel Cells—Kalte Verbrennung*, Steiner, Wiesbaden, 1962;
8a. Reference 8, p. 141.
8b. Reference 8, p. 199 ff.
9. G. Grüneberg, Dissertation, Braunschweig, 1956.
10. G. Grüneberg, M. Jung, E. Justi and H. Spengler, *Austrian Pat.* 199238 (1957).
11. K. Friese, E. Justi and A. Winsel, *Ger. Pat.* 1109752 (1957).
12. W. Vielstich, *Z. Physik. Chem.*, **15**, 409 (1958); J. A. Plambeck, J. P. Elder and H. A. Laitinen, *J. Electrochem. Soc.*, **113**, 931 (1966).

[*] Hydrogen containing CO or CO_2 as impurities could be used for this cell and the CO_2 content of air would cause no difficulty, since the electrolyte is acidic.

13. M. Jung and H. Kröger, *Ger. Pat. Appl.* 1137779 (1961).
14. E. Justi and A. Winsel, *J. Electrochem. Soc.*, **108**, 1075 (1961).
15. P. J. Clemm, *Electrochem. Tech.*, **1**, 351 (1963).
16. H. Krupp, R. McJones, H. Rabenhorst, G. Sandstede and G. Walter, *J. Electrochem. Soc.*, **109**, 553 (1962).
17. R. Ch. Burshtein, private communication.
17a. F. P. Dousek, J. Jansta and J. Riha, *Coll. Czech. Chem. Comm.*, **31**, 457 (1966).
17b. M. Jung and H. van Döhren, *5th Intern. Symp. Batteries, Brighton*, 1966, paper No. 23; M. Jung and H. van Döhren in *Aktuelle Batterieforschung*, (Ed. J. Euler), Varta A. G., Frankfurt, 1966, p. 583.
17c. K. Weidinger in *Aktuelle Batterieforschung*, (Ed. J. Euler), Varta A.G., Frankfurt, 1966, p. 113.
17d. D. Hahn in *Aktuelle Batterieforschung*, (Ed. J. Euler), Varta A.G., Frankfurt, 1966, p. 36.
18. E. Guth, H. G. Plust, C. G. Telschow and W. Vielstich, Hitherto unpublished. cf. A. R. Despić, D. M. Drazić, C. B. Petrović, and V. Lj. Vulcić, *J. Electrochem. Soc.*, **111**, 1109 (1964).
19. H. G. Plust, *Proc. Inst. Electr. Electron. Engr.*, **51(5)**, 790 (1963).
20. D. K. Worn and R. P. Perks, *Belg. Pat.*, 613223 (1962).
21. W. Fischer, unpublished.
22. R. Thacker, *Nature*, **206**, 186 (1965).
22a. S. Rundquist, *Acta Chem. Scand.*, **13**, 1193 (1959).
22b. H. Jahnke, *Proc. Journ. Intern. Étud. Piles Combust.*, Brussels, 1965, Ed. Serai Brussels, 1966, Vol. 4, p. 33.
22c. R. Jasinski, C. Mullen and L. Swette, *Proc. Journ. Intern. Étud. Piles Combust.*, Brussels, 1965, Ed. Serai Brussels, 1966, Vol. 4, p. 38.
22d. J. Lindholm, *Proc. Journ. Intern. Étud. Piles Combust.*, Brussels, 1965, Ed. Serai Brussels, 1966, Vol. 4, p. 27.
22e. W. J. Deiss and P. Blum, *Compt. Rend.*, **244**, 464 (1957).
22f. *ASEA-Zeitschrift*, **58**, 12 (1966), p. 171.
22g. J. M. Auclair, *Proc. Journ. Intern. Étud. Piles Combust.*, Brussels, 1965, Ed. Serai Brussels, 1966, Vol. 2, p. 45.
P. Dubois and P. Biro, *EDU-Meeting*, Essen, 1966.
23. M. Jung and H. Kröger, *Ger. Pat. Appl.* 1156768 (17.12.1960).
24. R. Paul, P. Buisson and N. Joseph, *Ind. Eng. Chem.*, **44**, 1006 (1952).
25. R. Jasinski, *Amer. Chem. Soc.*, Fuel Cell Symp., New York, 1963, p. 95 (publ. 1965); *Proc. 18th Ann. Power Sources Conf.*, Atlantic City, 1964, p. 9.
26. M. Pilkuhn and A. Winsel, *Z. Elektrochem.*, **63**, 1056 (1959).
27. W. Vielstich, unpublished.
28. E. Justi, W. Vielstich and A. Winsel, *Ger. Pat. Appl.* 1118843 (27.11.1959).
29. E. Guth and H. G. Plust, unpublished.
30. P. Ruetschi, J. C. Duddy and D. T. Ferell, *Proc. 3rd. Intern. Symp. on Batteries*, Bournemouth, 1962, Pergamon Press, 1963, p. 235.
31. P. Ruetschi, *Ger. Pat. Appl.* 1161965 (6.6.1960).
32. P. Ruetschi, *US Pat.* 3020827 (11.8.1959).
33. K. V. Kordesch in W. Mitchell, *Fuel Cells*, Academic Press, New York, London, 1963, p. 329.
34. K. V. Kordesch, *Proc. Inst. Electr. Electron. Engr. Part.*, **51** (5), 806 (1963).
35. G. E. Evans, *Lecture Amer. Inst. Chem. Eng.*, New Orleans, 1963.
35a. L. W. Litz and K. W. Kordesch, *Amer. Chem. Soc. Symp.* on Fuel Cells, New York, 1963, *Advances in Chemistry Series 47*, Amer. Chem. Soc., 1965, p. 166.

35b. M. C. Clark, W. G. Darland and K. Kordesch, *Proc. 18th Ann. Power Sources Conf.*, Atlantic City, 1964, p. 11; Washington Meet., *Electrochem Soc.*, 1964, Abstr. No. 24.
35c. K. V. Kordesch, *Proc. 19th Ann. Power Sources Conf.*, Atlantic City, 1965, p. 17 (1965).
35d. M. B. Clark, W. G. Darland and K. V. Kordesch, *Electrochem. Tech.*, **3**, 166 (1965).
35e. K. V. Kordesch, paper presented at II. Intern. Conf. EDU, Essen, 1966.
35f. K. V. Kordesch and G. E. Evans, *Proc. Journ. Intern. Étud. Piles Combust.*, Brussels, 1965, Ed. Serai, Brussels, 1966, Vol. 4, p. 93.
36. K. V. Kordesch, private communication.
36a. K. V. Kordesch in B. S. Baker, *Hydrocarbon Fuel Cell Technology*, Academic Press, New York, 1965, p. 17.
37. M. Barak, M. I. Gillibrand and J. Gray, *The Electrochem. Soc. Meeting*, Detroit, 1961, Ext. Abstr. No. 22.
37a. M. I. Gillibrand and J. Gray, *Brit. Pat.* 931 732 (1963).
37b. M. I. Gillibrand and J. Gray in D. H. Collins, *Batteries 2, Proc. 4th Intern. Symp.*, Brighton, 1964, Pergamon Press, 1965, p. 323.
37c. M. I. Gillibrand and J. Gray, *Proc. 5th Intern. Symp.*, Brighton, 1966 (paper No. 26).
38. C. E. Kent, R. R. Nilson and P. Moran, *Gen. Electr. Rep.* No. 63-WA-351, *Lecture Amer. Chem. Soc. Mech. Eng.*, Nov. 1963.
38a. K. R. Williams and D. P. Gregory, *Brit. Pat.* 874 283 (1961).
38b. K. R. Williams, J. W. Pearson and W. J. Gressler in D. H. Collins, *Batteries 2, Proc. 4th Intern. Symp.*, Brighton, 1964, Pergamon Press, 1965, p. 337.
     K. R. Williams, *An Introduction to Fuel Cells*, Elsevier Publ. Co. London, 1966.
38c. F. von Sturm, *Siemens-Zeitschrift*, **39**, 453 (1965).
38d. G. R. Frysinger, *Proc. Journ. Intern. Étud. Piles Combust.*, Brussels, 1965, Ed. Serai, Brussels, 1966, Vol. 2, p. 7.
39. W. T. Grubb, *US Pat.* 29113511 (1959).
40. W. T. Grubb and L. W. Niedrach, *J. Electrochem. Soc.*, **107**, 131 (1960).
41. L. W. Niedrach, *Proc. 13th Ann. Power Sources Conf.*, Atlantic City 1959, p. 120.
42. E. A. Oster, *Proc. 14th Ann. Power Sources Conf.*, Atlantic City, 1960, p. 59.
43. J. Perry, *Proc. 14th Ann. Power Sources Conf.*, Atlantic City, 1960, p. 50.
44. J. Bone, *Proc. 14th Ann. Power Sources Conf.*, Atlantic City, 1960, p. 62.
45. E. Joachim and W. Vielstich, *Electrochim. Acta (London)*, **3**, 244 (1960).
46. E. J. Cairns, D. L. Douglas and L. W. Niedrach, *A.I.Ch.E. Journal*, **7**, 551 (1961).
47. H. Hunger, *Proc. 14th Ann. Power Sources Conf.*, Atlantic City, 1960, p. 55.
48. L. W. Niedrach, *J. Electrochem. Soc.*, **109**, 1092 (1962).
49. L. W. Niedrach and W. T. Grubb, *Ion Exchange Membrane Fuel Cells*, in W. Mitchell, *Fuel Cells*, Academic Press, New York, 1963, p. 253–298.
50. W. Juda, C. E. Tirrell and R. N. Lurie, *Prog. Astron. Rocketry*, **3**, 445 (1961); R. M. Lurie, C. Berger and H. Viklund, *J. Electrochem. Soc.*, **110**, 1173 (1963).
51. D. Jahn, W. Vielstich and D. Wolf, unpublished.
52. H. Kröger and M. Jung, *Ger. Pat.* 1139558 (31.5.1960).
53. L. W. Niedrach, *French Pat.* 1324555 (8.5.1960).
53a. S. J. Krumbein and R. R. Russell, *Gen. Electr. Tech. Rept.*, No. 65-C-021 (1965).
53b. L. J. Lemaignen, *J. Intern. Étud. Piles Combust.*, Brussels, 1965; P. Blondeau and L. J. Lemaignen, in *Les Piles à Combustible*, Publ. de l'Institut Français du Pétrole Paris, 1965.
54. R. G. Haldeman, W. P. Colman, S. H. Langer and W. A. Barber, *Amer. Chem. Soc.*, Fuel Cell Symp., New York, Sept., 1963, p. 106 (Publ. 1965).
55. A. Dravnieks, D. B. Boies and J. I. Bregman, *Proc. 16th Ann. Power Sources Conf.*, Atlantic City, 1962, p. 4.

55a. C. Berger and M. P. Stries in B. S. Baker, *Hydrocarbon Fuel Cell Technology*, Academic Press, New York, 1965, p. 485.

56. R. A. Wynveen and T. G. Kirkland, *Proc. 16th Ann. Power Sources Conf.*, Atlantic City, 1962, p. 24.

57. R. Jasinski and T. G. Kirkland, *Fuel Cells—State of the Art*, Allis Chalmers, Milwaukee, Sept. 1963.

58. R. Jasinski and T. G. Kirkland, *Fuel Cells—State of the Art*, Allis Chalmers, Milwaukee, Dec. 1963.

59. R. Jasinski, private communication, Oct. 1963.

60. R. Jasinski, J. Huff and J. Endide, *The Electrochem. Soc. Meeting*, New York, Oct. 1963.

61. J. L. Platner and P. D. Hess, *Static Moisture Removal Concept for Hydrogen–Oxygen Capillary Fuel Cell*, Allis Chalmers, Milwaukee, Sept. 1963.

62. E. Baur and J. Tobler, *Z. Elektrochem.*, **39**, 169 (1933).

63. W. Ostwald, *Z. Elektrochem.*, **1**, 122 (1894).

64. W. W. Jacques, *Harper's Mag.*, **94**, 144 (Dec. 1896–May 1897); *Electr. Eng.* **21**, 261 and 497 (1896).

65. F. Haber and L. Bruner, *Z. Elektrochem.*, **10**, 697 (1904).

65a. H. Binder, A. Köhling, K. Richter and G. Sandstede, *Electrochim. Acta*, **9**, 255 (1964).

66. E. Justi, *Akad. Wiss. Lit. Mainz, Abh. Math. Naturwiss. Kl.* 1956, H. 1; K. Bischoff and E. Justi, *Ann. Mines Belg.*, **52**, 381 (1953); E. Justi and H. Spengler, *Ger. Pat.* 899212 (1953), *Ger. Pat.* 932026 (1955), *U.S. Pat.* 2830109 (1958).

67. E. Baur and H. Preis, *Z. Elektrochem.*, **43**, 727 (1937); **44**, 695 (1938); cf. also W. Schottky, *Wiss. Veröff. Siemens-Werke*, **14**, 1 (1935).

67a. A. N. Frumkin and B. I. Podlovchenko, *Dokl. Akad. Nauk*, **150**, 349 (1963); H. Wroblowa, B. J. Piersma and J. O'M. Bockris, *J. Electroanal. Chem.*, **6**, 401 (1963); J. W. Johnson, H. Wroblowa and J. O'M. Bockris, *J. Electrochem. Soc.*, **111**, 863 (1964); S. S. Beskorovainaya, Yu. B. Vasil'ev and V. S. Bagotskii, *Elektrokhimiya*, **2**, 44 (1966).

68. E. Gorin, *U.S. Pat.* 2570543 (1951), *U.S. Pat.* 2581650 (1952), *U.S. Pat.* 2581651 (1952) *U.S. Pat.* 3 108 857 (1963), *U.S. Pat.* 3 115394 (1963).

69. E. J. Cairns, A. D. Tevebaugh and G. J. Holm, *J. Electrochem. Soc.*, **110**, 1025 (1963); E. J. Cairns and A. D. Tevebaugh, *J. Chem. Eng.*, Data **9**, 453 (1964); K. Pohl and G. Martens, *Erdöl and Kohle*, **16**, 367 (1963); G. H. J. Broers and B. W. Treijtel, *Advan. Energy Conv.*, **5**, 365 (1965).

70. Y. L. Sandler, *J. Electrochem. Soc.*, **109**, 1115 (1962).

71. O. K. Davtyan, *Bull. Acad. Sci. URSS, Classe Sci. Tech.*, 1946, 107, 1946, 215; O. K. Davtyan, *Zh. Fiz. Khim.*, **22**, 1043 (1946); **37**, 1949 (1963); O. K. Davtyan, *Problema neprosredstvennogo prevrashcheniya khimicheskoi energii topliva v elektricheskuyu*, Izd. Akad. Nauk SSSR, Moscow, 1947.

72. E. Baur, W. D. Treadwell and G. Trümpler, *Z. Elektrochem.*, **27**, 199 (1921).

73. G. H. J. Broers, Ph.D. Thesis, University of Amsterdam, 1958; G. H. J. Broers, *Dechema Monogr.*, Vol. 38, No. 277, Verlag Chemie, Weinheim/Bergstr., 1960; G. H. J. Broers and J. A. A. Ketelaar, *Ind. Eng. Chem.*, **52**, 303 (1960); *Fuel Cells*, (Ed. G. J. Young), Vol. 1, p. 78, New York 1960; G. H. J. Broers and M. Schenke: *Fuel Cells*, (Ed. G. J. Young), Vol. 2, p. 6, New York 1963; G. H. J. Broers, *Fuel Cells*, CEP Technical Manual, p. 90, *Amer. Inst. Chem. Engs.*, New York 1963; G. H. J. Broers, M. Schenke and G. G. Piepers, *Advan. Energy Conv.*, **4**, 131 (1964); G. H. J. Broers and M. Schenke, Revue Energie Primaire, *Journ. Intern. Étud. Piles Combust.* **1**, 54 (1965); G. H. J. Broers and M. Schenke, *Hydrocarbon Fuel Cell Technology*, (Ed. B. S. Baker), p. 225, New York 1965.

74. Cf. A. M. Adams, Conf. Paper Nr. CP56-266, *Inst. Electr. Eng.*, 1956.

75.   H. H. Greger, *U.S. Pat.* 1963550 (1934), *U.S. Pat.* 2175523 (1939), *U.S. Pat.* 2276188 (1942); E. Gorin and H. L. Recht, *U.S. Pat.* 2,901,524 (1959).
75a. J. Millet and R. Buvet, *Hydrocarbon Fuel Cell Technology*, (Ed. B. S. Baker), p. 285, New York 1965.
76.   L. G. Marianowski, John Meek, E. B. Shultz jr. and B. S. Baker, *Proc. 17th Ann. Power Sources Conf.*, **72**, (1963); J. J. Millet and R. Buvet, Revue Énergie Primaire, *Journ. Intern. Étud. Piles Combust.*, **1**, 49 (1965).
77.   H. H. Chambers and A. D. S. Tantram, *Fuel Cells*, (Ed. G. J. Young), Vol. 1, p. 94, New York 1960; A. D. S. Tantram, A. C. C. Tseung and B. S. Harris, *Hydrocarbon Fuel Cell Technology*, (Ed. B. S. Baker), p. 187, New York 1965.
78.   D. L. Douglas, *Fuel Cells*, (Ed. G. J. Young), Vol. 1, p. 129, New York 1960.
79.   E. B. Shultz jr., K. S. Vorres, L. G. Marianowski and H. R. Linden, *Fuel Cells*, (Ed. G. J. Young), Vol. 2, p. 24, New York 1963.
80.   E. Gorin and H. Recht, *Fuel Cells*, (Ed. G. J. Young), Vol. 1, p. 109, New York 1960.
81.   G. H. J. Broers, private communication.
82.   D. Staicopoulos, E. Yeager and F. Hovorka, *J. Electrochem. Soc.*, **98**, 681 (1951).
83.   F. G. Will, *J. Electrochem. Soc.*, **110**, 145 and 152 (1963); M. Bonnemay, G. Bronoël and E. Levart, *C. R. hebd. Séances Acad. Sci.*, **257**, 3394 and 3385 (1963).
83a. G. J. Janz, F. Colom and F. Saegusa, *J. Electrochem. Soc.*, **107**, 581 (1960); G. J. Janz and F. Saegusa, *J. Electrochem. Soc.*, **108**, 663 (1961).
83b. G. G. Arkhipov and G. K. Stepanov, *Tr. Akad. Nauk SSSR*, Uralskii Filial, Sverdlovsk, Institut Elektrokhimii 6, Transl. Consultants Bureau, New York, 1966, p. 67; G. K. Stepanov and A. M. Trunov, *Tr. Akad. Nauk SSSR*, Uralskii Filial, Sverdlovsk, Institut Elektrokhimii 6, Transl. Consultants Bureau, New York, 1966, p. 73; M. V. Smirnov, L. A. Tsiovkina and V. A. Oleinikova, *Tr. Akad. Nauk SSSR*, Uralskii Filial, Sverdlovsk, Institut Elektrokhimii 6, Transl. Consultants Bureau, New York, 1966, p. 61.
83c. G. J. Janz and M. R. Lorenz, *J. Electrochem. Soc.*, **108**, 1052 (1961); G. J. Janz and F. Saegusa, *J. Electrochem. Soc.*, **110**, 452 (1963).
83d. M. Schenke, G. H. J. Broers and J. A. A. Ketelaar, *J. Electrochem. Soc.*, **113**, 404 (1966).
83e. G. V. Vorob'ev, S. F. Pal'guef and S. V. Karpachev, *Tr. Akad. Nauk SSSR*, Uralskii Filial, Sverdlovsk, Institut Elektrochimii 6, Transl. Consultants Bureau, New York 1966, p. 33.
83f. M. Eisenberg and B. Baker, *Electrochem. Tech.*, **2**, 558 (1964).
84.   M. L. Kronenberg, *J. Electrochem. Soc.*, **109**, 753 (1962).
85.   I. Trachtenberg, *J. Electrochem. Soc.*, **111**, 110 (1964); J. K. Truitt, *Fuel Cells*, CEP Tech. Manual, p. 1, *Amer. Inst. Chem. Engr.*, New York 1963; G. Frysinger, J. K. Truitt and G. Peattie, *Proc. 18th Ann. Power Sources Conf.*, Atlantic City, 1964, p. 14; C. G. Peattie, I. Trachtenberg and J. K. Truitt, *Proc. First Australian Conf. Electrochem.*, Sydney, Hobart, Australia, 1963, 683 (Pub. 1965); I. Trachtenberg, *Advan. Chem. Ser.*, **47**, 232 (1965); J. K. Truitt, Revue Énergie Primaire, *Journ. Intern. Étud. Piles Combust.*, **1**, 88 (1965); J. K. Truitt and F. L. Gray, *Proc. 19th Ann. Power Sources Conf.*, p. 49 (1965); I. Trachtenberg, *Hydrocarbon Fuel Cell Technology*, (Ed. B. S. Baker), p. 251, New York 1965; J. K. Truitt and F. L. Gray, *Chem. Eng. Prog.*, **62**, No. 5, 72 (1966).
85a. B. S. Baker, L. G. Marianowski, J. Zimmer and G. Price, *Hydrocarbon Fuel Cell Technology*, (Ed. B. S. Baker), p. 293, New York 1965.
85b. G. Frysinger and J. K. Truitt, *Proc. 18th Ann. Power Sources Conf.*, Atlantic City, 1964, p. 14.

86.   F. T. Bacon, *Fuel Cells*, (Ed. G. J. Young), Vol. 1, p. 51, New York 1960; A. M. Adams, F. T. Bacon and R. G. Watson, *Fuel Cells*, (Ed. W. Mitchell jr.), p. 130, New York–London 1963; F. T. Bacon, *Fuel Cells*, CEP Tech. Manual, p. 66, Amer. Inst. Chem. Engs., New York 1963.

87.   I. Taitelbaum, *Z. Elektrochem.*, **16**, 286 (1910).

88.   G. V. Elmore and H. A. Tanner, *J. Electrochem. Soc.*, **108**, 669 (1961).

89.   R. P. Tischer, unpublished.

90.   E. Baur, *Z. Elektrochem.*, **16**, 300 (1910).

91.   W. T. Grubb and L. W. Niedrach, *Proc. 17th Ann. Power Sources Conf.*, p. 69 (1963); W. T. Grubb and L. W. Niedrach, *J. Electrochem. Soc.*, **110**, 1086 (1963); W. T. Grubb, *Nature (London)*, **201**, 699 (1964); W. T. Grubb, *J. Electrochem. Soc.*, **111**, 1086 (1964); W. T. Grubb and C. J. Michalske, *J. Electrochem. Soc.*, **111**, 1015 (1964); W. T. Grubb, *J. Electrochem. Soc.*, **113**, 191 (1966).

91a. L. W. Niedrach and H. R. Alford, *J. Electrochem. Soc.*, **112**, 117 (1965).

91b. R. P. Tischer, *Z. analyt. Chem.*, **224**, 93 (1967).

91c. R. P. Hamlen and E. J. Szymalak, *Electrochem. Tech.*, **4**, 172 (1966).

92.   R. Jasinski, J. Huff, S. Tomter and L. Swette, *Ber. Bunsenges. physik. Chem.*, **68**, 400 (1964).

92a. W. T. Grubb and C. J. Michalske, *Proc. 18th Ann. Power Sources Conf.*, Atlantic City, 1964, p. 17.

93.   D. P. Gregory, paper presented at UCLA course 6, Aug. 1963; D. P. Gregory and H. Heilbronner, *Hydrocarbon Fuel Cell Technology*, (Ed. B. S. Baker), p. 509, New York 1965.

94.   G. Grüneberg, M. Jung and H. Spengler, *Ger. Pat.* 1 146 562 (18.1.1958).

95.   G. Grüneberg, Thesis, Braunschweig, 1958.

96.   E. J. Cairns and D. I. Macdonald, *Electrochem. Tech.*, **2**, 65 (1964); E. J. Cairns and D. C. Bartosik, *J. Electrochem. Soc.*, **111**, 1205 (1964).

96a. E. J. Cairns, *Hydrocarbon Fuel Cell Technology*, (Ed. B. S. Baker), p. 465, New York, 1965; E. J. Cairns, *Nature*, **210**, 161 (1966).

96b. E. B. Butler, C. B. Miles and C. S. Kuhn, jr., *Ind. Eng. Chem.*, **38**, 147 (1946).

96c. D. C. Grahame, *Chem. Rev.*, **41**, 441 (1947); A. N. Frumkin, *Advan. Electrochem. and Electrochem. Eng.*, (Ed. P. Delahay). **5**. 3, 1963.

96d. R. V. Windsor and G. H. Cady, *J. Amer. Chem. Soc.*, **70**, 1500 (1948); G. H. Cady, *J. Amer. Chem. Soc.*, **56**, 1431 (1934); E. B. R. Prideaux and K. R. Webb, *J. Chem. Soc.*, 2 (1937) and 111 (1939).

97.   H. A. Liebhafsky and E. J. Cairns, *GE Res. Lab. Rept.*, No. 63-RI-3480 C(1963).

98.   W. Schottky, *Wiss. Veröff. Siemens-Werken*, **14**, 1 (1935).

99.   W. Nernst and W. Wald, *Z. Elektrochem.*, **7**, 373 (1900); W. Nernst, *Z. Elektrochem.*, **6**, 41 (1899).

100.   J. L. Weininger and P. D. Zemany, *J. Chem. Phys.*, **22**, 1469 (1954).

101.   K. Kiukkola and C. Wagner, *J. Electrochem. Soc.*, **104**, 379 (1957).

102.   A. Dietzel and H. Tober, *Ber. Dtsch. Keram. Ges.*, **30**, 47 (1953).

103.   E. Koch and C. Wagner, *Z. Physik. Chem.*, **B38**, 295 (1937).

104.   E. Zintl and U. Croatto, *Z. Anorg. Chem.*, **242**, 79 (1939).

105.   O. Ruff and F. Ebert, *Z. Anorg. Chem.*, **180**, 19 (1929).

106.   R. E. Carter, *J. Amer. Ceram. Soc.*, **43**, 448 (1960).

107.   J. M. Dixon, L. D. LaGrange, U. Merten, C. F. Miller and J. D. Porter, *J. Electrochem. Soc.*, **110**, 276 (1963).

108.   R. E. Carter and W. L. Roth, *GE Res. Lab. Rept.*, No. 63-RL-3479 M (1963).

108a. W. H. Rhodes and R. E. Carter, *Gen. Electr. Rept.*, No. 65-C-004, Sept. (1965).

109.   C. Wagner, *Naturwissenschaften* **31**, 265 (1943).

110.   F. Hund, Z. *Physik. Chem.*, **199**, 142 (1952).
      F. Hund, Z. *Elektrochem.*, **55**, 362 (1951).
111.   W. D. Kingery, J. Pappis, M. E. Doty and D. C. Hill, *J. Amer. Chem. Soc.*, **42**, 394 (1959).
112.   T. Y. Tien and E. C. Subbarao, *J. Chem. Phys.*, **39**, 1041 (1963).
113.   H. Schmalzried, Z. *Elektrochem.*, **66**, 572 (1962).
114.   F. Trombe and M. Foëx, *C.R. hebd. Séances Acad. Sci.*, **236**, 1783 (1953).
115.   E. J. Cairns and H. A. Liebhafsky, *The Thermodynamics of the Complete Fuel Cell*, GE Res. Lab. Rept. No. 63-RL-339C (1963).
116.   D. H. Archer and E. F. Sverdrup, *Proc. 16th Ann. Power Sources Conf.*, Atlantic City, 1962, p. 34.
116a.   S. V. Karpachov, A. T. Filyayev and S. F. Palguyev, *Electrochimica Acta*, **9**, 1681 (1964).
116b.   M. Kleitz, J. Besson and C. Deportes, *Proc. Journ. Intern. Étud. Piles Combust.*, Brussels, 1965, Ed. Serai, Brussels, 1966, Vol. 3, p. 35.
116c.   G. Robert, J. Besson and C. Deportes, *Proc. Journ. Intern. Étud. Piles Combust.*, Brussels, 1965, Ed. Serai, Brussels, 1966, Vol. 3, p. 5.
116d.   H. Tannenberger, H. Schachner and P. Kovacs, *Proc. Journ. Intern. Étud. Piles Combust.*, Brussels, 1965, Ed. Serai, Brussels, 1966, Vol. III, p. 19.
116e.   T. Takahashi, K. Ito and H. Iwahara, *Proc. Journ. Intern. Étud. Piles Combust.*, Brussels, 1965, Ed., Serai, Brussels, 1966, Vol. III, p. 42.
117.   K. Bischoff and E. Justi, *Jb. Akad. Wiss. Lit. Mainz., Abh. Math. Naturwiss., Kl.* 1955, p. 250.
118.   D. Korda, *Elektrotechnik Z.*, 16, 272 (1895).
119.   E. Baur, A. Petersen and G. Fulleman, Z. *Elektrochem.*, **22**, 409 (1916).
120.   J. Weissbart and R. Ruka, *J. Electrochem. Soc.*, **109**, 723 (1962).
121.   J. Weissbart, and R. Ruka, *Rev. Sci. Instruments*, **32**, 593 (1961).
122.   J. Weissbart and R. Ruka, Paper presented at the Electrochemical Society Fall Meeting, Detroit, Mich. (Oct. 2–5, 1961). Ext. Abstr. No. 44, (Battery Division).
122a.   D. H. Archer, J. J. Alles, W. A. English, L. Elikan, E. F. Sverdrup and R. L. Zahradnik, *145th ACS Meeting, Division of Fuel Chemistry*, Vol. 7, No. 4, 8 (Sept. 1963).
122b.   D. H. Archer, R. L. Zahradnik, E. F. Sverdrup, W. A. English, L. Elikan and J. J. Alles, *Proc. 18th Ann. Power Sources Conf.*, Atlantic City, 1964, p. 36, and *Chem. Eng. Progr.*, **60**, 64 (1964).
122c.   D. H. Archer, L. Elikan and R. L. Zahradnik in B. S. Baker, *Hydrocarbon Fuel Cell Technology*, Academic Press, New York, London, p. 51 (1965).
123.   R. E. Carter, W. A. Rocco, H. S. Spacil and W. E. Tragert, *Chem. Eng. News*, **47**, Jan. 14 (1963).
124.   R. E. Carter, private communication.
124a.   D. W. White, *Proc. Journ. Intern. Étud. Piles Combust.*, Brussels, 1965, Ed. Serai, Brussels, 1966, Vol. III, p. 10.
124b.   D. W. White, *Gen. Electr. Rept.* No. 66-C-132 (April, 1966).
125.   H. Binder, A. Köhler, H. Krupp, K. Richter and G. Sandstede, *Electrochim. Acta* (*London*), **8**, 781 (1963).
125a.   H. H. Möbius and B. Rohland, *Proc. Journ. Intern. Étud. Piles Combust.*, Brussels, 1965, Ed. Serai, Brussels, 1966, Vol. III, p. 27.
125b.   H. Schaohner and H. Tannenberger, *Proc. Journ. Intern. Étud. Piles Combust.*, Brussels, 1965, Ed. Serai, Brussels, 1966, Vol. III, p. 49.
125c.   E. F. Sverdrup, D. H. Archer, J. J. Alles and A. D. Glasser, in B. S. Baker, *Hydrocarbon Fuel Cell Technology*, Academic Press, New York, London p. 311 (1965).
126.   V. S. Daniel-Bek, M. Z. Mints, V. V. Sysoeva and M. V. Tikhonova, *Conf.*, Moscow, 1956, p. 794–800.

127. V. S. Daniel-Bek and co-workers, *J. Appl. Chem. USSR*, **32**, 679 (1959). Translated from *Zh. Prikl. Khim.*, **32**, 649 (1959). Fuel Abstr., 1960, 2208 (Orig.).

128. S. F. Palguev and Z. S. Vlochenkova, *Tr. Inst. Khim. Akad. Nauk. SSSR, Ural. Filial,* 1958, No. 2, p. 183–200; *Chem. Abstr.*, **54**, 9542h (1960).

129. E. Baur and R. Brunner, *Z. Elektrochem.*, **43**, 725 (1937)

130. E. Justi and A. Winsel, Ref. 8, p. 46 *Ger. Pat.* 1071175 (9.2.1956). *U.S. Pat.* 2925454.

131. K. Kordesch and A. Markov, *Österr. Chemiker-Ztg.*, **52**, 125 (1951).

132. J. F. Yeager, *The Electrochem. Soc. Meeting*, Indianapolis, 1961, Ext. Abstr., No. 109.

133. J. N. Murray and P. G. Grimes, *Rept. Res. Div.*, Allis Chalmers, 1963.

133a. G. Wolf, *Aktuelle Batterieforschung*, Varta AG, Frankfurt/Main, 1966, p. 138.

133b. Demonstration Varta AG, Kelkheim, November 1966.

133c. K. R. Williams, J. W. Pearson and W. J. Gressler, *3rd Intern. Symp. Batteries,* Brighton, 1964.

133d. K. R. Williams, M. R. Andrew and F. Jones in K. R. Williams, *Introduction to Fuel Cells,* Elsevier Publ. Comp., Amsterdam, London, New York, 1966.

133e. B. L. Tarmy and C. E. Heath, *Proc. 15., 16., 17., 18., 19., Ann. Power Sources Conf.,* Atlantic City.

134. H. Spengler and G. Grüneberg, *Dechema-Monogr.* Vol. 38, Nr. 579–599 (1961); G. Grüneberg, J. Kubisch and H. Spengler, *Ger. Pat. Appl.* 1163412 (1958).

135. W. Vielstich, *Proc. 4th Intern. Symp. on Batteries*, Brighton, 1964, Pergamon Press, 1964.

136. E. Guth, H. J. Haase, H. G. Plust and W. Vielstich, *Proc. 7th Intern. Conf. on Lighthouses and other Aids to Navigation,* Rome, 1965.

136a. H. G. Plust, *Brown Boveri Mitt.*, **53**, 5 (1966).

136b. *Funkschau*, **38**, 138 (1966).

136c. W. Vielstich in B. S. Baker, *Hydrocarbon Fuel Cell Technology*, Academic Press, New York, 1965, p. 79.

136e. W. Vielstich and U. Vogel, *Amer. Chem. Soc. Meeting*, Chicago, 1967.

136d. P. G. Grimes and H. H. Spengler in B. S. Baker, *Hydrocarbon Fuel Cell Technology,* Academic Press, New York, 1965, p. 121.

136f. G. Grüneberg and H. Barth, unpublished.

136g. H. Schmidt and W. Vielstich, *Z. anal. Chemie*, **224**, 84 (1957).

137. M. I. Gillibrand and G. R. Lomax, *Proc. 3rd Intern. Symp. on Batteries,* Bournemouth, 1962, p. 221, Pergamon Press, 1963; M. Barak, *C.E.P. Techn. Manual,* Amer. Inst. Chem. Engs., p. 74 (1963); M. I. Gillibrand, *Brit. Pat.* 963254 (1964); Ashworth, Gray and Gillibrand, *Brit. Pat. Appl.* 14590/62 (1962).

138. M. Barak, private communication, October 1963.

139. S. S. Tomter and A. P. Antony, *Hydrazine Fuel Cell System*, Res. Div., Allis Chalmers, 3 July, 1963.

140. R. Jasinski, *Catalysts for Hydrazine Fuel Cell Anods*, Res. Div., Allis Chalmers, Sept. 1963.

141. E. J. Cairns and D. C. Bartosik, *The Electrochem. Soc. Meeting*, Toronto 1964, Ext. Abstr. No. 214.

142. O. Bloch, P. Degobert, M. Prigent and J. C. Balaceanu, *6 Welt Petroleum Conf.,* Frankfurt, 1963, Sect. VI—paper 3—PD 10.

143. P. G. Grimes, B. Fiedler and J. Adams, *Proc. 15th Ann. Power Sources Conf.,* Atlantic City, 1961, p. 29, *Brit. Pat.* 948 984, 948 985, 948 986 (1961); cf. also K. Hamann, W. Vielstich and U. Vogel, *Chemie in unserer Zeit*, **1**, 62 (1967).

144. W. Vielstich, *Z. Instrumententechnik*, **71**, 29 (1963).

145. C. E. Heath, *Proc. 17th Ann. Power Sources Conf.*, Atlantic City, 1963, p. 96; J. A., Shropshire and B. L. Tarmy, *Fuel Cell Systems*, (Ed. G. J. Young and H. R. Linden), Advances in Chemistry Series 47, p. 153, Washington, D.C. 1965.

146. *Monsanto Res. Co.*, USAF Contract No. AF 33(616)-7735, Rep. ASD-TDR-62-42 (1962).

147. E. Baur and A. Glässner, *Z. Elektrochem.*, **9**, 534 (1903).

148. A. M. Posner, *Fuel*, **34**, 330 (1955).

149. B. R. Stein, *Status Report on Fuel Cells*, U.S. Army Research Office, Rep. available from OTS; PB 151 804 (1959).

150. W. Carson and M. Feldmann, *Proc. 13th Ann. Batt. Res. a. Dev. Conf. USASC Lab.*, p. 120 (1959); M. Feldmann, A Redox Type Fuel Cell, *Proc. 13th Ann. Power Sources Conf.*, Atlantic City, 1959, p. 111.

151. H. H. Geissler, *Proc. 17th Ann. Power Sources Conf.*, Atlantic City, 1963, p. 75

151a. H. H. Geissler and L. E. Goodman, *Proc. 18th Ann. Power Sources Conf.*, Atlantic City, 1964, p. 28.

151b. J. Meek, B. S. Baker and A. C. Allen in B. S. Baker, *Hydrocarbon Fuel Cell Technology*, Academic Press, New York, London, 1965, p. 25.

151c. M. A. Vertes and A. J. Hartner, *Proc. Journ. Intern. Étud. Piles Combust.*, Brussels, 1965, Ed. Serai, Brussels, 1966, Vol. I, p. 63.

152. G. Sandstede, *Brennstoffzellen für die elektrochemische Verbrennung*, Battelle-Institut, Frankfurt, 1966.

153. W. T. Grubb and C. J. Michalske, *Nature (London)*, **201**, 287 (1964).

154. Demonstration Cell of the Battelle-Institut Frankfurt, exhibited at ACHEMA Frankfurt 1964; H. Binder, A. Köhler and G. Sandstede, *Rev. Énergie Primaire, Journ. Intern. Étud. Piles Combust.*, **1**, 74 (1965); J. A. Shropshire and H. H. Horowitz, *J. Electrochem. Soc.*, **113**, 490 (1966).

155. V. S. Bagotzky, *Electrode Processes in New Electrochemical Sources of Current*, in *Proc. 4. Electrochem. Conf.* Moscow, 1956, p. 737–743; cf. also O. Pavela, *Suomen Kem.*, **30B**, 240 (1957).

156. J. Mrha and W. Vielstich, *Abh. sächs. Akad. Wiss. Leipzig, Math. Naturw. Kl.*, **49** (5), 71 (1968).

157. G. Grüneberg and F. Weddeling, *Paper at AGARD-Meeting*, Liège, 1967.

157a. G. Grüneberg and F. Dünger, unpublished.

158. M. Eisenberg, *Proc. 17th Ann. Power Sources Conf.*, Atlantic City, 1963, p. 97.

159. P. Terry, J. Gallagher, R. Salathe and J. O. Smith, *Proc. 20th Ann. Power Sources Conf.*, Atlantic City, 1966, p. 39.

160. P. Terry, *U.S. Govt. Report* AD 629 220, Feb. 1966.

161. E. A. Gills, *Proc. 20th Ann. Power Sources Conf.*, Atlantic City, 1966, p. 41.

162. K. V. Kordesch, *Abh. sächs. Akad. Wiss. Leipzig, Math. Naturw. Kl.*, **49** (5), 47 (1968).

163. K. R. Williams, *An Introduction to Fuel Cells*, Elsevier Publishing Comp., 1966, pp. 147 and 298.

164. F. Goebel, W. Nippe and W. Vielstich, unpublished.

165. cf. ref. 157.

166. B. Warszawski, *Entropie*, No. 14 (1967), p. 33.

167. K. Hamann, W. Vielstich and U. Vogel, *Chemie in unserer Zeit*, **1**, 62 (1967).

CHAPTER V

Special Types

A series of cells are first discussed that, like classical fuel cells, employ oxygen or air electrodes as cathodes. Metals in liquid (amalgams, gallium) or solid form (zinc, aluminium, magnesium) are used as fuels in the wider sense, and no catalyst electrode is required to bring them into reaction. Barium[1], lithium and calcium[1a] have also been proposed for use as anode materials. As in primary cells (e.g. the dry cell), the anode is consumed in use, and the metal concerned must be either continuously or periodically replenished.

A zinc–oxygen cell with liquid alkaline electrolyte and hydrophobic, air-diffusion electrode has been commercially available for several decades[2], and a magnesium–air cell has been known even longer. Fuel-cell technology, however, has so greatly improved the performance of oxygen and air electrodes that previously normal current densities have been increased ten-fold, with electrodes of minimal volume and weight. Metal–air cells have therefore recently received much attention in relation to use for traction. The possibility of rechargeable anodes is of particular interest in this connexion (section V.3).

Cells using an amalgam as fuel are far advanced technically. Compact batteries of rating more than 10 kW have been built (section V.1). Cells of this kind could lend themselves to application in association with chlorine–alkali electrolysis, since they perform the amalgam process more economically.

Under the heading of cells with self-consuming anodes four types are described which are as yet in the stage of laboratory development. The hydrogen–chlorine cell has not yet attracted technical interest because of the objectionable nature of the oxidant. Biochemical cells have been the subject of particularly intensive study.

V.1 THE AMALGAM–OXYGEN CELL

The principle of the amalgam–oxygen cell developed by Yeager[3] has been briefly described in chapter I. Sodium is added to the mercury circulation system in an amalgamation vessel outside the cell (Figure I.7). The liquid amalgam, on entering the cell, runs down a vertical steel plate which is completely wetted by the amalgam. Oxygen is taken to the opposite face of a cathode of porous carbon or porous metal. The reaction product, sodium hydroxide, is rejected from the cell. The anode reaction is the oxidation of the sodium in the amalgam:

$$Na \rightleftharpoons Na^+ + e^-$$

The standard potential for sodium at 25°C is -2.714 volt; for sodium amalgam it is -1.957 volt. The concentration dependence is given by

$$\phi_0 = -1.957 + 0.059 \log(a_{Na^+}/a_{Na})$$

Use of a vertical electrode allows concentration and charge-transfer over-potentials to be neglected, even at current densities as great as 1 A cm^{-2}. To attain equally satisfactory mass transfers at a horizontal electrode, the solution would have to be so strongly stirred as to ripple the mercury surface.

The current efficiency of the amalgam anode is close to 100%. Thanks to the high hydrogen overpotential, the spontaneous rate of reaction of the amalgam in strongly alkaline solution is very small (with 4 mole% sodium amalgam in 4N KOH at 25°C, it corresponds to a current density of less than 1 mA cm^{-2}). Carbon particles are well known to catalyse the amalgam reaction very strongly. In this respect, the use of metal electrodes on the oxygen side is safer than that of porous carbon.

Sodium dissolves in mercury very vigorously. The solubility limit for liquid amalgams must not be exceeded, otherwise blockage of the feed to the cell (see Figures V.2 and 3) will occur.

Eidensohn[4] has proposed to utilize the energy difference between sodium and sodium amalgam in a preceding galvanic cell. The *sodium–sodium amalgam cell* in the form Na|NaX|Na(Hg) can be operated either at medium or at high temperatures. A water-free electrolyte is of course essential. For high temperatures, a melt of 76% sodium hydroxide, 10% sodium boride and 14% sodium iodide at about 230°C is suitable. For temperatures below 100°C, a solution of sodium iodide in ethylamine is appropriate. The emf of the cell is between 0.8 and 0.9 volt.

V.1.1 The Laboratory Cell of Yeager[3]

For investigation of the oxygen side of his cell Yeager used a double-layer electrode of porous carbon. The carrier layer was a 5 mm thick plate of porous graphite; the 1.5 mm thick active layer was made of active carbon (Nuchar C115, West Virginia Pulp and Paper Co.) with 15% of silver and 10% of binder (e.g., polyethylene). Copper tubes built into the graphite carrier (see

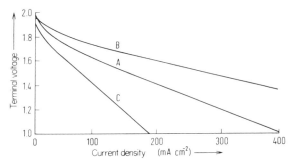

Figure V.1. Current–voltage curves of an amalgam–oxygen cell at 25°C (after Yeager[3]). Curve A, experimental cell; Curve B, as curve A, but with ohmic voltage drop eliminated; curve C, average terminal voltage of single cells of a 5-cell battery, 0.52 wt % Na.

Figure V.3) served the purposes of oxygen supply and of electrical contact, to the electrode. The oxygen pressure was so adjusted that minimal gassing was observed on the electrolyte side. Glass capillaries delivered the amalgam to the vertical steel plate; at the moment of wetting the 'self-reaction' of the amalgam is at first quite brisk.

Curves A and B of Figure V.1 were obtained with this experimental cell (electrode area, 25 cm^2; distance between electrodes, 0.3 mm). The sodium concentration in the amalgam was 0.52%, the oxygen pressure 1 atm in excess of atmospheric and the electrolyte was 5N NaOH. Before these curves were taken, the cell had been delivering 150 mA cm^{-2} for 30 hours. Ohmic voltage drop has been eliminated from curve B. The fall in terminal voltage with increasing current density is almost entirely attributable to the oxygen electrode. Electrolytes prepared with sea water gave identical results. At 52°C, a current density of 250 mA cm^{-2} was obtained at a terminal voltage of 1.5 V.

Similar results have been obtained with use of potassium amalgam in KOH solution. The analogous lithium system provided an emf as large as 2.3 volt, but the spontaneous 'self-reaction' of the lithium amalgam was considerably faster.

V.1.2   The 5-cell Battery

Figure V.2 and V.3 show the construction of this battery built by Yeager[3]. The oxygen electrode is in four sections, each of about 100 cm^2 area, mounted on a perforated steel plate held in a plastic frame. The sections are made of active carbon (Union Carbide Consumer Products, $8 \times 12 \times 0.6 \text{ cm}^3$). As in the laboratory cell, the oxygen is distributed through copper pipes.

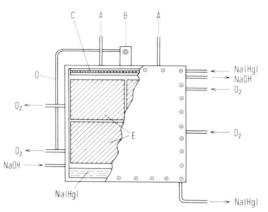

Figure V.2. Sketch of the front view of an amalgam–oxygen battery (after Yeager[3]). Surface area of a four-section cathode, 400 cm². A, vents; B, side-bar of steel plate; C, amalgam delivery tube; D, electrical connection; E, carbon.

The steel plates for the anodes are about 1.6 mm thick, and the back of each is in contact with the plastic frame of a cathode (see Figure V.3). A set of jets (diam. 0.4 mm) distributes the amalgam at the head of each anode plate, the rate of amalgam flow being typically 3 cm³ sec⁻¹. If the sodium content of the amalgam falls to 0.1 wt% in the battery (this corresponds to a current density of

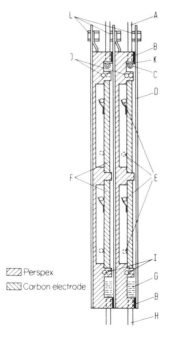

Perspex
Carbon electrode

Figure V.3. Section through the amalgam–oxygen battery (after Yeager[3]). Compare Figure V.2. A, vent; B, packing; C, amalgam jets; D, steel plate covered with flowing amalgam; E, copper tubes, welded to the perforated steel plates backing the carbon electrodes (for oxygen distribution and electrical contact); F, carbon electrodes; G, amalgam sump; H, amalgam outflow.

150 mA cm^{-2}), the rate of flow must be at least 1.3 cm^3 sec^{-1}. The electrodes are fixed 0.5 cm apart. The voltage drop in the electrolyte of course affects the current–voltage curve of the cell (curve C, Figure V.1); it amounts to 0.2 volt at 25°C, with 3N NaOH as electrolyte, at a current density of 150 mA cm^{-2}. In the experiment to which curve C relates, there was a flow of electrolyte of some few cm^3 sec^{-1} per cell. For a flow rate of 0.7 cm^3 sec^{-1} of initially 2.5N NaOH, and a current density of 150 mA cm^{-2}, the effluent electrolyte is 3.5N NaOH.

The weight of each cell is 3.1 kg (750 g amalgam, 250 g electrolyte).

The arrangement of the associated apparatus is shown in Figure V.4. The sodium storage vessel is kept at 120°C, and liquid sodium is added dropwise to the amalgam stream by an injector, operated by a relay controlled by an ancillary circuit which contains the cell

$$\text{Amalgam } (0.3\% \text{ Na}) \,|\, 3\text{N NaOH} \,|\, \text{amalgam } (X\% \text{ Na}).$$

The electrical *isolation* of the cells from the common amalgam circulation is accomplished by means of a motor-driven valve system, which opens for 0.5 sec each 2.5 sec for each cell. A similar device is used for the amalgam leaving the cells.

Nitrogen is used as a protective gas in the liquid sodium store. The electrolyte is stored in an 18 l container and is pumped by means of a stainless steel centrifugal pump to a 1 l glass vessel at a higher level, with an overflow returning to the container. The hydrostatic pressure provided by this constant-head device serves to maintain a uniform flow of electrolyte to each cell, independently

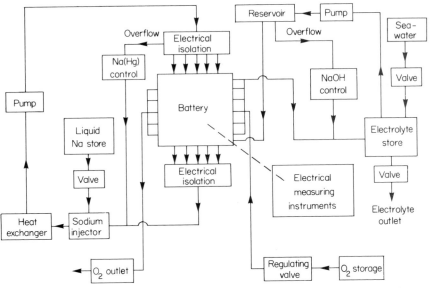

Figure V.4. Block diagram of the amalgam–oxygen battery and its associated equipment (after Yeager[3]).

of pumping rate. The concentration of the NaOH is controlled in a manner similar to that of the amalgam, by use of an ancillary cell

$$\text{Amalgam} \mid 3\text{N NaOH} \mid \text{satd KCl} \mid \text{X}_\text{N} \text{ NaOH} \mid \text{Amalgam.}$$

Sea water is drawn from an 18 l container for dilution of the electrolyte in the circulation system, and simultaneously a corresponding quantity of electrolyte is rejected. All the cells are included in the same electrolyte flow circuit. Isolation is unnecessary because the resistance of the conducting paths along the electrolyte delivery tubes is so large that the loss of energy concerned is negligible. Experiments lasting longer than 24 hours have not been carried out with this battery.

In developing the Yeager cell further, the M. W. Kellogg Company[5] have constructed a 16 kW unit (Figure V.5). This battery is built up of five modules, each module consisting of 37 cells in parallel, with a gross electrode area of 0.97 m^2. A departure from the Yeager design was made in the adoption of a cylindrical arrangement of the electrodes (Figure V.6). A porous silver tube from the Electric Storage Battery Co. (cf. section IV.1.1.1 and reference 30 and 32 of chapter IV) was used for the oxygen electrode. Isolation of each cell from the amalgam circuit (both entry and exit) was effected by a dropping flow.

The emf of the battery amounted to about 10 volts, and the maximum rating of 16 kW was attained at a current density of 400 mA cm^{-2}. The estimated working life at 100 mA cm^{-2} output was 5000 hr. The volume of the battery was approximately 0.3 m^3.

Figure V.5. Sodium amalgam–oxygen battery of 16 kW rating by the M. W. Kellogg Company.

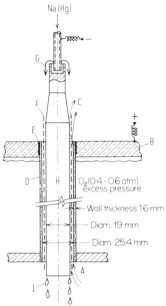

Figure V.6. Cylindrical construction of the Kellogg cells (after Jasinski and Kirkland[5]). A, electrolyte inlet; B, copper busbar; C, electrolyte outlet; D, silver cathode; E, soldered joint; F, porous polyethylene separator; G, overflow; H, steel anode; I, dropping mercury.

The amalgam–oxygen battery provides a compact energy source of high rating, capable of supplying large current densities at a cell voltage exceeding 1 volt, at relatively low temperatures. Storage of the fuel is easy and requires no great volume. In spite of its CO_2 content, air can be used without difficulty as oxidant, because electrolyte is rejected during operation.

A disadvantage is that large quantities of water are required for dilution of the electrolyte and also for cooling. The high cost of sodium ($4.50 per kg) and the great weight and poisonous nature of mercury are further adverse factors. The complex regulation system required would not be economic for installations of less than 10 kW rating. A possible technical application of these cells is discussed in the next section in relation to chlorine–alkali electrolysis. A prerequisite would appear to be the production of an oxygen electrode with a working life of a few years. The silver electrodes of the Kellogg battery have attained 7500 hr in half-cell use (cf. section IV.1.1.1).

V.1.3   Use of Amalgam–oxygen Cells as Secondary Cells in Chlorine–alkali electrolysis

In the usual industrial process the amalgam formed in the primary chlorine–alkali electrolysis cell is catalytically decomposed, with formation of hydrogen and caustic alkali, in an attached 'pile' by the action of water in presence of pieces of graphite or iron (Figure V.7a). This is a local-action corrosion process and the energy of the reaction is lost as heat.

If, now, an oxygen electrode is brought into the secondary amalgam-decomposer cell, the latter is converted into a galvanic cell, namely, the amalgam–oxygen cell discussed above (Figure V.7b).

The voltage of this cell can be considered in relation to that required to perform the primary electrolysis of the brine. 30 to 35% of the electrolysis voltage (3.6 to 4 V) can indeed be provided by an amalgam–oxygen cell (1.3 to 1.5 V at 4000 A m^{-2}) connected in series, and the energy of the amalgam decomposition can be utilized*. It would, of course, be necessary to forego hydrogen as an electrolysis product.

If, instead of the oxygen-dissolution electrode, a hydrogen-evolution electrode is used in the secondary cell (Figure V.7c) then there is still a galvanic cell with a terminal voltage that could contribute to the primary electrolysis. The potential of the reversible hydrogen electrode is about 950 mV positive to the electrode potential of the amalgam. Table V.1 shows the results of laboratory experiments with an *amalgam–hydrogen cell* of this kind. The amalgam concentration was 0.2 wt% of sodium, the electrolyte was 7 to 8% NaOH solution, and the distance between the electrodes was 1 cm. A magnetic stirrer was used to maintain uniformity of composition of the horizontal amalgam electrode. The working temperature was 70°C. The results show that the catalytic activity of the skeleton-nickel electrode is adequate to attain practical current densities for the hydrogen deposition[6] (terminal voltage about 0.6 V).

Difficulties can arise in the continuous operation of primary and secondary cells in series for purposes of saving energy in the amalgam process. Spontaneous 'self-reaction' of the amalgam has

Table V.1. Current and voltage values of an amalgam–hydrogen cell (after Vielstich, Justi and Winsel[8]).

Current (A)	Current density (A m^{-2})	Terminal voltage (mV)	
		Nickel foil	DSK-Raney-nickel (after Justi)
0	0	830	950
0.05	42	690	942
0.2	167	570	920
0.4	335	470	885
0.8	666	270	810
1.2	1000	60	745
1.6	1333		675
2.0	1665		605

* J. A. LeDuc, J. G. Kourilo and Ch. Lurie, *J. Electrochem. Soc.*, **116**, 546 (1969).

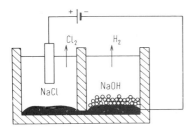

(a) Catalytic decomposition of amalgam with graphite or iron in the pile.

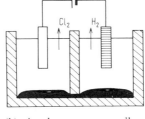

(b) Amalgam–oxygen cell as secondary cell, with sacrifice of hydrogen.

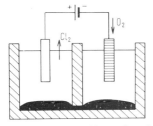

(c) Electrochemical formation of hydrogen and alkali from amalgam and water, with recovery of part of the amalgam energy for use in electrolysis of the alkali chloride solution; amalgam–hydrogen cell as secondary cell.

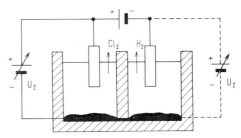

(d) Circuit for current adjustments in primary and secondary cells.

Figure V.7. Electrochemical utilization of the amalgam decomposition in chlorine–alkali electrolysis (after Vielstich[6]).

the consequence that less current is required for the secondary than for the primary cell. If the same current were passed, premature consumption of all the sodium in the amalgam would lead to a collapse of voltage and the formation of mercuric oxide. A number of suggestions have been made of methods to keep primary and secondary cells in step[7]. Two examples are shown in Figure V.7d. An ancillary voltage source is connected in parallel with one or other of the cells, and so adjusted as to compensate for the spontaneous reaction of the amalgam.

A pilot plant combining electrolysis and amalgam–oxygen cells has recently been set up by the Kellogg Company[7a], and has operated satisfactorily at 3500 A, with 2% of self-reaction in the amalgam cell. The solution to the problem of coupling the cells to attain 98% utilization of the amalgam has not yet been published. It is hoped within two years to establish commercial cells working at 30 000 to 35 000 A, involving energy economy of 33 to 35%. Costs for the amalgam cell (capital, operation, amortization) should amount to 0.43 cent per kW hr (in 1965), not including working materials (amalgam and air).

V.2 GALLIUM–OXYGEN CELL

The possible use of gallium as the negative electrode in primary or secondary cells has recently been discussed by Jahn and Plust[9]. From the electrochemical standpoint, gallium in equilibrium with $H_2GaO_3^-$ ions has the very negative standard potential of -1220 mV at 25°C. Gallium metal also has a high hydrogen overpotential. Its low melting point (29.8°C) and its low equivalent weight (23 g per faraday) are of advantage for its use in galvanic cells. A decisive advantage is that the reaction

$$Ga + OH^- \rightleftharpoons H_2GaO_3^- + H_2O + 3e^-$$

proceeds rapidly with small charge-transfer overpotential.

In 8N $HClO_4$ the overpotential at 50°C and 220 mA cm^{-2} is only 100 mV. In sulphuric acid a film of $Ga_2(OH)_4SO_4$ is formed so that in 2N H_2SO_4 at 50°C with an overpotential of about $+600$ mV only 20 mA cm^{-2} can be attained.

The anodic behaviour in 6N KOH has been studied in more detail because silver-activated oxygen electrodes of similar capacity can easily be set up in this electrolyte. Figure V.8 shows current–potential curves for the electrodes of such a cell. At 100 mA cm^{-2} and 80°C the current efficiency is near to 100%. The overpotential changes insignificantly with gallium content of the electrolyte up to 300g Ga l^{-1}. On open circuit gallium dissolves spontaneously at a rate equivalent to a corrosion current of about 8 mA cm^{-2}. The authors estimate that by using oxygen stored under pressure a gallium-consuming cell could be made of weight about 20 kg kW hr^{-1}.

A suitable cell construction, taking the normally liquid state of gallium into account, has not yet been developed. The high cost of the metal is to be considered in relation to practical applications.

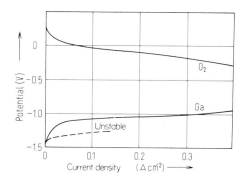

Figure V.8. Single electrode potentials (relative to standard hydrogen electrode) of a gallium–oxygen cell with 6N KOH as electrolyte, at 80°C (after Jahn and Plust[9]).

V.3   ZINC–OXYGEN CELLS

Cells based on the reaction of zinc with atmospheric oxygen in alkaline electro-lytes have been known for a long time. They have been used in radio stations, for signalling installations, flashing warning lights, operation of safety devices in mines, illumination in remote areas, and many other purposes. Figure V.9 shows an American design of cell.

In these cells, a hydrophobic carbon electrode dips into 20% NaOH as electrolyte; the other electrode is of amalgamated zinc. The emf is 1.45 V. The capacity of the cells is limited by the reserve of zinc, which cannot, of course, be completely consumed. The anodically formed zinc hydroxide goes into solution as zincate. After the recharging of a cell with fresh electrolyte and the renewal of the zinc electrode the carbon–air electrode can be repeatedly reused. Up to 10 000 hours of operation have been achieved. Figure V.10 shows the discharge characteristics of a French-built alkaline zinc–air cell (CIPEL, type 608A). Table V.2 contains data important for practical applica-tion relating to two zinc–air cells (with liquid, alkaline electrolytes), compared with a dry cell which also works with zinc and atmospheric oxygen. The dry cell employs a semi-solid ammonium chloride–zinc chloride electrolyte, and a cathode of compressed, active carbon powder; at higher discharge rates its terminal voltage falls away relatively rapidly (Figure V.11, curve A)*.

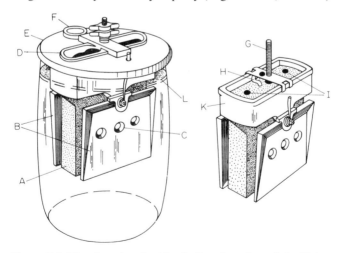

Figure V.9. The zinc–air cell with alkaline electrolyte of the Union Carbide Corporation (Eveready Air Cell) (from Ullmanns Enzyklo-pädie der technischen Chemie, ed. W. Foerst, vol. 7, p. 769, 1956). A, carbon electrode; B, zinc electrodes; C, indicator for state of charge; D, air entry; E, porcelain lid; F, filler for oil layer; G, securing bolt; H, connecting cable; I, ventilation holes; K, plastic supporting frame; L, oil layer.

* New developments in the dry-cell field have, however, effected considerable improvements[9a].

Table V.2. Comparative data for two zinc–air cells with liquid electrolytes and a zinc–air dry cell. I.E. = Internationale Entladungsvorschrift (international discharge specification): 5 min on load, 10 min rest.

	Dimensions (cm)	Weight (kg)	Discharge resistance or current	Capacity (A hr)	E, average (V)	(kW hr)	(kg kW hr^{-1})	Price (1964)
Zinc–air cell, CIPEL (type 608A)[26]	21.5, 21.5, 28.0 (12.9 l)	14.5	400 mA 1000 mA	2500 2100	1.22 1.13	3.05 2.37	4.75 6.12	NF40.
Zinc–air cell, Union Carbide Co.[27] (Eveready Air Cell), Type AC 500 B, NaOH	8.8$^2\pi$, 30 (7.2 l)	8.75	600 mA	500	1.20	0.60	14.6	$9.91
Zinc–air cell, Union Carbide Co.[27] (Eveready Air Cell), Type AC 750, KOH	8.8$^2\pi$, 30 (7.2 l)	8.75	2000 mA	750	1.30	0.98	8.90	$9.91
Zinc–air dry cell, Carbone[26] (Type 513)	16.0, 16.0, 18.0 (4.6 l)	5.2	2.5 Ω 2.5 ΩI.E.	732 702	0.96 1.02	0.702 0.715	7.42 7.27	DM25.

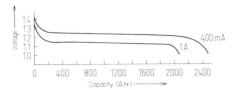

Figure V.10. Discharge curves of a zinc–air cell
with alkaline electrolyte of the firm CIPEL,
type 608A ; continuous discharge at 400 mA and
1A[26].

Zinc–oxygen cells have attracted renewed interest in relation to fuel-cell
research[10]. The Electric Storage Battery Company has demonstrated a zinc
cell involving the *continuous* feed of zinc powder to the cell, and circulation of
the electrolyte through a regenerator in which the zinc-containing solution is
continuously renewed by addition of calcium carbonate*. The somewhat
complex equipment required, particularly that associated with the zinc feed,
has so far been a barrier to the production of a technically developed cell.

The same Company has, however, tested on the semi-technical scale a zinc–
oxygen cell with compact cylindrical zinc electrodes[5]. The oxygen electrodes,

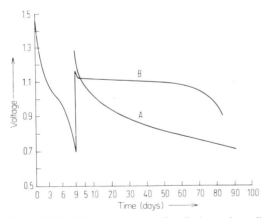

Figure V.11. Discharge curves of a Carbone dry cell,
Type 513, (after catalogue of Carbone AG, Frankfurt-
Bonames). Curve A: periodical load, 3 min at 1.35A
through 0.6 ohm, 12 min rest; capacity 589 A hr. Curve
B: continuous discharge at 1.02 A (220 A hr), followed
by discharge at 0.2 A (405 A hr)—total capacity 625 A hr.

* The General Dynamics Corporation[9b] has recently proposed an electrochemical regeneration of
zinc from the zinc hydroxide formed by means of an electrolyte circulation system.

porous silver tubes of 2.5 cm diameter (cf. section IV.1.1.1 and references 30 and 32 of chapter IV), are supplied with oxygen under pressure from steel cylinders. The electrolyte is 30% KOH. The cell gives a terminal voltage of 0.9 V at 80 mA cm^{-2}.

An 18-cell battery, driving a 12 V dc motor, has been developed as a power unit for a small truck. The rest voltage of the cell is 25 V, and it will provide a starting current of 80 A. A maximum speed of about 12 mph is attained at 30 A and 12 V; this corresponds to approximately $\frac{1}{2}$ hp. With one charge of zinc the truck will travel about 75 miles in 6 hr.

The carbon-air electrodes used in fuel cells have been applied to the development of a zinc–air cell of dry-battery type. Aqueous KOH solution, with added zinc oxide to repress corrosion, is used as electrolyte. Vielstich and Vogel[10a] have made plastic cased cells, of monocell size, constructed similarly to the formate–air battery of Figures IV.85 and IV.85a; the fuel electrode is merely replaced by a zinc cylinder of 1 mm wall thickness. The air electrode secured to the lid is of active carbon hydrophobically treated with polyethylene. Figure V.11a shows several discharge curves at various temperatures.

This cell has clear advantages over normal dry cells. Their capacity is greater. and, with the quoted thickness of the zinc electrode, the cell can be renewed

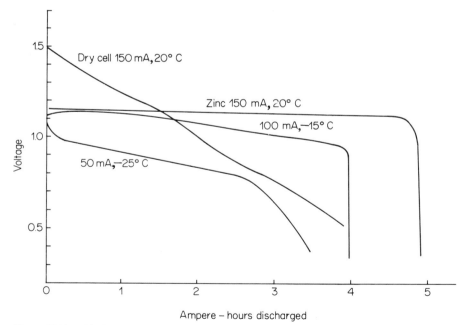

Ampere – hours discharged

Figure V.11a. Discharge curves of zinc–air cells with aqueous electrolyte (after Vielstich and Vogel[10a]); commercial zinc sheet, 10M KOH + 36 g l^{-1} ZnO. For comparison, dry cell IEC R 20, 150 mA, 20°C.

three times by replacement of the electrolyte (20–22 cm^3). The discharge voltage is approximately constant, and, despite the liquid electrolyte, the cell remains serviceable at temperatures below − 20°C. Crystallization occurs in Leclanché cells below − 10°C, leading to increase of internal resistance[10b]. The higher conductance of KOH solutions gives a significant advantage for continuous discharge at higher current densities.

A disadvantage is that gas-tight closure of the cell involves danger of penetration of electrolyte through the carbon electrode, particularly as a result of temperature fluctuation. Use of a vent for pressure equalization restricts the cell to upright use.

Schwabe and co-workers have developed an even smaller cell (a button cell), with a polythene–carbon air electrode prepared by Grüneberg and Jung's method (section IV.3.1.2 and Figure IV.83). A thin polythene membrane over the air electrode situated in the lid makes the cell effectively gas- and liquid-tight in use. Terminal voltage is between 1.2 and 1.0 V at 60 mA (20 mA cm^{-2}) at room temperature.

Consideration has recently been given to rechargeable zinc–air cells in relation to traction[10f,g]. For cells of lower rating recharging presents few problems. A membrane, similar to that usually used in the silver–zinc accumulator, is placed between the zinc and air electrodes. It is expedient, however, not to develop the carbon-diffusion electrode for purposes of charging, but to use a third electrode for this purpose, e.g., a nickel screen[10d] inserted between the zinc and air electrodes. This has the advantage that selective catalysis keeps oxygen evolution overvoltage at a low value during charging.

For batteries of high rating, there are problems apart from those of recharging. The major ones are passivation at low temperatures (− 10 to − 40°C), due to formation of a zinc oxide film on the anode, corrosion of the anode at high temperatures, and flooding of the air cathode on prolonged standing[10e]. The air electrodes mainly used are hydrophobically treated mixtures of platinum or silver (and for active carbon) with Teflon on a conductive screen. Charkey and Dalin[10e] have found that an electrode with a content of hydrophobic material adjusted for optimum catalytic activity is unsatisfactory in resisting flooding while standing on open circuit, and the difficulty is worsened at elevated temperatures. Incorporation of a permselective membrane adjacent to the air electrode considerably reduces this tendency, both at room temperature and at 52°C. For example, an air electrode–membrane combination can stand immersion in KOH for 200 hr at room temperature and 50 hr at 52°C without substantial loss in performance; without membrane protection, the corresponding periods are 36 hr and 10 hr respectively. An interesting electrochemical technique has been found which delays flooding for up to two months. Zinc–air cells with permselective membranes that were subjected to a continuous 'parasitic' drain of about 5×10^{-4} of their normal load had not flooded after a month.

Mechanical recharging is of interest for zinc–air cells of higher rating. Chodosh, Rosansky and Jagid[10f,g] have developed a 24 V, 25 A hr battery of this type for servicing radio equipment. It weighs less than 6 kg (cf. 8.5 kg. for an equivalent Ag–Zn battery) and, designed to operate at 50–70 mA cm^{-2}, it provides 60 W hr lb^{-1}. A hydrophobic air cathode and a KOH electrolyte are used. The removable anode of each cell is made of porous zinc, with a permanently attached double paper separator, and is mounted in the cell by means of a leak-proof O-ring seal. After fabrication, each anode is soaked in 36 wt% KOH solution and then dried; this process leaves solid KOH uniformly dispersed throughout the porous structure, stiffening it and facilitating handling during recharging. Battery activation is effected simply by inserting the anodes so prepared into the water-filled cell compartments.

V.4   ALUMINIUM– AND MAGNESIUM–OXYGEN CELLS

Aluminium and magnesium are similar in behaviour, but aluminium is the better in activity, magnesium in ease of handling.

V.4.1   The Aluminium–oxygen Cell

In the last few years it has been proposed to replace zinc by aluminium in alkaline metal–oxygen cells[11–13]. Aluminium is superior to zinc and magnesium in relation to its equivalent weight, its negative electrode potential and its electrochemical activity. It has not hitherto come into use because of the low current efficiency arising from its rapid corrosion, with evolution of hydrogen, in alkaline solutions. The corrosion can, however, be largely retarded by alloying and by adjustment of working conditions[12].

In alkaline electrolytes, aluminium undergoes the anodic reaction

$$Al + 3OH^- \rightarrow Al(OH)_3 + 3e^- ; \qquad \phi_{00} = -2.31 \text{ V}$$

In concentrated alkali hydroxide solution ($>2N$), aluminate is the main product:

$$Al + 4OH^- \rightarrow H_2AlO_3^- + H_2O + 3e^-$$

and the calculated standard potential becomes -2.35 V. Observed rest potentials are, however, only -1400 to -1700 mV relative to the saturated calomel electrode.

Foust[12], using such electrodes, obtained current densities of 20 to 300 mA cm^{-2} at temperatures from 20 to 65°C, with working electrode potentials (SCE) between -700 and -1600 mV. KOH electrolyte, 3N and 10N, was used, and current efficiencies varied between 35 and 99%. Corrosion loss increased rapidly with rising temperature, but decreased with rising current density. At 65°C and 300 mA cm^{-2} it was thus possible to attain almost 100% efficiency.

Foust,[12] has constructed a 20-cell battery (2 × 10 cells in parallel) that delivers 10 A at 12 V terminal voltage. The positive electrodes are of porous nickel which, supplied with air, operate at 20 mA cm^{-2}. The size of a cell is 14 × 23 × 0.55 cm^3. The aluminium in each cell is adequate for 50 hr working but the electrolyte needs hourly renewal.

Zaromb[11] quotes an energy–weight ratio of 200 W hr kg^{-1} for an aluminium–air cell; he obtained an efficiency of 80% at 1.3 V and 10 mA cm^{-2}.

To summarize, the aluminium–air cell with alkaline electrolyte is limited in usefulness by self-discharge and excessive gas and heat evolution. Whether or not these undesirable effects can be effectively suppressed by addition of chromate to the electrolyte[13a], it is certain that significant technical applications of the alkaline aluminium–oxygen cell are limited to special cases. Only very limited power densities can be obtained with use of neutral or acid electrolytes.

V.4.2   The Magnesium–air Cell

The magnesium–air cell has the advantage over its aluminium analogue of a higher theoretical emf (3.09 V, as compared with 2.71 V) but encounters similar operational difficulties—although the difficulty of self-discharge is substantially less. Higher terminal voltages are attainable with suitable alloy electrodes, but these are in general so active in self-dissolution as to lead to a pronounced rise in temperature. In practice a compromise has to be made between higher rating and shorter life on the one hand, and lower rating with slower self-discharge on the other. High terminal voltages are obtained with amalgamated magnesium electrodes, but for low and medium rating cells, ternary magnesium alloys are mostly used, containing small proportions of aluminium and zinc.

The General Electric Company has recently developed a magnesium battery repeatedly regenerable by replacement of the magnesium electrodes and the electrolyte[13b,c]. Figure V.11b shows a module of five such cells. The anodes are of magnesium alloy AZ 31 B (3% Al, 1% Zn), and the cathodes are platinum-catalysed air electrodes, as used in alkaline fuel cells. The electrolyte is 7% NaCl but other salts may be used and the battery has even been successfully operated with available river or seawater. The air electrodes (two per anode) are integral with the casing, and the anodes fit into guides which facilitate their removal and replacement. Thirty working cycles with the same air electrodes have been attained, corresponding to 400 hr working life; 1000 to 2000 hr operation are regarded as practicable for batteries of this kind.

Figure V.11c shows the current–voltage curve for such a cell; the rest overpotential is essentially due to the magnesium electrode, whereas the steep

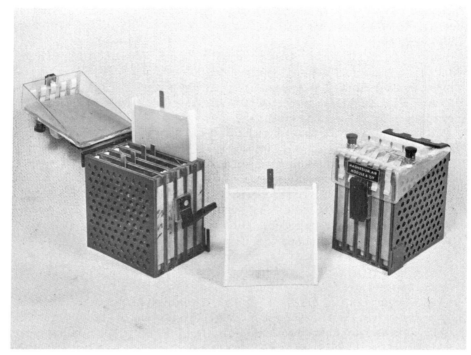

Figure V.11b. Magnesium–air module (after Kent and Carson[13b]).

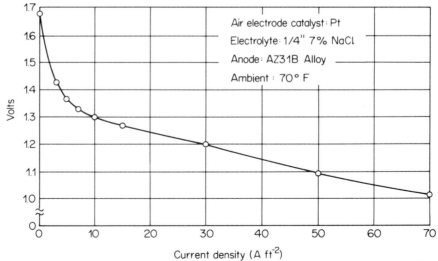

Air electrode catalyst: Pt
Electrolyte: 1/4" 7% NaCl
Anode: AZ31B Alloy
Ambient: 70° F

Current density (A ft⁻²)

Figure V.11c. Typical performance of a magnesium–air cell (after Kent and Carson[13b]).

initial fall of the curve and the linear section which follows are attributable to the air electrode and ohmic drop. The battery works at 12 W and 6.0 V, giving 15 A hr per charge. The compact construction gives the favourable energy density of $71\,W\,hr\,lb^{-1}$ per charge, increasing to $500\,W\,hr\,lb^{-1}$ over 60 operational cycles.

A larger unit of 23 cells has since been developed, giving at least 264 W at 24 V[13d]. It has been successfully tested over 30 recharging cycles.

The Yardney Electric Corporation has carried out experiments on the use of other electrolytes for the purpose of extending the working range of the magnesium–air cell to lower temperatures[10e]. The aspect of efficiency is favourable, since the proportion of 'self-reaction' decreases with falling temperature. The best results have been achieved with the cryohydrate formed by 4N $NaClO_4$ and 4N $LiClO_4$ with 7% LiCl which crystallizes at $-30°C$.

V.5   HYDROGEN–CHLORINE CELLS

The electrochemical reaction of hydrogen and chlorine to form hydrogen chloride

$$
\begin{array}{ll}
\text{Anode:} & H_2 \longrightarrow 2H^+ + 2e^- \\
\text{Cathode:} & Cl_2 + 2e^- \longrightarrow 2Cl^- \\
\hline
& H_2 + Cl_2 \longrightarrow 2HCl
\end{array}
$$

had already been studied in some detail by Foerster[14] in 1923. His object was to find whether the *production of hydrochloric acid* could be carried out more economically by electrochemical means than by the usual chemical process. The electrochemical preparation of hydrochloric acid is, however, especially expensive in regard to apparatus, and requires high capital investment (platinum must be used as catalyst on the hydrogen side).

In contrast to the hydrogen–oxygen cell, for example, the hydrogen–chlorine cell involves the difficulty that the product, HCl, must be removed. Further, the use of a diaphragm is obligatory because the anode material is catalytically active to chlorine. The overpotential at the anode must not be allowed to exceed 200 to 300 mV, otherwise the platinum used as catalyst is strongly corroded. The emf of the cell corresponds fairly closely with the thermodynamic value. The following cell emf was found by Foerster[14]:

$$1N\ HCl,\ 25°C,\ E° = 1.366\ volt$$

$$11.5N\ HCl,\ 30°C,\ E° = 1.000\ volt$$

Because a diaphragm must be used to separate anode and cathode compartments, the chloride ion concentration in the latter increases markedly as the

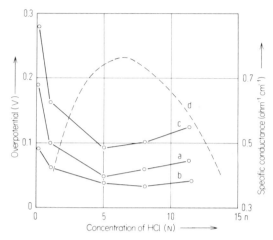

Figure V.12. Influence of HCl concentration on over-potential and conductance (after Yoshizawa, Hine, Take-hara and Kanaya[15]). Current density $= 10\,\text{mA cm}^{-2}$, 30°C. a, Overpotential of the Cl_2 electrode; b, overpotential of the H_2 electrode; c, sum of a and b; d, conductance of HCl.

cell operates. This affects the emf. Thus, for the cell[14]

$$H_2 \mid 3.5\text{N HCl} \mid 12.0\text{N HCl} \mid Cl_2$$

at 20°C, $E° = 1.14$ volt. The conductance of the electrolyte is optimal between 4 and 10N HCl (cf. Figure V.12).

Foerster[14] used plates of Acheson graphite ($12 \times 5 \times 2.5\,\text{cm}^3$) in his experiments. No additional catalyst was needed for the reduction of chlorine (cf. section III.7). The hydrogen electrodes were platinized at 2 A. Anode and cathode compartments were separated from each other by an earthenware diaphragm 2 mm thick. Both gases were passed through the electrolyte in excess—no gas-diffusion electrodes were used. The hydrogen electrode dipped only 0.5 cm into the electrolyte, but the chlorine electrode was completely immersed; both were corrugated to increase the active surface area.

A battery of three such cells, with 2N HCl as electrolyte and delivering 0.2 A at room temperature, had a terminal voltage of 2.93 volt when supplied with $1.5\,\text{l hr}^{-1}$ of chlorine and $12\,\text{l hr}^{-1}$ of hydrogen. The voltage hardly changed over 12 hours of operation. Reduction of the chlorine flow to $0.4\,\text{l hr}^{-1}$, with current at 0.3 A, reduced the terminal voltage only to 2.6 V. The hydrogen consumption under these conditions was only 2%; that of the chlorine, because of its greater solubility, was 91%.

A single cell was allowed to pass 0.5 A for 72 hours. The terminal voltage decreased from 0.78 to 0.67 volt and the HCl concentration in the electrolyte increased by about 60% (electrolyte 300 cm^3, initial concentration 230 g l^{-1} HCl).

Increase of the working temperature of the cell diminishes the terminal voltage, especially above 80°C, because, at the higher temperatures, the lesser solubility of chlorine has more effect than the facilitation of the charge-transfer step.

Yoshizawa, Hine, Takehara and Kanaya[15] in 1962 studied a laboratory cell with *gas-diffusion* electrodes of platinized carbon (H$_2$) and of carbon (Cl$_2$), each in the form of discs 1.3 cm thick and of area 5 cm^2.

Figure V.12 shows the influence of HCl concentration on the overpotentials of the electrodes at 10 mA cm^{-2} and 30°C; chlorine was supplied at 15 cm^3 cm^{-2} min^{-1}. The variation in overpotential is much greater than the concentration dependence of emf.

In Figure V.13, the potentials of the individual electrodes in 5N HCl at 50°C are shown as functions of current density. In calculating terminal voltages from these curves, voltage drop in electrolyte and diaphragm must be taken into account. The limit of current density lies at about 40 mA cm^{-2}; if this should be exceeded, the terminal voltage falls so rapidly that there is a risk, mentioned earlier, of loss of platinum from the hydrogen electrode.

Bianchi[16] has also recently reported measurements on a lower-temperature hydrogen–chlorine cell. A terminal voltage of 1 volt was obtained at 50 mA cm^{-2}

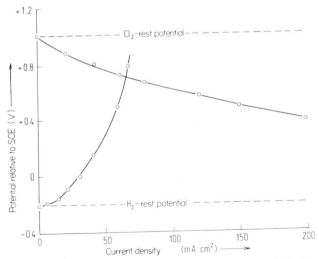

Figure V.13. Current–potential curves for the electrodes of a hydrogen-chlorine cell (after Yoshizawa, Hine, Takehara and Kanaya[15]); 5N HCl, 50°C, 84 cm^3 of H$_2$ cm^{-2} min^{-1} and 14 cm^3 of Cl$_2$ cm^{-2} min^{-1}.

and 45°C. Even at 100 mA cm^{-2} a steady voltage of 0.65 V should be obtainable.

Yoshizawa and co-workers[17] have investigated a hydrogen–chlorine cell with a *molten salt* electrolyte; KCl/LiCl at 450 to 500°C. Porous carbon (Cl$_2$) and platinized porous carbon (H$_2$) were used as electrode materials. The emf was 1.55 V and 200 mA cm^{-2} was obtained at a terminal voltage of 550 mV.

In comparison with an oxygen electrode in an acid electrolyte, the lower overpotential of the chlorine electrode would seem to give the hydrogen–chlorine cell an advantage at temperatures up to about 80°C. However, for practical purposes the difficulties of corrosion, the poisonous nature of the oxidant, the necessity of the diaphragm and the problem of getting rid of the reaction product, HCl, are disadvantageous.

V.6   ALKALI BOROHYDRIDE AS FUEL

The electrochemical reaction of alkali borohydride as fuel in cells with alkaline electrolytes proceeds, as in the case of hydrazine, by way of a preliminary dehydrogenation (cf. section III.5.2). Jasinski[18], using a laboratory apparatus similar to that illustrated in Figure IV.88, has studied the properties of an alkali borohydride–oxygen cell at a temperature of 70°C. The electrodes, 15 × 15 cm^2, were of palladized nickel for the fuel electrode, and silver-activated nickel for the oxygen electrode. The electrolyte, a 2% solution of KBH$_4$ in 6N KOH, was circulated. Anode and cathode compartments were separated by a membrane; without the membrane, the oxygen electrode showed a mixed potential, displaced by some 120 mV. Figure V.14 shows a current–voltage curve obtained after 40 hours operation of the cell.

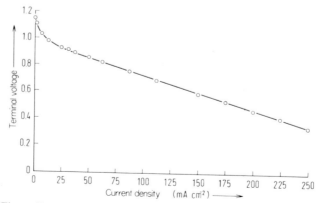

Figure V.14. Current–voltage curve of a potassium borohydride–oxygen cell with 15 × 15 cm^2 electrodes at 70°C (after Jasinski[18]).

Use of *nickel boride*, Ni_2B, instead of palladium as catalyst for the anode improved the terminal voltage; at $100\ mA\ cm^{-2}$ under otherwise similar conditions, it lay above 0.8 V.

This cell has the advantages of high terminal voltage and ease of storage of the fuel. Disadvantages associated with technical application are, beside the cost of the fuel, the accumulation of borate in the electrolyte and the loss of part of the hydrogen contained in the fuel. Loss of gaseous hydrogen occurs if the potential of the electrode is negative with respect to that of the hydrogen electrode in the same solution (cf. section III.5.2).

V.7 AMMONIA–OXYGEN CELL WITH ALKALINE ELECTROLYTE

Wynveen[19] has examined a *gaseous* ammonia–oxygen cell, using a cell construction similar to that of the Allis-Chalmers hydrogen–oxygen cell illustrated in Figure IV.26. Both electrodes were of platinized carbon, separated by an asbestos diaphragm 0.2 cm thick, soaked in KOH solution. Both gases were passed in excess through the electrode chambers.

The use of ammonia in solution as a fuel is disfavoured by its much reduced solubility in alkaline electrolytes; in the present case this effect is put to advantage by the use of a high concentration of KOH, which minimizes the diffusion of ammonia to the cathode.

Figure V.15 shows current–voltage curves obtained at different temperatures. A three-fold excess of ammonia over that electrochemically consumed was passed—$20\ cm^3\ min^{-1}$ of ammonia are required for a current of 4 A. In spite of the KOH-permeated membrane, the oxygen electrode shows a mixed

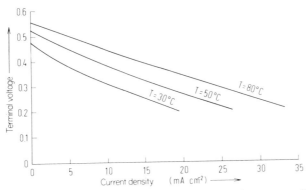

Figure V.15. Current–voltage curves of an ammonia–oxygen cell (after Wynveen[19]), with KOH-soaked asbestos diaphragm as electrolyte, platinized carbon electrodes, both gases at atmospheric pressure.

potential; use of a silver cathode raises the emf to 0.7 V. If the fuel gas is circulated, the accumulation of nitrogen must be taken into account, although a nitrogen content of less than 50% has a comparatively small effect.

In continuous operation at 80°C the volume of nitrogen produced, as well as the electrical output, corresponded with a current efficiency of 95 to 100%, calculated on the basis of three electrons per molecule (cf. equation III.35). No hydrogen was found in the exhaust gas.

Advantages for practical application are, first, the lesser explosion hazard with NH_3-O_2 mixtures as compared with H_2-O_2; secondly, that ammonia is easily liquefied (1 cylinder of NH_3 is equivalent to 9 of H_2); thirdly, that operating costs are relatively low.

V.8 BIOCHEMICAL FUEL CELLS

The idea of biochemical fuel cells is not new. In 1911, Potter[19a] set up a battery which generated small amounts of electricity from the activity of bacteria and yeast extracts. Similar work was done by Cohen[19b] in 1931, but neither of these was followed up until recently, when, after a suggestion by Sisler[19c,19d] some cells were made by the Magna Corporation[20-23] and other companies[22-25] in the U.S.A. In these cells, a great variety of substances, including waste products such as humus, faeces, or compost, as well as marine plant and animal wastes, were acted upon by biological catalysts (enzymes, yeasts, bacteria) to generate substances of greater electrochemical activity, which were then used as reactants at one or other electrode of a fuel cell. Of the substances present in waste products, those most suitable as fuels in biochemical fuel cells are hydrocarbons, fatty acids, alcohols, carbohydrates and urea. Lately, cells have been constructed using the juices of fruits or coconuts as biological fuels.

A detailed study of the theoretical relationships between the biological reactions and the electrochemical reactions in biochemical fuel cells has been made in the U.K. by Gray Young[20a], from which he concludes that the least severe theoretical limitations would be encountered by using biological agents in a redox type of cell in preference to any other type of fuel cell.

V.8.1 Biochemical Anodes

If the biochemical process is the generation of a substance suitable for electrochemical use as a fuel, then it may be allowed to occur in the anode compartment of a cell—the cathode being a more or less conventional oxygen electrode. This arrangement constitutes a direct biochemical fuel cell. The biochemical processes must be allowed to occur under natural conditions, i.e., at normal temperatures, in nearly neutral media, and with only small concentrations of salts and biological substances. As the biological solution must also be used as electrolyte, it follows that, in direct biochemical cells, the current densities can attain only small values, of the order of 1 mA cm^{-2}. However, as the biological electrolyte solution will contain many substances, some of which may be poisonous to electrochemical catalysts, and since it is difficult to match bio-chemical and electrochemical working conditions, an alternative arrangement may be preferable. This is to arrange for the fuel suitable for electrochemical use to be generated in a separate biochemical reactor, and then to be used in a fuel cell of the more usual kind. Because of the difficulty of separating liquid or soluble reaction products from the biochemical reactor, this arrangement is practically limited to producing gases such as H_2, CH_4, NH_3 or H_2S for use in the electrochemical part of the fuel cell.

The modes of action with glucose and with urea are given as typical examples of biochemical anode reactions.

1. Glucose
   Direct use:

$$C_6H_{12}O_6 \xrightarrow[\text{Electrode}]{\text{Enzyme}} C_6H_{10}O_6 + 2H^+ + 2e^-$$

(the reaction proceeds only at the electrode surface).

Indirect use:

$$C_6H_{12}O_6 \xrightarrow[\text{Clostridium butyricum}]{\text{Bacteria}} H_2 + CO_2 + \text{other products}$$

$$H_2 \xrightarrow[\text{catalyst}]{\text{Electrode}} 2H^+ + 2e^-$$

(note that an additional catalyst is required at the electrode for the ionization of the hydrogen).

Figure V.16 shows current–potential curves for two bioanodes. The over-potentials observed in the absence of the biochemical catalyst are plotted for comparison.

2. Urea

In both the direct use and the indirect use, ammonia is formed as a stable intermediate product in the decomposition of urea. In the direct use it is formed at the anode and utilized immediately; in the indirect case it is formed in a separate biochemical reactor and then brought to the anode of a fuel cell, such as that of Wynveen[19], where it is electrochemically oxidized using another catalyst.

$$CO(NH_2)_2 + H_2O \xrightarrow{\text{Enzyme}} 2NH_3 + CO_2 \text{ (biological)}$$

$$2NH_3 + 6OH^- \xrightarrow[\text{catalyst}]{\text{Electrode}} N_2 + 6H_2O + 6e^- \text{ (electrochemical)}$$

Rohrback, Scott and Canfield[22] have used a direct urea anode with an air cathode of platinized carbon. The maximal emf was 0.8 V on open circuit, and on short circuit 4 mA cm^{-2} was measured.

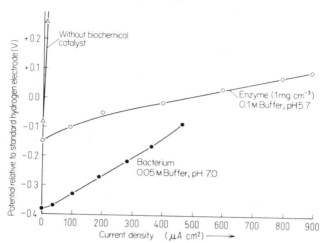

Figure V.16. Bioanodes with glucose as fuel (after Rohrback, Scott and Canfield[22]). Electrolyte, phosphate buffer + 0.1M glucose; platinized platinum electrodes; temperature 25°C.

Taylor and co-workers[22a] have built an indirect 20 W biocell based on urea/urease which gives $2 \, mA \, cm^{-2}$ at a terminal voltage of 28 V (0.4 V per cell). The battery has an energy density of $103 \, W \, hr \, lb^{-1}$ and can deliver its 20 W for about two weeks with consumption of $0.0376 \, lb \, hr^{-1}$ of urea and $0.030 \, lb \, hr^{-1}$ of oxygen.

Brake, Momeyer, Cavallo and Silverman[23] have used the bacterium *Bacillus pasteurii* as a biochemical catalyst; $0.026 \, mole \, min^{-1}$ of NH_3 per g catalyst, was obtained. The enzyme urease gives $1.43 \, mole \, NH_3 \, min^{-1} \, g^{-1}$. The enzyme catalyst is, however, expensive and short-lived.

Extensive studies of hydrogen evolution from natural products have been made by Melpar, Inc.[23a,b]. Very many vegetables, fruits and grains have been investigated with the use of the enzyme from *Chlostridium Welchii*. Whilst in earlier trials it had been found that there was a lag of two to three hours before hydrogen evolution began, immediate hydrogen production has since been achieved by use of a large inoculum of freshly harvested 18-hour cells. The best results have been obtained with bananas and cucumbers. The fact that cucumbers, with the lowest carbohydrate content of all the substrates examined, gave a particularly large yield of hydrogen, suggests that other substances than carbohydrates play an important part. It seems probable that natural products which are rich in readily available, fermentable sugars but give only small yields of hydrogen, contain other substances having a hindering effect on gas evolution. Studies in this field are continuing.

Tests of a microbiological 10 l hydrogen generator, coupled to a hydrogen–oxygen fuel cell by the General Electric Company have given, without lag,

1 W for 8 hr without adjustment of pH or substrate concentration in the fermentation unit. Automatic regulation of these factors should improve efficiency considerably. The rate of gas evolution appears to be limited by the concentrations of H_2 and CO_2 in solution.

The performance of cells using biochemically produced hydrogen is limited by the processes of collecting and concentrating the hydrogen. Palladium electrodes, used for this purpose, have given 10 mA cm^{-2} on intermittent, and 1.5 mA cm^{-2} on continuous load[23c].

The extensive investigations of the Magna Corporation[23c] suggest that formic acid is the most promising biochemically-produced fuel. Sugar, fruits, coconuts, yams, etc. are possible primary materials; from them, with use of *Aeromonas formicans* or *E. Coli*, more than a mole of formic acid per mole of glucose has been produced—together with a series of other electrochemically active substances. A current density of 4 mA cm^{-2}, with overpotential 200 mV, has been attained at an 'integrated bioanode' working with coconut milk at practically 100% current efficiency. Separation of the formate anode

Figure V.16a. Demonstration model of a biochemical fuel cell produced by the Magna Corpn. (after Brake[23c]).

from the biological system allows $40 \, \text{mA cm}^{-2}$ to be reached at ambient temperature.

Figure V.16a shows a demonstration battery. It is of plastic construction with separate compartments for anode, cathode and biological culture. The anode space is separated from the culture by a cellophane membrane and from the cathode by an exchange membrane. Anode and cathode are palladized palladium and platinized platinum respectively. The electrolyte is a 0.2M phosphate buffer of pH 7. Trials with the milk of *one* coconut and air operated a transistor radio intermittently for 50 hours over 45 days. The catalyst (*Aer. formicans*) can be stored in a lyophilized state indefinitely over extreme ranges of temperature, and can be easily transported under almost any conditions.

It should be noted that an electrochemical catalyst is still needed in most of these cells after the biochemical catalyst has generated the intermediate (electrochemical) fuel.

V.8.2   Biochemical Cathodes

Biochemical processes can also be used on the cathode side of a fuel cell. For instance, oxygen can be generated from CO_2 photosynthetically by algae

$$CO_2 + H_2O \xrightarrow{\text{algae}} (CH_2O)_x + O_2$$

and can then be used directly at an electrode or transferred to the cathode of a separate oxygen-breathing fuel cell (where air or oxygen is not otherwise available). A direct cathode of this type was investigated by Reynolds and Konikoff.

Alternatively many microorganisms in the absence of oxygen will reduce sulphates, nitrates or carbonates whilst oxidizing a variety of substances ranging from H_2 to carbohydrates. Bacteria of this type can be used as catalysts for sulphate ion reduction at a cathode, and this may be accompanied by the production of hydroxyl ions:

$$SO_4^{2-} + 4H_2O + 8e^- \xrightarrow[\text{D. desulphuricans}]{\text{Bacteria}} S^{2-} + 8OH^-$$

or by the consumption of hydrogen formed in a preceding reaction.

$$2H_2O + 2e^- \longrightarrow H_2 + 2OH^- \ (H_2 \text{ evolution})$$

$$SO_4^{2-} + 4H_2 \xrightarrow[\text{D. desulphuricans}]{\text{Bacteria}} S^{2-} + 4H_2O$$

Figure V.17 shows a current–potential curve for a sulphate biocathode at 25°C. The electrode is of sand-blasted platinum and the electrolyte is seawater ($0.27\% \, Na_2SO_4$, pH 6.8).

The Magna Corporation, commissioned by the U.S. Navy, has developed a battery that uses magnesium as fuel in the form of a sacrificial anode, whilst

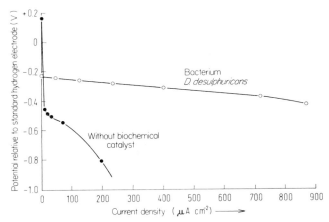

Figure V.17. Sulphate biocathode in sea-water at 25°C (after Rohr-back, Scott and Canfield[22]).

the hydrogen discharged at the cathode is consumed by bacteria in the course of reducing sulphate ions in the electrolyte, which is seawater. The overvoltage of hydrogen discharge at the cathode is said to be reduced by these means. Figure V.18 shows a unit of several watts rating.

Figure V.18. Magnesium–sulphate biocell (after Rohrback, Scott and Canfield[22]).

V.8.3   Biochemically Regenerated Redox Electrodes

Many common bacteria can use redox substances, such as potassium ferricyanide, benzoquinone, or redox indicators, as oxidants for the complete oxidation of fuel substances to CO_2 and H_2O, and, in so doing, the oxidant is reduced in exact equivalence to the extent of the oxidation reaction. For example, the common bacterium *Escherichia coli* reacts with glucose and ferricyanide:

$$C_6H_{12}O_6 + 24Fe(CN)_6^{3-} + 6H_2O \xrightarrow[E.coli]{Bacteria} 6CO_2 + 24H^+ + 24Fe(CN)_6^{4-}$$

The reduced solution of ferrocyanide can then be used as the electrolyte at a redox anode of a fuel cell, being continually regenerated by the biological reactions, which can take place either in the anode compartment or in a separate vessel. Gray Young[27a] has constructed anodes of this type*, and reports that current densities of 40 mA cm^{-2} have been maintained at an anode potential about 300 mV positive to the hydrogen electrode; however, at present, difficulties are being encountered in coupling such an electrode to a conventional air-depolarized cathode.

V.8.4   General Comments

Biocells have several intrinsic disadvantages in comparison with conventional fuel cells. These are essentially due to the difficulties of cell design imposed by the nature of the biological systems, and the apparently conflicting requirements of biological and electrochemical reactions, as far as environment is concerned. In direct cells the solution must contain only low concentrations of salts and of hydrogen and hydroxyl ions but must also be used as electrolyte— so large ohmic losses and only small current densities are to be expected. Moreover, for the propagation of the microorganisms, the solution must also contain nutrient substances and will also contain side-products of biological processes, all of which make it difficult to avoid poisoning of the electrode catalysts. Because of these limitations, it appears unlikely that a practical direct biochemical cell can be constructed. In indirect cells the problem is in two parts:

(a) The design of a biological reactor for the economic generation of an electrochemical fuel from waste or other low grade substance. The electrochemical fuel will probably have to be a gas, because of the difficulty of separating liquid or soluble substances which will be present only in low concentration in the effluent from the biochemical reactor. The problems of this are outside the scope of this book.

(b) Design of an electrochemical fuel cell to use the fuel produced at (a)—but this will be a conventional fuel cell, probably constrained to work at near atmospheric temperature.

* Plain graphite is used as the electrode material, no electrode catalyst being needed.

The biologically regenerated redox fuel cell appears to offer some advantages over the above partly because no additional electrode catalyst is needed, but there are troubles in marrying the biological part to the electrochemical. If these troubles can be overcome, the biological redox anode may be of some use in being able to utilize unconventional fuels.

REFERENCES

1. J. A. LeDuc and C. Lurie, *4th Intern. Sympos. Batteries (IDCB)*, Brighton 1964, Abstract No. 33.
1a. H. G. Oswin, 29th Panel Meeting AGARD, Liège 1967.
2. Zinc–Air Electrolyte Activated Cell of CIPEL S.A. (France); Eveready Air Cell of Union Carbide Corp. (USA).
3. E. Yeager, The Sodium Amalgam–Oxygen Continuous Feed Cell, in W. Mitchell: *Fuel Cells*, Academic Press, New York and London, 1963.
4. S. Eidensohn, *Ger. Pat. Appl.* 1156458; *U.S. Pat.* 3,057,946 (9.10.62).
5. R. Jasinski and T. G. Kirkland: *Fuel Cells—State of the Art*, Allis-Chalmers, Milwaukee, September 1963.
6. W. Vielstich, *Chemie-Ing.-Techn.* **34**, 346 (1962).
7. W. Vielstich, *Chemie-Ing.-Techn.* **33**, 75 (1961).
7a. J. A. LeDuc, M. W. Kellogg Co., New Market, N.J., U.S.A., personal communication.
8. W. Vielstich, E. Justi and A. Winsel, *Ger. Pat.* 1 094 723 (22.12.1958).
9. D. Jahn and H. G. Plust, *Nature (London)* **199**, 806 (1963).
9a. R. Huber, *Elektrotechn. Z.* **18**, 864 (1966).
9b. P. R. Shipps, *Proc. 20th Power Sources Conf.*, Atlantic City, 1966, p. 86.
10. *Fuel Cell, Power for the Future*, Fuel Cell Research Associates, P.O. Box 157, Cambridge 38, Mass.
10a. W. Vielstich and U. Vogel in B. S. Baker, '*Fuel Cell Systems II*', Advan. Chem. Ser., 90, 341 (1969).
10b. F. Jolas, *Elektrotechn. Z.* **18**, 869 (1966).
10c. Fuel Cell Meeting, Dresden, April 1967.
10d. H. Schmidt and W. Vielstich, *Chem. Ing. Techn.* **39**, 761 (1967); *Austrian Pat.* 176 889, 1953; *U.S. Pat.* 3219486 (23.11.1965).
10e. A. Charkey and G. A. Dalin, *Proc. 20th Power Sources Conf.*, Atlantic City, 1966, p. 79.
10f. S. M. Chodosh, M. G. Rosansky and B. F. Jagid, *Proc. 21st Power Sources Conf.*, Atlantic City, 1967, p. 103.
10g. A. M. Moos, N. J. Palmer, *Proc. 21st Power Sources Conf.*, Atlantic City, 1967, p. 51.
11. S. Zaromb, Fuel Cell Progress, July 1963, Fuel Cell Corporation St. Louis, U.S.A.; *J. Electrochem. Soc.* **110**, 253 (1963).
    R. A. Foust, The Electrochem. Soc., Meeting Boston 1962, Extended Abstract No. 49.
13. G. R. Drengler, M. B. Clark, R. E. Stark and T. R. Beatty, *Brit-Pat.* 875,977, U.S. Notification 16.6.1959.
13a. R. von Schorlemer, H. Schmidt and W. Vielstich, unpublished.
13b. C. E. Kent and W. N. Carson, General Electric Rep. No. 66-C-233 (1966).
13c. C. E. Kent and W. N. Carson, *Proc. 20th Power Sources Conf.*, Atlantic City, 1966, p. 76.
13d. C. E. Kent, General Electric Rep. No. 67-C-284 (1967).

14.   F. Foerster, *Z. Elektrochem.*, **29**, 64 (1923).
15.   S. Yoshizawa, F. Hine, Z. Takehara and Y. Kanaya, *J. Electrochem. Soc. (Japan)* **30**, E 10 (1962).
16.   *Fuel Cell Progress*, Vol. II No. 6, 1964, Fuel Cell Corporation St. Louis, U.S.A.
17.   *Fuel Cell Progress*, Vol. II No. 4, 1963, Fuel Cell Corporation St. Louis, U.S.A.
18.   R. Jasinski, The Electrochem. Soc., Meeting Pittsburgh 1963.
19.   R. A. Wynveen, Sympos. Amer. Soc., Chicago 1961, B 49.
19a.   M. C. Potter, *Proc. Roy. Soc., London* (Sec. B) **84**, 260 (1911).
19b.   B. Cohen, *J. Bacteriol.*, **21**, 18 (1931).
19c.   F. D. Sisler, *New Scientist*, **12**, 110 (1961).
19d.   F. D. Sisler, *J. Wash. Acad. Sci.*, **52**, 181 (1962).
20.   E. A. De Zubay and E. B. Shultz Jr., *Ind. Res.* No. 14, 19 (1961).
20a.   T. Gray Young, L. Hadjipetrou and M. D. Lilly, *Biotech. and Bioeng.*, **8**, 581 (1966).
21.   R. Witkin, Battery with bacteria catalysts could use sea water as fuel, *The New York Times*, 25.10.1961.
22.   G. H. Rohrback, W. R. Scott and J. H. Canfield, *Proc. 16th Annual Power Sources Conf.*, Atlantic City, 1962, p. 18.
22a.   J. E. Taylor, N. Fatica and G. H. Rohrback (1963) Progress Report.
23.   J. Brake, W. Momyer, J. Cavallo and H. Silverman, *Proc. 17th Annual Power Sources Conf.*, Atlantic City, 1963, p. 56.
23a.   P. S. May, G. C. Blanchard and R. T. Foley, *Proc. 18th Power Sources Conf.*, Atlantic City, 1964, p. 1–3.
23b.   G. C. Blanchard and C. R. Goucher, Biochemical Fuel Cell, 11th Quarterly Progress Report, Contract Nr. DA 36-039-SC-90878 (1965).
23c.   J. M. Brake, Biochemical Fuel Cells, Final Report, Contract No. DA 36-039-SC-90866 (1965).
24.   M. Shaw, *Proc. 17th Annual Power Sources Conf.*, Atlantic City, 1963, p. 53.
25.   G. C. Blanchard, R. T. Foley and M. J. Allen, *Proc. 17th Annual Power Sources Conf.*, Atlantic City, 1963, p. 59.
26.   Carbone AG, Frankfurt-Bonames, private communication.
27.   Union Carbide Co., U.S.A., private communication.
27a.   T. Gray Young, private communication (1967).

CHAPTER VI

The Electrochemical Heat Converter*

Of the physical methods of converting heat into electrical energy, the Seebeck effect in the thermoelectric converter, and the Richardson effect in the thermionic converter, have already been exploited on a technical scale. Application of the Hall effect in streaming plasma is the subject of intensive research towards development of the magnetohydrodynamic converter. Besides other diverse physical methods the possibilities of electro-chemical heat conversion have recently come under investigation. The electrochemical processes which have been discussed in this connection can be divided into four groups:

1. *The electrochemical system is subjected to a periodic temperature variation.*
2. *The products from a galvanic cell are thermally decomposed outside the cell to regenerate the original reactants.*
3. *The aqueous electrolyte of a concentration cell is regenerated outside the cell by thermal fractionation.*
4. *The electrodes of an electrochemical system are situated in regions which differ in temperature.*

These methods are illustrated by means of specific examples, and, finally, are compared with the physical methods of heat conversion.

---

* By Dr. G. Grüneberg, Arbeitsgruppe der Fraunhofer-Gesellschaft, Aachen, W. Germany.

VI.1. INTRODUCTION

Heat, as the unorganized kinetic energy of random molecular motion, or as the energy of long-wave electromagnetic radiation, is the less specialized and less useful form of energy the smaller the average value of a single quantum. The second law of thermodynamics indicates that heat can be converted into more specialized and useful forms of energy, such as mechanical, electrical and chemical energy only with an efficiency of less than 100%, quite independently of what conversion method is used. Even Carnot efficiency is not attained in practice because of irreversible losses inseparable from the operation of real processes, and of technical imperfections in the practical methods of carrying them out.

The conventional route for the conversion of heat into electrical energy involves the intermediate stages of the kinetic energy of a flowing medium and of mechanical energy. The direct conversion avoids either both of these intervening stages or, at least, the second. The electrochemical method replaces the first stage by that of chemical energy.

The aim of modern technical developments has been to devise contrivances that have heat as the sole input and electrical energy as direct output.

VI.2 PHYSICAL METHODS

The physical methods which are at the present time being applied[1,2] are based on effects which have long been known. Their practical exploitation, however, is only now becoming feasible because of growing command of technological problems.

This is the case in the application of the Seebeck effect in the *thermoelectric generator*. The development of this depends on the interrelation of two well-known basic phenomena:

1. The contact potential difference between two different materials at the same temperature, and
2. the absolute thermoelectric power—the electrical potential difference ($\mu V \ ^\circ K^{-1}$) that is established by a temperature difference within a homogeneous material.

In thermojunctions between metals, only those electrons with energy in excess of the Fermi level contribute to the thermoelectric effect. The high concentration of electrons in metals causes their behaviour to deviate from that governed by classical Boltzmann statistics. This is called degeneracy and is in accordance with Fermi statistics. Because of degeneracy the energy distribution of metallic electrons changes only slightly over the temperature range accessible to thermoelectric converters ($< 2000^\circ C$), and, in consequence, the thermoelectric power is very small. In thermojunctions between n- and p- semiconductors, on the other hand, there is a very much lower concentration of charge-carriers. These charge-carriers behave in conformity with Boltzmann statistics, having, like gas molecules, an average thermal energy. For this reason and because of the dependence of their average kinetic energy, $\frac{3}{2}kT$, on temperature, they collectively contribute to the thermoelectric effect, and give rise to a thermoelectric power greater by orders of magnitude than that due to metallic electrons.

The *thermionic converter* is, in principle, based on an ordinary diode with a hot cathode and a cold anode, and exploits the Richardson effect. The hottest electrons surmount the potential barrier at the surface of the cathode. The useful potential difference generated by a system of this kind is determined by the difference in the work functions of cathode and anode.

The extremely small efficiency of the common diode can be brought to a technically useful level by reducing the extent of the space charge which builds up in front of the anode. This can be done (a) in the high-vacuum thermionic converter by considerably reducing the distance between the electrodes, and (b) in the plasma thermionic converter by the use of caesium vapour.

The *magnetohydrodynamic converter* (MHD generator) is based on the same fundamental physical law, known to Faraday, as the steam turbogenerator, namely that an electric charge moving through a magnetic field in a direction at right angles to the field experiences a force acting at right angles to the direction of motion and to the field. In the turbogenerator the kinetic energy of a medium in motion (steam) is converted into mechanical energy. In the MHD generator on the other hand, the medium in motion is a plasma, of which the charge, in virtue of the Hall effect, is deflected in a direction at right angles to that of the original motion and to the direction of the applied magnetic field towards a collecting electrode.

Whereas the first two methods named above are in a stage of technical development and experimental application, the MHD generator is still at the stage of intensive fundamental research. There are, in addition, some other physical principles on which the conversion of heat into electrical energy can be based.

In the *thermodielectric converter* (also called the ferroelectric generator)[3], use is made of the high temperature-dependence of the static dielectric constant of certain ferroelectric materials above the Curie temperature.

Heat converted into the kinetic energy of a stream of caesium vapour can be transformed into electrical energy in the *caesium vapour generator*[4]. A gas (caesium vapour) with an ionization potential less than the thermionic work function of a heated metal (tungsten) with which it comes into contact, loses electrons and becomes ionized. If the ionized gas (Cs^+ ions) now encounters a metal surface of work function lower than the ionization potential, the ionized gas is discharged.

VI.3   ELECTROCHEMICAL METHODS

In reviews of the subject of 'Heat Conversion', electrochemical methods are frequently neglected, although they are fundamentally equivalent to the physical processes discussed in this connexion. Several publications[5,6] give the erroneous impression that the only electrochemical cells that can be used for electro-chemical heat conversion are fuel cells with reactants that can be thermally regenerated. The electrochemical heat converters described in these publica-tions are not fuel cells as customarily defined—involving the reaction of a fuel with oxygen or other oxidant—but special galvanic cells involving cell reactions that can be thermally reversed.

VI.3.1 Electrochemical Systems Subjected to Alternating Changes of Temperature; the Sn–CrCl₃ Cell

A simple galvanic cell suitable for the conversion of heat into electrical energy was discovered at the end of the last century by Case[7]. It is the tin–chromium (III) chloride cell which was improved by Skinner[8] who replaced pure tin by tin amalgam electrodes. Its distinguishing characteristic is that the cell reaction is reversible with its preferred direction depending on the temperature. Thus:

$$Sn(Hg) + 2CrCl_3 \text{ (aq)} \underset{15°C}{\overset{95°C}{\rightleftharpoons}} Hg + SnCl_2 \text{ (aq)} + 2CrCl_2 \text{ (aq)}$$

At lower temperatures (T, ~ 15 to $20°C$) the equilibrium lies predominantly to the left; at higher temperatures ($T + \Delta T$, ~ 90 to $100°C$) predominantly to the right. If the temperature of this system, initially in equilibrium at T, is raised quite rapidly to $T + \Delta T$, it will seek to establish the new equilibrium appropriate to the higher temperature by a purely *chemical* reaction. This, however, pro-ceeds relatively slowly. If, now, the negatively charged tin is connected via an external circuit, R (Figure VI.1), with an inert electrode dipping into the same solution, the new equilibrium will be established considerably faster, with

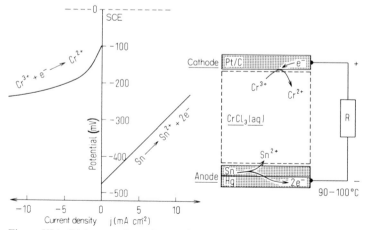

Figure VI.1. Right: reaction scheme of the Sn–CrCl₃ cell. Left: experimental potential-current curves for this cell.

provision of electrical energy. It can be seen that the inert electrode will operate as a reversible Cr(III), Cr(II) redox electrode, and will initially be positive in potential because of the high Cr^{3+} concentration which at first prevails. At the inert, positive cathode, of platinum or porous graphite, Cr(III) is electrochemically reduced to Cr(II), whilst the tin goes into solution anodically as Sn(II).

When this reaction has proceeded so far towards completion that the cell begins to fail in its electrical output, the cell may be disconnected and its temperature reduced to T. Then, in a purely *chemical* reaction

$$2Cr^{2+} + Sn^{2+} \rightarrow 2Cr^{3+} + Sn\downarrow,$$

Cr(III) and metallic tin are reformed. The tin is precipitated in finely crystalline form and dissolves in the mercury at the bottom of the cell to regenerate the amalgam. After the equilibrium appropriate to the lower temperature T has been reestablished the cell is ready for renewed operation, as before, when its temperature is restored to $T + \Delta T$.

Thus by periodical temperature change, a periodical supply of electrical energy is available, derived by the cell from the heat content of its surroundings at the higher temperature, via the chemical energy of the cell system. Since the reaction

$$Sn(Hg) + 2CrCl_3 (aq) \rightarrow Hg + SnCl_2 (aq) + 2CrCl_2 (aq)$$

proceeds endothermally, the enthalpy change accompanying the reaction, ΔH, is positive. Since, further, the Gibbs free energy change, ΔG, must be negative for extensive spontaneous reaction, the entropy change of the system, ΔS, must be positive, and additional heat must in consequence be drawn from the surroundings. The system therefore works like a steam engine. In the case of the steam engine the increase of entropy which accompanies absorption of heat is due to the transition from the liquid to the gaseous phase. In either case regeneration occurs on cooling.

More detailed examination of the electrode processes shows that, besides tin oxidation, some electrochemical reduction of Cr^{3+} ions occurs at the tin electrode. Fortunately, this process is strongly hindered, and causes a positive displacement of potential not more than 60 to 80 mV. The Sn^{2+} ions also tend to pull the potential of the cathode in a negative direction, but only to a tolerable extent because of their low concentration.

In principle, the Sn–$CrCl_3$ cell could be adapted for continuous operation, the electrolyte enriched in Sn^{2+} and Cr^{2+} being led to a regenerator at a lower temperature. Means would have to be found, of course, for continuously returning the precipitated tin to the anode of the cell.

VI.3.2   Recovery of Reactants by Thermal Decomposition of Reaction Products in an External Regenerator; the Lithium–hydrogen Cell

For the galvanic cell described in the previous section, regeneration of the reactants occurred at a temperature *lower* than the operating temperature of the cell. As an electrochemical heat converter it could be seen to be thermodynamically equivalent to a steam engine. Since, however, one of the essential reactants is tin, which is a solid, continuous operation of such a cell would be difficult to manage.

If a cell and a regenerator can be operated at a temperature at which the reaction products and one reactant are liquid, and the other reactant is gaseous, continuous operation can be conveniently achieved. In addition, to attain the highest possible Carnot factor, there should be the widest possible difference of temperature between the cell and the regenerator. The electrochemical hydrogenation of alkali and alkaline-earth metals appears to be fundamentally suitable. The hydrides decompose thermally at temperatures below 1000°C into pure metal (liquid reactant) and molecular hydrogen (gaseous oxidant).

An electrochemical heat converter based on the lithium–hydrogen system and designed for continuous operation in this way has been constructed and tested by Shearer and co-workers[9,10]. The reaction scheme and flow diagram are shown in Figure VI.2.

The reactor must work at or above the decomposition temperature, T_D, of the hydride[11], whereas the cell must have a working temperature, T_W, above, but not far from, the melting point, T_M, of the alkali or alkaline-earth metal, and, of course, above that of the electrolyte.

Table VI.1 shows the temperatures T_D and T_M and also the thermodynamic standard potentials (at 25°C) of metal/hydride electrodes calculated from standard Gibbs free energies of formation of the hydrides.

From this table it can be seen that, of the alkaline earth hydrides, BaH_2 is excluded because $T_M > T_D$: for SrH_2, although $T_M < T_D$, the temperature difference is too small, but CaH_2 is suitable, providing that there is an electrolyte, liquid at temperatures above 810°C, in which it is sufficiently

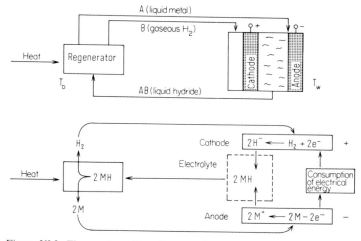

Figure VI.2. Flow and reaction diagrams for a thermally regenerable metal hydride cell.

Table VI.1. Characteristic data for alkali and alkaline earth hydrides.

Hydride	Hydride decomposition temperature, T_D (°C) $p_{H_2} = 1$ atm	Melting point of the metal, T_M (°C)	Standard potential at 25°C ϕ_{00} (V)
LiH	850	186	0.725
NaH	425	97.5	0.40
KH	432	66.3	0.23
CsH	479	28.5	0.32
CaH$_2$	985	810	0.77
SrH$_2$	840	800	0.7
BaH$_2$	730	850	0.68

soluble. Of the alkali-metal hydrides, NaH, KH and CsH have such low decomposition temperatures, T_D, that, so far, no suitable non-aqueous electrolyte has been found with a melting point low enough. For LiH, on the other hand a eutectic mixture of LiCl (80%) and LiF (20%), melting at 485°C can be used. This eutectic is prepared, according to the directions of Laitinen, Ferguson and Osteryoung[12], by melting the salts together in a stream of HCl.

The rest voltage theoretically to be expected from this heat converter, with the cell at 570°C and the regenerator at 850°C

$$2Li + H_2 \quad \underset{\text{regenerator at 850°}}{\overset{\text{cell at 570°}}{\rightleftharpoons}} \quad 2LiH,$$

is, according to the graphical presentation of the thermodynamic data in Figure VI.3, only about 250 mV. In practice, the considerably higher voltage of 720 mV is established.

This signifies that the cathode, which has not been described in detail, must come to a potential 470 mV positive with respect to the reversible hydrogen potential for the same electrolyte, since there is no doubt about the establishment of the proper Li, Li$^+$ redox potential at the liquid lithium anode. The authors, Werner, Shearer and Ciarlariello[9], remark on this, to quote:

'The difference is probably due to the solvent action of the electrolyte in lowering the partial pressure of the hydrogen in the electrolyte. This increases the differential hydrogen pressure and results in a higher voltage. This voltage would fall to a lower value when the electrolyte became saturated with the hydride.'

This explanation is open to doubt, since variation in the solubility of hydrogen could hardly lead to such a great positive displacement of the cathode potential.

At a cell voltage of 0.3 V, current densities of 250 to 450 mA cm^{-2} were attained. Chemical attack of the cathode is a problem not yet explained.

VI.3.3　Regeneration of the Electrolytes of a Concentration Cell by Thermal Fractionation in an External Regenerator; the H$_2$, H$_2$SO$_4$ Concentration Cell

Instead of thermally regenerating the reactants of a galvanic cell, heat can be used partially to separate the components of an inert, aqueous electrolyte and so maintain a high concentration difference between two reversible electrodes of the same kind. In concluding a general discussion on galvanic cells with

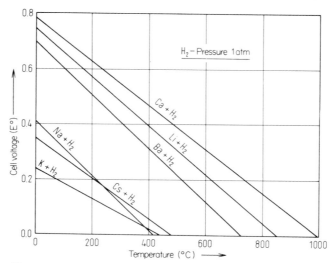

Figure VI.3. Thermodynamic cell voltages, $E°$, of alkali and alkaline earth hydride–hydrogen cells, in dependence on temperature.

reaction products susceptible to thermal decomposition, King, Ludwig and Rowlette[13] proposed a thermally regenerable sulphuric acid concentration cell, and provided the construction diagram shown in Figure VI.4.

Two porous platinum electrodes are situated in the upper part of the cell, furthest removed from the heat source. The electrolyte (aqueous H_2SO_4) flows out of the space between the electrodes (a) to the region of the cell (b) where fractionation occurs. The evaporated water dilutes the electrolyte at the anode, whilst the hot, more concentrated, H_2SO_4 returns, via the heat exchanger (c) to the cathode.

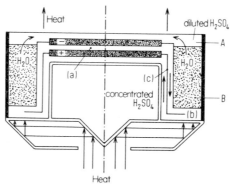

Figure VI.4. Schematic constructional sketch of the thermally regenerable sulphuric acid concentration cell (after King, Ludwig and Rowlette[13]). A, porous layer; B, fractionator

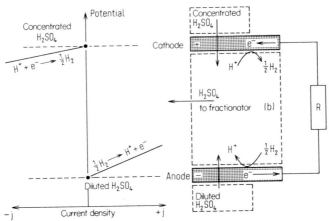

Figure VI.5. Reaction scheme of the hydrogen-operated sulphuric acid concentration cell, and potential–current density curves.

Presumably, as illustrated in Figure VI.4, the electrolytes of differing concentration must percolate through the porous electrodes. It is not clear from this sketch how the hydrogen necessary for reversible operation of the electrodes is supplied. Without it, as the essential vehicle for the charge-transfer reaction, the scheme outlined by the authors is not workable, since platinum electrodes are themselves inert in sulphuric acid. The appropriately improved reaction scheme is represented in Figure VI.5.

In fact, by using *hydrogen*, the authors[13] obtained rest voltages of 0.7 to 0.8 V, and outputs of 6 mW cm^{-2}.

Theoretically, for an acid concentration cell with transport,

$$E° = -2t^- \frac{RT}{nF} \ln(a_{\mathrm{dil}}/a_{\mathrm{conc}})$$

where t^- is the transport number of the negative ion. Since $t^- \sim 0.2$, the observed rest voltage is surprisingly high.

The working life of the cell was stated to be 18 hours. It was further reported that hydrogen brought into circulation was oxidized by the concentrated acid.

The sulphuric acid concentration cell could be made practicable by placing a diaphragm between the electrodes, as indicated in Figure VI.6, and by circulating the catholyte 'used up by dilution' through a fractionator, the water removed in the fractionator being returned to the anolyte.

For the diaphragm a porous material is required, for example porous graphite, capable of conducting away the heat of dilution and allowing sufficiently unhindered passage through it, from anolyte to catholyte, of as much diluted H_2SO_4 as water added to the anolyte from the fractionator. If the cell is supplied only with an initial charge of hydrogen, the hydrogen deposited at the cathode must be quantitatively transferred to the anode gas compartment.

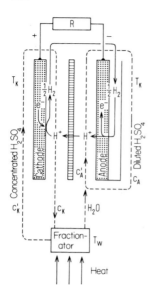

Figure VI.6. Mode of operation of a functional H_2, H_2SO_4 concentration cell.

VI.3.4　Electrochemical Systems with Electrodes at Different Temperatures; the Iodine Thermocell

The potential difference between two chemically identical, reversible half-cells (I) and (II), containing the same homogeneous electrolyte and at the same temperature, T_0, is zero:

$$\phi_0 = \phi_0(\text{I}) = \phi_0(\text{II}) = \phi_{00} + \frac{RT_0}{nF} \ln\left(\frac{a_{\text{ox}}}{a_{\text{red}}}\right)$$

If the temperature of half-cell (II) is raised to the temperature $T = T_0 + \Delta T$, the potential of this half-cell is changed to

$$\phi_0(\text{II}) = \phi_0(\text{I}) + \frac{\mathrm{d}\phi_0}{\mathrm{d}T} \cdot \Delta T$$

where

$$\frac{\mathrm{d}\phi_0}{\mathrm{d}T} = \frac{\mathrm{d}\phi_{00}}{\mathrm{d}T} + \frac{R}{nF} \cdot \ln\left(\frac{a_{\text{ox}}}{a_{\text{red}}}\right) + \frac{RT}{nF} \cdot \frac{\mathrm{d}\ln(a_{\text{ox}}/a_{\text{red}})}{\mathrm{d}T}$$

is the thermal temperature coefficient. It is positive if the potential of the warmer electrode is positive to that of the cooler electrode.

In the definition of the galvanic thermo emf,

$$E_{\text{th}} = \phi_0(\text{II}) - \phi_0(\text{I}) = \frac{\mathrm{d}\phi_0}{\mathrm{d}T} \cdot \Delta T,$$

two of the four effects which, in general, enter are to be considered: the *electrode temperature effect* (1) and the *homogeneous thermoelectric effect* (2) at (in the initial state) the junction between two electrolytes of the same concentration but at different temperatures (thermal liquid junction potential, tljp).

In general the sum of effects (1) and (2) is two to three orders of magnitude greater than the *thermoelectric effect* (3), (see section VI.2, Seebeck effect) arising at the junction in the external circuit between the identical electrode metals differing only in temperature. This can usually be neglected in relation to galvanic thermocells.

After the liquid electrolytes at a different temperature have been in contact for some time, a *thermal diffusion effect* (4) comes into operation, in the sense that the cooler electrolyte increases in concentration (Soret effect). The formation of a concentration gradient in this way can generally be avoided by stirring or convection; the effect does not occur in the case of solid electrolytes in which only one kind of ion is mobile.

A galvanic thermocell with a solid electrolyte was studied in 1957 by Weininger[14]; it is described in detail below as a practical example.

VI.3.4.1   *The Weininger Galvanic Thermocell*

The silver halides are solid electrolytes of particularly high conductance. Weininger[14] used α-AgI, which is the stable form of silver iodide over the temperature range 145 to 552°C. It is a pure cation-conductor.

The construction of the cell is indicated in Figure VI.7. A slice of α-AgI is held between two inert gas-diffusion electrodes of carbon. One side of the cell is supplied with heat, and is at 550°C; the other side is held at about 350°C. The whole system is enclosed and contains an atmosphere of I_2 vapour; there is one common gas compartment. Iodine vapour is evolved at the cooler electrode (negative terminal) and is absorbed at the hotter electrode (positive terminal).

Figure VI.8 shows Weininger's experimental cell—the layer of solid electrolyte between the two carbon electrodes can be seen.

The experimental thermoelectric power of the cell lies between 1.2 and 1.4 mV °K^{-1}; at 523°K the measured value was 1.36 mV °K^{-1}, as compared with a theoretical value of 1.45 mV °K^{-1}.

At the higher temperature electrode, iodine combines, as iodide ions, with silver ions migrating towards the electrode; electrons are used up:

$$I + e^- \rightarrow I^- ; \qquad I^- + Ag^+ \rightarrow AgI$$

At the lower temperature electrode silver ions migrate away and iodide ions are discharged, forming elementary iodine; electrons are liberated:

$$AgI \rightarrow Ag^+ + I^- ; \qquad I^- \rightarrow I + e^-.$$

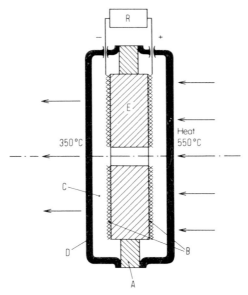

Figure VI.7. Constructional diagram of the iodine thermocell (after Weininger[14]). A, annular ceramic mounting; B, porous gas-diffusion electrodes; C, iodine vapour; D, casing; E, solid α-AgI.

In the net cell reaction, iodine is neither consumed nor generated (Figure VI.9).

This description of events does not explain what is the source of the driving force, i.e., how the potential difference between the hot cathode and the 'cold' anode is spontaneously established. M. v. Stackelberg[15] has suggested the following mechanism of operation of this cell.

For the cell to work iodine vapour must flow from the cooler to the warmer side. The equilibrium pressure of iodine vapour must therefore be smaller at the higher than at the lower temperature. It follows that the reaction

$$\tfrac{1}{2}I_2(\text{gas}) + e^-(\text{metal}) \rightarrow I^-(\text{in AgI}) \tag{VI.1}$$

must be endothermic, since at the higher temperature the activities of I_2 and of electrons are lower.

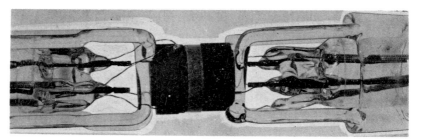

Figure VI.8. I_2 thermocell for laboratory experiments with α-AgI slice and carbon gas-diffusion electrodes.

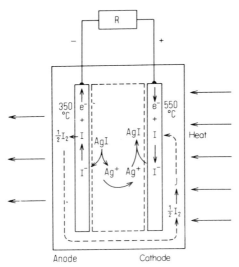

Figure VI.9. Reaction diagram for the iodine thermocell (after Weininger[14]).

Then there is an attraction, to the hot side, of (a) I_2 through the gas phase, (b) electrons through the external circuit, and (c) silver ions through the solid electrolyte.

The process

$$I(gas) + e^-(metal) \rightarrow I^-(in\ AgI) \qquad (VI.2)$$

is probably exothermic. For $\frac{1}{2}I_2(gas) \rightarrow I(gas)$, the enthalpy of dissociation is about 18 kcal per mole of I. Since $I_2(gas)$ is about 1 % dissociated at 550°C, but only about 0.01 % dissociated at 350°C, the partial pressure of I atoms in the hotter region of the gas space is about two orders of magnitude greater than in the cooler region. Because $p_{I,hot} > p_{I,cold}$, it follows from equation (VI.2) that (a) the electron activity in the hot electrode is smaller than that in the cold, and (b) the I^- activity in the hot side of the electrolyte is greater than that in the cold side. (a) gives rise to the emf of the cell, and (b) causes the flow of Ag^+ in the AgI to the hot side.

The silver ion does not migrate in response to a concentration gradient, since this can only be of insignificant magnitude, but in response to an electrical potential gradient in the AgI which arises from (b), i.e., as a result of the affinity of the reaction

$$I + e^- \rightarrow I^- \qquad (VI.3)$$

Weininger obtained a rest voltage of about 260 mV for a temperature difference of 200°C. In spite of the low electrolytic resistance of the α-AgI slice (thickness 1–2 mm), the cell would provide only a few mA cm^{-2}. This was due to the inadequacy in catalytic properties of the carbon diffusion electrodes used. Improvement of the electrodes is the first essential for the development of a technical cell.

A thermocell of the type Ag|α-AgI|Ag has provided about 100 mA cm^{-2}. Practical application of this cell, however, does not seem feasible because of rapid growth of dendrites at the surface of the cathode.

The theoretical conversion efficiency of the iodine cell is in the region of 6.5 %, and is therefore comparable with the best electronic thermocells. Application is conceivable in relation to the conversion of solar energy or of waste heat.

VI.4 A COMPARATIVE ASSESSMENT OF HEAT CONVERTERS, THEIR STATE OF DEVELOPMENT, AND TECHNICAL DIFFICULTIES

It is necessary to take various viewpoints into account in attempting to compare electrochemical heat converters with each other, as well as with those based on physical effects that have not been discussed in any detail in this chapter. The possible or desirable objectives of their application are of primary significance.

If it were desired to apply a heat converter to use the extremely hot exhaust gases of a first-stage rocket motor to supply a large quantity of electrical energy to operate control mechanisms, a working life not exceeding a few minutes would be adequate, but any proposed method would have to be assessed for its high electrical output in relation to size and weight, its insensitivity to shock and extreme temperature fluctuation, and its readiness for instant operation. On the other hand, a converter to deal with a large supply of moderate temperature heat would need to have an extraordinarily long working life with minimal attention and control; only cheap materials should be required for its construction, because its power density can only be quite small.

The most important data and characteristics of the presently known physical and electrochemical heat converters are assembled in Table VI.2; it is, of course, unavoidable that relative assessments of the methods can be based only on present knowledge of their states of development.

The *thermoelectric heat converter*, based on the combination of semiconductors, takes the predominant place at the present time. Its overriding advantage stems from the fact that it is a purely solid-state device and that the many combinations that can be made to suit alternative applications and various ranges of temperatures allow optimum working efficiency to be attained. It is therefore not surprising that this type of converter, as the most promising of success, has already been considerably further developed than its possible competitors. The disadvantages are the temporal variation of the properties of semiconducting materials, and their relatively high cost.

The *thermionic converter* is by no means so widely adaptable since it requires a high cathode temperature, and, in addition, the provision of a vacuum or an atmosphere of caesium vapour. The higher loading of 18 watt cm^{-2} attained is paid for by the short working life, limited by evaporation of the cathode material or failure of the vacuum. All its characteristics suggest that its use is practically confined to smaller, portable units, and its application may be of advantage when short periods of high load with minimal attention and control are called for.

Table VI.2. Comparison of physical and electrochemical methods of converting heat into electrical energy.

Heat converter	Temperature range (upper temperature) (°C)	Present working life	Cost	Installation	Manipulation and control	Rating (mW cm^{-2}) (as yet)	Remarks	Difficulties and irreversible processes
Physical methods								
(1) Thermoelectric, semiconductor	100/1000	Months	Variable	Portable small and large	Very simple	5 kW, with ca. 20–150 mW cm^{-2}		Evaporation of alloy additives
(2) Thermionic converter	1300	Days/weeks	High	Portable small and large	Simple	18000		Maintenance of vacuum
(3) MHD generator	2500	Minutes	Very high	Fixed large	Complex	Whole plant 1.5 mW	Cycle not yet complete	Evaporation of electrodes; Curie-point of magnets; suitable working material
Electrochemical methods	Lower/upper temperature							
(4) Sn–CrCl$_3$ cell	20/100	Year	Very small	Fixed small and large	Very simple	1	Only periodic operation possible	Not yet known
(5) Li–H$_2$ cell	570/850	Days	High	Fixed large	Difficult	100		Corrosion of cathode
(6) H$_2$–H$_2$SO$_4$ concentration cell	20/150	Hours/day	High	Portable small and large	Simple	6		Oxidation of the working gas, H$_2$
(7) I$_2$ thermocell	350/550	Weeks	Very small	Portable small and large	Very simple	1	Working efficiency about 6.5 per cent	Improvement of the properties of the iodine electrodes.

The working medium required by the *MHD generator* is a plasma which can be produced only at very high temperatures. If this device is to be used as a heat converter, the working material must be capable of traversing the temperature-controlled cycle liquid–vapour–gas–plasma–gas–vapour–liquid reversibly. As yet, only 'open' systems (as distinct from closed systems involving a cycle) for the conversion of chemical into electrical energy in this way are known. The high temperatures (of more than 2500°C) required are not easy to attain and introduce considerable technological difficulties—such as those associated with evaporation of the electrodes and the ease with which the Curie-point of the magnetic material may be exceeded. For these reasons the undoubtedly very great load densities which can be reached can, as yet, be maintained for no longer than minutes. Because of the complexities of application and control, such physical systems are likely to find application for heat conversion only in large, fixed installations. There may be, however, good prospects for application of the MHD generator as an 'open' converter of chemical into electrical energy for short-period, heavy-load operation (rockets).

Electrochemical heat converters have hitherto received but little attention, either in research or development, although they require less investment and control than, for example, the thermionic converter or the MHD generator.

Particularly modest in these respects is the *Sn–CrCl$_3$ cell*, which converts 'reserve heat' in the temperature range of 20 to 100°C. Its load density, at 1 mW cm^{-2} is certainly very small, but the cell can be made from the cheapest materials, has a very long working life, and could be made the basis of large, fixed installations. As an example consider its possible application to the utilization of waste heat: during the day the converter (as a periodically working cell) could provide electrical energy while absorbing heat (acting as an electrical cooling device), and regeneration in the resting state could take place at the lower temperature prevailing at night.

The *Li–H$_2$ cell*, with a hundred-fold greater output, must work between the temperatures of 570 and 850°C and requires quite accurate maintenance of these temperatures. The main technological problem associated with this cell is to find an insoluble cathode material; it is analogous to the corresponding problems of thermionic converters and MHD generators. The use of a regenerator external to the electrochemical system always represents a loss of electrical output, since the liquid reaction product will not flow of its own accord—it must be mechanically pumped.

In the present scheme of construction of the *H$_2$SO$_4$ concentration cell*, working over the temperature range of 20 to 150°C, the circulation of the electrolyte, regenerated by thermal fractionation, takes place by thermo-syphoning. This considerably simplifies management and control. The disadvantages of this, in principle, promising type of converter are the expensive electrode materials needed (platinum and/or palladium) and the irreversible oxidation of the working gas, hydrogen, in the closed system, with simultaneous reduction of H_2SO_4. In an improved form these cells might have a prospect of application, either in larger, fixed installations, or in smaller, mobile units.

The galvanic I_2 *thermocell* is the electrochemical counterpart of the thermo-electric converter. For the latter, use of alternative semiconducting materials has enabled the optimum working efficiency to be attained for a given temperature range. It is possible that intensive research will perform a similar service for the iodine thermocell.

REFERENCES

1. J. Kaye and J. A. Welsh, *Direct Conversion of Heat to Electricity,* J. Wiley & Sons, New York, 1960.
2. N. W. Snyder, (a) *Energy Conversion for Space Power*, Vol. 3; (b) *Space Power Systems*, Vol. 4. *Progress in Astronautics and Rocketry*, Academic Press, New York, 1961.
3. (a) *Electronics* **32**, No. 51, 88 (Dec. 1959); (b) W. Ackmann, *Ferroelektrika als Energiequellen für Satelliten*, VDI-Nachr. **16**, No. 1; 7 (1962).
4. S. Krapf, *Energie* **4**, 297 (1952).
5. W. S. Caple and E. L. Shriver, Thermally Decomposable inorganic Compounds, ACS–1, Allison Division, General Motors Corporation, Indianapolis, Ind. 1960.
6. R. E. Henderson, B. Agruss and W. S. Caple, Allison Division, General Motors Corporation, in N. W. Snyder, *Energy Conversion for Space Power*, Academic Press, New York, 1961.
7. W. E. Case, *Jb. d. Elektrochemie* **2**, 56 (1896).
8. S. Skinner, *Proc. Physic. Soc. London* **13**, 477 (1895); *Philos. Mag.* **39**, 444 (1895).
9. R. C. Werner, R. E. Shearer and P. A. Ciarlariello, *Proc. 13th Annual Power Sources Conf.*, Atlantic City, New York, 1959, 122, Resumé of Thermally Regenerative Fuel Cells.
10. R. E. Shearer, J. W. Mausteller, P. A. Ciarlariello and R. C. Werner, Regenerative Metal Hydride System, Proc. 14th Annual Power Sources Conference, 1960.
11. D. T. Hurd, *An Introduction to the Chemistry of the Hydrides,* J. Wiley & Sons, Inc., New York, 1952.
12. H. A. Laitinen, W. S. Ferguson and R. A. Osteryoung, *J. Electrochem. Soc.* **104**, 516 (1957).
13. J. King, F. A. Ludwig and J. J. Rowlette, General Evaluation of Chemicals for Regenerative Fuel Cells, in N. W. Snyder, *Energy Conversion for Space Power*, Vol. **3**, 387, Academic Press, New York, 1961; F. A. Ludwig and J. J. Rowlette, *U.S. Pat.* 3,231,426 (1966).
14. J. L. Weininger, Thermocell, *U.S. Pat.*, 2,890,259 (1959); *J. Electrochem. Soc.* **111**, 769 (1964).
15. M. v. Stackelberg, private communication.

Chapter VII

Electrochemical Conversion of Nuclear Energy*

Most of the reviews on the conversion of nuclear into electrical energy, avoiding of the intermediate stages of mechanical energy and heat, are confined to physical methods.

The available information on electrochemical methods of effecting this conversion is discussed in this chapter. These methods are based on the intermediate transformation of nuclear into chemical energy by the radiolysis of water or of oxygen, and utilization of the radiolysis products as reactants in specially designed galvanic cells. The most direct and obvious application is the use, in a fuel cell, of the hydrogen and oxygen which are formed radiolytically in water-moderated nuclear reactors.

The results of radiochemical investigations allow the theoretical conversion efficiencies of various types of electrochemical converters to be estimated.

The 15 to 25% efficiency of the oxygen–ozone cell with radiolytic regeneration offers particularly good technical prospects, and the radiolytically regenerated hydrogen–iron (III) cell could be superior to the existing physical conversion methods.

An outline is given of the possibilities of developing an electrochemical, homogeneous ^{235}U reactor.

VII.1 INTRODUCTION

Nuclear reactors, which provide electrical energy on the largest technical scale, derive their energy chiefly by the controlled fission of the uranium isotope ^{235}U by thermal neutrons. The primary energy carriers (heavy, charged nuclei,

* By Dr Gerhard Grüneberg, Arbeitsgruppe der Fraunhofer-Gesellschaft, Aachen, W. Germany.

neutrons, γ-quanta) coming directly from the fission process are very diverse in nature and in the quantities of energy they carry. Interaction with a suitable material medium levels out these heterogeneous energies, degrading them to the technically utilizable form of heat. According to the type of reactor concerned, however, one or other technically utilizable form of chemical energy is also produced. In the reactors at present in use, however, this chemical energy (e.g. in the hydrogen and oxygen produced by radiolysis of moderator water) is undesired, and is degraded to heat by suitable means. The very large output of electrical energy from nuclear fission is obtained in the conventional way by the thermomechanical route involving the steam turbine.

The mechanical stage in these energy transformations is undesirable in many applications of nuclear energy. It involves noise, vibration and the maintenance of complex mechanical units. It is the present concern of science and technology to devise and perfect unconventional methods of direct conversion to eliminate the mechanical energy stage.

For the conversion of heat as a form of energy there are physical methods— thermoelectric with use of semiconductors, thermionic and magnetohydro- dynamic—and also electrochemical devices, such as galvanic cells (chapter VI).

In the case of nuclear energy conversion, however, it is senseless to proceed via the most degraded form of energy, heat, with the Carnot limitation on efficiency which this imposes. It is of fundamental advantage to apply the energy liberated in fission processes, without dispersion and levelling to the form of heat, in the direct production of electrical energy.

Physical methods, to be briefly reviewed in section VII.3, allow only a very small fraction of the fission energy (201 meV per nucleus) to be directly converted, namely, the energy stored, after fission, in longer-lived radioactive substances. Such substances may be radioactive fission products, or artificial, radioactive elements produced by neutron-capture reactions.

Electrochemical methods for the direct conversion of nuclear energy have been investigated, as yet, only sparsely. The treatment of the subject which follows seeks to classify all the information at present available, and to survey the existing stage of development of electrochemical nuclear energy conversion, together with a number of associated problems.

VII.2   TYPES OF NUCLEAR RADIATION AND APPLICABLE RADIO- ACTIVE ISOTOPES

The 201 MeV liberated per act of ^{235}U fission in thermal reactors is distributed as follows:
kinetic energy;

(a) fission products		83.5%
(b) secondary neutrons		2.5%
(c) fission product β-radiation		3.5%

quantized energy;

(d) primary γ-radiation	2.5%
(e) fission product γ-radiation	3.0% and
the unavailable neutrino energy from β-decay	5%

This inhomogeneous radiation mixture can be regarded as 'raw' nuclear energy, the use of which in direct conversion involves rather large technical difficulties. The situation is not dissimilar to that of using coal in galvanic fuel cells, or of crude oil for internal combustion engines. By physical methods the 'raw' nuclear energy as a whole can be converted into electrical energy via heat; by electrochemical methods, only via chemical energy (Figure VII.1).

Homogeneous nuclear energy, either selectively derived from the inhomogeneous energy of nuclear fission, or by artificial means, is more easily dealt with. An analogous situation is again found in the use of pure H_2 and O_2 in a fuel cell, or of petrol in an engine.

Nuclear energy in a 'refined' form is stored in those radioactive fission products which have technically serviceable half-lives, τ (e.g. from a month to 25 years), and a sufficiently high specific activity. Such products are listed in Table VII.1. The entries in the first column, in %, relate to the mass of uranium from which they are derived.

When it is considered that

1. Only 3.5% of the energy of a thermal reactor is stored in such radioactive fission products,

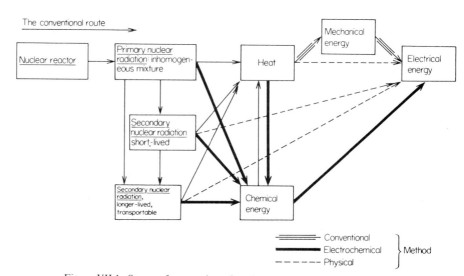

Figure VII.1. Stages of conversion of nuclear energy into electrical energy.

Table VII.1. Technically serviceable, separable fission products.

%	ε(MeV)	τ
1.3 ^{85}Kr $\rightarrow$ ^{85}Rb $+ \beta^-$	0.67	10.6 yr
5.6 ^{90}Sr $\rightarrow$ ^{90}Y $+ \beta^-$	0.54	25 yr
5.8 ^{91}Y $\rightarrow$ ^{91}Zr $+ \beta^-$	2.24	65 hr
5.7 ^{144}Ce $\rightarrow$ ^{144}Pr $+ \beta^- \rightarrow$ ^{144}Nd $+ \beta^-$		285 days
2.4 ^{147}Pm $\rightarrow$ ^{147}Sm $+ \beta^-$	0.22 (100%)	2.6 yr

2. the usable fraction of these amounts only to a few % and
3. the extraction of individual fission products itself demands a great deal of energy,

it becomes understandable that this form of nuclear energy is very expensive, and that its application will be practicable only in very special circumstances.

If thermal neutrons from a reactor are used to make, by (n, γ) processes, certain α-, β- or γ-active isotopes, this is equivalent to storing 'artificial' nuclear energy, of the greater potential value the more homogeneous the radiation of the artificial isotope concerned. The high costs are, in this case, bound up with the separation of the isotope initially required. Table VII.2 lists such (n, γ) processes and the radioisotopes they provide, suitable to be considered for the purpose of direct energy conversion.

Isotopes of particular scientific and technical interest, because of their homogeneous radiation and relative ease of separation are γ-active ^{60}Co, β-active ^{147}Pm, and α-active ^{210}Po.

Table VII.2. Technically serviceable artificial radio-isotopes produced by (n, γ) reactions.

Kind of radiation	ε(MeV)	τ
^{34}S (n, γ) ^{35}S $\rightarrow$ ^{35}Cl $+ \beta^-$	0.167 (100%)	87.6 days
^{44}Ca (n, γ) ^{45}Ca $\rightarrow$ ^{45}Sc $+ \beta^-$	0.25 (100%)	164 days
^{59}Co (n, γ) ^{60}Co $\rightarrow$ ^{60}Ni $\begin{cases} + \gamma \\ + \beta^- \end{cases}$	1.33, 1.17 0.308	5.24 yr
^{112}Cd (n, γ) ^{113}Cd $\rightarrow$ ^{113}In $+ \beta^-$	0.58 (100%)	5.1 yr
^{184}W (n, γ) ^{185}W $\rightarrow$ ^{185}Re $+ \beta^-$	0.43 (100%)	74 days
^{203}Tl (n, γ) ^{204}Tl $\rightarrow$ ^{204}Pb $+ \beta^-$	0.76	3 yr
^{209}Bi (n, γ) ^{210}Bi $\quad\quad \beta^- \downarrow 5$ days $\quad\quad\quad ^{210}$Po $\rightarrow$ ^{206}Pb $\begin{cases} + \alpha \\ + \gamma \end{cases}$	5.3 (100%) 0.8 (0.0012%)	138 days

VII.3   PHYSICAL METHODS FOR CONVERSION OF NUCLEAR RADIATION

At the present time, as far as published information goes, five physical methods are available for the direct conversion of homogeneous nuclear radiation into electrical energy. Each of them serves for only one kind of radiation because of the vastly different characteristics of α-, β- and γ-rays, and of neutrons. In most cases there is an additional requirement that the single kind of radiation concerned shall be confined to a sufficiently narrow range of primary energies.

VII.3.1   The High-voltage Nuclear Battery (Direct Charging Converter)

An electrode, carrying on its surface an isotope emitting α- or β-particles, assumes a negative or positive charge, respectively, relative to a collecting electrode, from which it is insulated by means of a vacuum or a suitable dielectric (Figure VII.2). The high potential difference that is built up is limited either by the efficiency of insulation or by the level of the primary kinetic of the monoenergetic, particulate radiation. In a vacuum system of this kind 250 mc of β-active ^{90}Sr, for example, has yielded 0.2 mW output at 365 000 volt. This represents a conversion efficiency of about 15 to 20%. Use of a solid, synthetic dielectric reduced the efficiency to a few % and the terminal voltage to a few thousand volts.

VII.3.2   The Neutron Converter

Certain isotopes (e.g. β-active ^{104}Rh) have the property of giving rise to α- or β-activity of short half-life under slow neutron irradiation. These, carried by

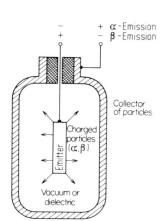

Figure VII.2. Diagram of a high-voltage nuclear battery.

an electrode of the high-voltage nuclear battery mentioned above, provide the basis of direct conversion of the energy of neutrons into electrical energy.

VII.3.3   Gas Ionization Nuclear Battery (Contact Potential Converter)

There is a contact potential difference between two different, spatially separated metals, equal to the difference between the work functions of the metals. If such a pair of electrodes is situated in an atmosphere of a heavy noble gas (e.g. argon) under pressure, and if the gas is ionized by means of α- or β-radiation, the positive ions are discharged by electron-uptake at the electrode of higher work function (the cathode, e.g. steel with a PbO-treated surface). The negative ions are discharged by electron-release at the electrode of lower work function (the anode, e.g., Mg). In a cell of this kind a voltage of about 1 volt has been realized with a current output of 2×10^{-9} A, the radioactive material (e.g. ^{90}Sr or tritium as β-emitter) introduced into the cell having an activity of 10 mc. The relevant efficiency amounts to 1 to 1.5%.

VII.3.4   The Semiconductor Nuclear Battery (Semiconductor Junction Battery)

If nuclear radiation—in practice, less energetic β-rays alone if structural damage is to be avoided—is allowed to act on semiconducting material, the kinetic energy of the β-particles is mainly used in the promotion of electrons from the lower energy levels of atomic cores and covalent bonds, to the higher energy levels corresponding to free electrons. These free electrons, and the positive holes left at the sites of ionization, can function as charge-carriers. The charge-carriers produced in this way at or near a p–n contact are separated by the internal field of the contact, so that the n- and p- semiconductors assume negative and positive charges, respectively. If an external, metallic circuit is completed, electrons flow through it from the n- to the p- semiconductor. With a source of β-radiation of ca 50 mc, a potential difference of several tenths of a volt has been observed, with currents of a few μA*.

VII.3.5   The Photoelectric Converter (Scintillation Converter)

If a material which emits light on excitation (e.g. a phosphor) is placed between conventional photocells and is irradiated with β- or γ-rays, nuclear energy can be converted to electrical energy through light as an intermediate stage.

* This kind of nuclear battery should not be confused with a semiconductor thermoelectric converter, the operation of which is due simply to the heat arising from degradation of the kinetic energy of the nuclear radiation. Such a device should be classified with heat converters (chapter VI).

VII.4   THE RADIOLYSIS OF AQUEOUS SOLUTIONS

In the interests of the intelligibility of the sections which follow, it seemed desirable to extract from the wealth of literature on the radiolysis of water such principles and experimental results as are fundamental to the electrochemical methods of converting nuclear energy.

VII.4.1   The Process of Energy Transfer

Fundamental chemical transformation of radiolysed matter—in this case, water—occurs only in the direct neighbourhood of the paths of charged particles. The following are radiochemically active:

(a) *Hard photons*, such as γ- or X-ray quanta, only after the transfer of their energy, in one or a few processes, to photoelectrons, Compton electrons or ion-pair forming electrons;

(b) *fast neutrons*, by transfer of their kinetic energy in elastic collision with nuclei, with loss of extranuclear electrons by neutron impact (recoil nuclei);

(c) *slow neutrons*, in so far as they give rise to short-lived α- or β-radiators, or nuclear fission products.

The kinetic energy of *charged* particles, in contrast to that of photons, suffers progressive dissipation by numerous elastic and inelastic collisions with atoms or molecules in, or close to, the paths they traverse. This does not occur in such a way that a given molecule, in collisional interaction with a large packet of energy, acquires directly only the energy required for the breaking of a chemical bond. On the contrary, lower-lying extranuclear electrons are ejected from the constituent atoms, which have energies far greater than chemical bonding energies, and themselves initiate further radiochemical reactions.

The paths of the *secondary* electrons so produced (also called δ-rays) branch off from that of the primary particle, and their energies attain several thousands of electron volts, the maximum energy being strongly dependent on the mass of the primary particle producing them.

The stopping power of a material (here only water is being considered) also increases with the mass of the charged particle; this can be seen from Figure VII.3, in which a comparison can be made of different particles of equal energy[1]. The heavier a particle, and the higher its charge, the more energy it dissipates per unit length of path, but the shorter its range.

This can be illustrated by the following approximate ranges in water of various particles, all of 2 MeV initial energy: electrons, 9000 μm; protons, 74 μm; deuterons, 48 μm; helium nuclei, 12 μm.

Since the number of radiochemical primary reactions (ionizations and excitations in about equal proportion) which occur along these distances is approximately the same in each case, it follows that the so called specific ionization is different from case to case. The more kinetic energy a particle loses to atoms and molecules along its path, the slower it becomes. At the same time, however, the chances of an encounter increase. According to Magee[2], the time of flight of a primary electron of

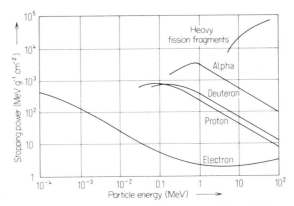

Figure VII.3. Stopping power of water for different kinds of
fast charged particles as a function of particle energy.

1 MeV energy through a molecule is 10^{-18} sec; that of a 0.025 eV thermal electron is about 10^{-15} sec. Consequently, the specific ionization power of the particle increases with decreasing kinetic energy; this can also be seen from Figure VII.3.

Feng[3] has pointed out that, with electrons of energies within the range of radiochemical interest (keV to MeV) about half of the energy is ultimately dissipated by secondary electrons of average energy less than about 60 eV.

VII.4.2   Stages of Radiolysis

The overall process of radiolysis begins with the exposure of a system to energetic radiation, and ends with the reestablishment of thermodynamic equilibrium. It can be divided into three characteristic stages[4], as follows:

In the *physical step* the large energy packet associated with the primary charged particle is dissipated in the target material in less than 10^{-13} sec. The first collisions eject secondary electrons from the molecules involved with accompanying generation of positive ions. When the secondary electrons are sufficiently retarded by further encounters and ionizations they are captured by molecules with formation of negative ions. In the case of water H_2O^+ ions are formed along a cylindrical core following the particle track and H_2O^- ions in an annular region closely surrounding it.

In the *physicochemical* step these unstable ions are transformed within about 10^{-11} sec, before any other chemical reaction can take place, into ions H_3O^+ and OH^-, appropriate to the medium, and into the free radicals OH and H:

$$H_2O \xrightarrow{10^{-11}\ sec}$$
$$H_2O^+ + aq \rightarrow H^+ . aq + OH \qquad\qquad (VII.1)$$
$$+$$
$$e^- + H_2O \rightarrow OH^- + H$$

Table VII.3. Average ion densities (ions μm^{-1}) of different radiations[6].

Light particle radiation:	
Smallest theoretical ion density	6.3
γ- or β-rays, 20 to 30 MeV	8.5
γ-rays from radium	11
γ-rays, 30 to 180 MeV	100
Heavy particle radiation:	
Radiation from ^{210}Po, α, 5.3 MeV	4500
Uranium fission products (ca. 85 MeV)	130 000

The cylindrical core along the particle track therefore consists essentially of OH radicals at a volume concentration considerably higher than that of the hydrogen atoms around it. The number of free radicals is approximately proportional to the linear ion-density of the radiation, and depends on the energy of the radiation used, cf. Table VII.3.

The linear ion-density provides an important experimental means of varying the local concentration of free radicals. An approximate calculation shows that the OH concentration in the core of an α-particle track (its radius amounts to about 20 Å) can be for a short time ($< 10^{-4}$ sec) as much as 1 molar. For high-energy electrons or γ-rays, however, the local concentration is very small, and at medium intensities can be considered as uniform over the whole irradiated volume.

In addition to ionized molecules it is assumed that there are also *excited* molecules[8,9] along the particle track, capable of direct, non-ionogenic dissociation:

$$\left. \begin{array}{ll} H_2O & H_2O \rightsquigarrow H_2O^* \\ {\Large\updownarrow} & \\ H_2O^* \rightarrow H + OH & H_2O \rightsquigarrow H_2O^* \end{array} \right\} \rightarrow H_2 + H_2O_2 \qquad (VII.2)$$

The *chemical step* requires first the diffusion of the OH and H radicals away from the particle track, and then their chemical reaction with reactive constituents in the aqueous solution. In the case of heavy-particle irradiation, however, the specific ion-density along the track is so large that there is some reaction between the radicals

$$H + H \rightarrow H_2 \text{ or } OH + OH \rightarrow H_2O_2 \qquad (VII.3)$$

or, with liberation of heat, recombination to form water molecules,

$$H + OH \rightarrow H_2O$$

Then, 10^{-8} sec after the passage of energetic particles, the constituents that may be considered as the primary radiolysis products of water are the radicals H and OH, and the molecules H_2 and H_2O_2.

VII.4.3   The Radiation Yield

There is no stoichiometric equivalence between the radicals on the one hand, or between the molecules on the other hand, produced by radiolysis, i.e., between $g(H)$ and $g(OH)$ or between $g(H_2)$ and $g(H_2O_2)$, although the following stoichiometric equation must be satisfied:

$$g(-H_2O) = 2g(H_2) + g(H) = 2g(H_2O_2) + g(OH) \qquad (VII.4)$$

where $g(H_2O)$ is the number of radiolysed water molecules per $100\,eV$ of radiation energy absorbed.

The magnitude of the g-value depends on the specific ionization-density of the radiation used and also somewhat on the nature and pH of the aqueous solution. The data in Table VII.4 apply to air-free, $0.8N\ H_2SO_4$[10]. It can be seen that the molecular radiolysis products increase, whereas the radicals correspondingly decrease, with increasing specific ionization density.

Table VII.4. g-Values for water radiolysis with various radiations (O_2-free, aqueous solutions).

Kind of radiation	$g(H)$	$g(OH)$	$g(H_2)$	$g(H_2O_2)$	$g(-H_2O)$
$^{60}Co - \gamma$	3.65	2.95	0.45	0.80	4.55
$18\ MeV - D^+$	2.39	1.75	0.71	1.03	3.81
$8\ MeV - D^+$	1.71	1.45	1.05	1.17	3.80
$32\ MeV - He^{++}$	1.28	1.06	1.14	1.25	3.56
$^{10}B(n, \alpha)^7Li$	0.23	0.41	1.66	1.57	3.55

The primary radiolysis products take part in secondary reactions such as

$$
\begin{aligned}
OH + H_2 &\rightarrow H_2O + H \\
H + H_2O_2 &\rightarrow H_2O + OH \\
H_2O_2 + OH &\rightarrow H_2O + HO_2 \\
HO_2 + HO_2 &\rightarrow H_2O + O_2
\end{aligned}
\qquad (VII.5)
$$

Such reactions lead finally to the disappearance of all radicals, only the stable radiolysis products being left. Particularly in the case of alkaline aqueous solutions, these are invariably molecular hydrogen and oxygen in $2:1$ stoichiometric ratio, since any H_2O_2 formed undergoes base-catalysed decomposition.

VII.5   PRINCIPLES AND MODE OF ACTION OF THE ELECTRO-CHEMICAL NUCLEAR ENERGY CONVERTER

Electrochemical conversion processes differ from the physical processes that also operate via an intermediate—although very short-lived—chemical stage (e.g., the gas-ionization nuclear battery) in the following characteristics. They do not utilize the ion mixture produced by the action of nuclear radiation on a gas, but, instead, either

(a) the 'ordered' chemical energy in spatially-separated chemical reactants, or
(b) the 'unorganized' chemical energy in the form of a mixture of reducible and oxidizable radicals or molecules produced by the radiolysis of suitable liquids or gases.

In the appropriately designed galvanic cell the original materials subjected to radiolysis are regenerated with liberation of electrical energy.

The optimum *efficiency*, η_{NE}, of an electrochemical nuclear energy converter is the product of two efficiencies:

(a) η_{NC}, for the transformation of nuclear energy into the chemical energy of such radiolysis products as the galvanic cell is specially designed to utilize, and
(b) η_{CE}, for the transformation of the chemical energy of these radiolysis products into electrical energy.

η_{NC} is the g-value for the radiolysis product electrochemically active in the galvanic cell, multiplied by an appropriate factor in eV, representing the maximum theoretical conversion.

η_{CE} is the ratio of the terminal voltage to the theoretical emf. The fraction $1 - \eta_{NE}$ is lost as heat.

In the case of *direct coupling* of a chemical nuclear reactor with a galvanic cell, the g-values, expressing the purely radiochemical results, cannot be used to predict the optimum efficiency to be expected without consideration of other factors likely to be significant. There is always the possibility that the radiolysis products, in relation to rates of formation and secondary reactions, may be considerably modified in behaviour by the presence of the electrodes in the electrochemical system—particularly in technical assemblies of compact construction. The electrode materials may act as sources of radiolytically active secondary radiations, or they may consume or transform reactive radiolysis products, depending on their specific properties, and also on whether they are operating as anodes or cathodes.

VII.5.1   Utilization of the Molecular Products of the Radiolysis of Water

As already mentioned, heavy-particle irradiation, because of the high ionization density, produces a predominance of the molecular products H_2 and H_2O_2 in aqueous solutions. This is apparent from the g-values of Table VII.4. The hydrogen escapes from the aqueous solution whilst the H_2O_2 becomes enriched to a limit set by its decomposition to water and oxygen under the influence of catalysts or radiation. Oxyhydrogen gas always collects in homogeneous reactors containing uranyl sulphate solutions or suspensions of uranium dioxide, and also over the water in which atomic waste (burnt-out nuclear fuel, radioactive residues) is kept. After establishment of a steady state, i.e. when the H_2O_2 concentration in the aqueous solution remains constant, the gas evolved is a stoichiometric mixture, $H_2 + \frac{1}{2}O_2$.

According to Cap[10a], a homogeneous reactor of 100 kW (thermal) capacity (30 kW electrical) produces 28–50 l min^{-1} of oxyhydrogen gas; used in a fuel cell, this would give an additional 3 kW, increasing electrical output by 10%. At present this gas is regarded as an unavoidable and undesirable by-product, and its formation is minimized. If, on the contrary, reactors were designed to produce as much of it as possible, electrical output could be further increased.

The following two methods are available for the electrochemical use of the molecular products of radiolysis.

VII.5.1.1   The Radiolytically Regenerated Hydrogen–oxygen Fuel Cell

Rosenblum and English[11], from the NASA Lewis Research Centre in Cleveland, Ohio, constructed a closed electrochemical system, presumably corresponding with that shown diagrammatically in Figure VII.4, which can be considered as a radiolytically regenerated hydrogen–oxygen fuel cell. It consists of a nuclear reactor, and a separate, conventional fuel cell, supplied separately with hydrogen and oxygen. The hydrogen comes directly from the chemical reactor, in which an aqueous solution, probably dilute sulphuric acid, is radiolysed by α-rays from

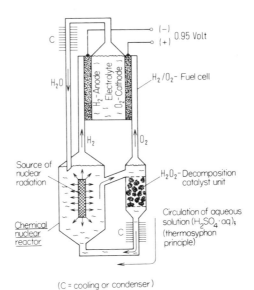

(C = cooling or condenser)

Figure VII.4. Constructional diagram of a radio-lytically regenerated hydrogen–oxygen fuel cell.

^{210}Po. The aqueous solution in which H_2O_2 is generated, but is not allowed to attain the limiting concentration, circulates through a vessel charged with catalyst for H_2O_2 decomposition. The oxygen so formed passes to the cathode gas compartment of the fuel cell. When current is drawn from the cell water is regenerated from its elements and is returned as condensate to the reactor. The efficiency attained (output of electrical energy/radiation energy absorbed) amounted to less than 1 %.

VII.5.1.2   *The Radiolytically Regenerated Oxyhydrogen Cell*

Circulation of the radiolysed aqueous solution can be obviated by using in the nuclear reactor either an alkaline solution, or one containing, in suspension, a catalyst for the decomposition of hydrogen peroxide. The catalyst must be inert to hydrogen. The only molecular products of radiolysis in the stationary state are then hydrogen and oxygen, mixed in the stoichiometric ratio, $H_2 + \frac{1}{2}O_2$, i.e. oxyhydrogen gas.

The problem of how to make direct electrochemical use of electrolytic gas in a galvanic cell had already been considered by Justi[12] in 1956. It was solved by Grüneberg, Wicke and Justi[13] in 1956, using the cell construction shown in Figure VII.5.

The gas mixture, $H_2 + \frac{1}{2}O_2$, flows into an elongated gas compartment of an oxygen-diffusion electrode of special design, made only of carrier and catalyst materials active to oxygen, but inert to hydrogen, i.e. of active carbon, silver or gold. The cathodic current passed by the oxygen electrode is continuously controlled, according to the rate of flow of the gas mixture into the cell, so that the gas issuing from the end of the long gas compartment of the electrode

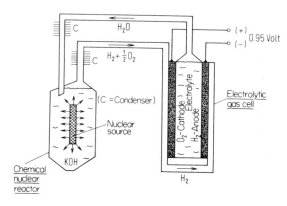

Figure VII.5. Constructional diagram of a radiolytically regenerated oxyhydrogen gas cell.

contains as little residual oxygen as possible—below the inflammability limit of 5% of O_2 in H_2. The gas then passes to the hydrogen-oxidation electrode of the cell.

Goebel, Struck and Vielstich[13a] have recently operated a normal hydrogen–oxygen cell with oxyhydrogen gas (91% H_2, 9% O_2) at room temperature using hydrophobic sintered-metal electrodes of area 8 cm², 3–4 mm apart and free electrolyte. The hydrogen–oxygen mixture was supplied at the same composition to each electrode chamber at an excess pressure of 0.08–0.1 atm. The oxygen electrode was of silvered, sintered nickel, and the hydrogen anode was catalysed by 4 mg cm⁻² of Pd, Pt (4:1). Under load the cell attained a steady terminal voltage within minutes, which remained constant for the several hours duration of the experiment. Electrode potentials as functions of current density and gas composition are shown in Figure VII.5a. An increase in operating temperature considerably improves the electrical characteristics.

If the oxygen content of the gas mixture is very small, it is advantageous to use air for the oxygen electrode.

VII.5.1.3   *Estimation of Optimum Efficiencies*

Comparison of the g-values given in Table VII.4 for the molecular products of the radiolysis of water indicates that, for α-irradiation by ^{210}Po, appropriate values are $g(H_2) = g(H_2O_2) = 1.1$. Under these circumstances, for the conversion of nuclear energy into usable chemical energy,

$$\eta_{NC} = \frac{1.1 \times 2 \times 1.23^*}{100} = 2.7\%$$

Assuming a terminal voltage of 0.95 V for the hydrogen–oxygen cell, the efficiency of conversion of chemical into electrical energy is

$$\eta_{CE} = \frac{2 \times 0.95}{2 \times 1.23} = 77\%$$

Hence the overall efficiency,

$$\eta_{NE} = \eta_{NC} \times \eta_{CE} \sim 2.1\%$$

According to Table VII.4 an electrochemical nuclear energy converter that utilizes only the molecular products of the radiolysis of water can hardly exceed an efficiency $\eta_{NE} = 4\%$.

* See Table II.3, p. 28.

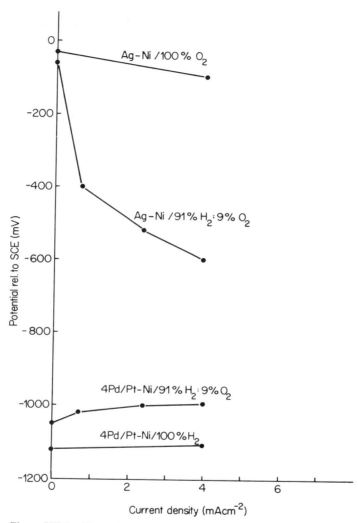

Figure VII.5a. Electrode potentials versus current density for noble-metal activated sintered nickel electrodes supplied with hydrogen, oxygen and hydrogen–oxygen mixture (91:9); electrolyte 6N KOH; total gas pressure 1.1 atm (after Goebel, Struck and Vielstich[13a]).

VII.5.2 Utilization of the Radical and Molecular Products of Water Radiolysis

The devices described in the preceding section made no use of the radicals generated by radiolysis. They could, on the contrary, undergo undesirable secondary reactions (cf. equations VII.5) with the molecular products, and in any practical application this could reduce the effective g-values for H_2 and H_2O_2 below those to be expected from radiochemical investigations.

Because radicals are short-lived, special precautions are necessary in attempting to make electrochemical use of them. Two suitable procedures are available:

(a) Either the galvanic cell electrodes are brought directly into the solution undergoing radiolysis, and efforts are made to involve as many radicals as possible in electrochemical reaction by adjustment of electrode design or other working conditions, or

(b) an electrochemically usable redox system is added to the solution to act as a radical trap, for example, Ce^{4+} as a trap for hydrogen atoms:

$$Ce^{4+} + H \rightarrow Ce^{3+} + H^+$$

or Fe^{2+} as a trap for OH radicals:

$$Fe^{2+} + OH + H^+ \rightarrow Fe^{3+} + H_2O$$

VII.5.2.1   *Electrodes Selective for Conversion of H and OH Radicals*

Investigations of the behaviour of catalytically selective pairs of electrodes in oxygen-free, aqueous electrolytes undergoing γ-irradiation, were first carried out by Zalkind, Veselovsky and Gochaliev[14]. They observed a rest voltage of 0.95 V between platinum and gold electrodes. Under the influence of H atoms and H_2 molecules the platinum assumed the potential of the reversible hydrogen electrode, whereas the gold came to a mixed potential mainly determined by OH radicals and H_2O_2.

It can be assumed from this that the radical constituents were active; on the one hand they are considerably more reactive than the molecular constituents; on the other hand their g-values are higher for radiation of lower ionization density (cf. Table VII.4).

Zalkind and Veselovsky[15] have studied the dependence of the anodic current carried by a platinum electrode, at a constant potential of $+ 400\,mV$ with respect to the reversible hydrogen electrode, on the duration of irradiation (Figure VII.6).

In this experiment a cylindrical platinum electrode ($10\,cm^2$) was arranged coaxially around a cylindrical ^{60}Co γ-source (80 Curie). Oxygen-free $0.8N\,H_2SO_4$ was used as electrolyte. The average energy absorbed was $2.5 \times 10^{15}\,eV$ $cm^{-3}\,sec^{-1}$. The constant current of about $13\,\mu A$ showed that the platinum had the greater activity in relation to the reducing constituents produced by radiolysis.

At overpotentials of more than $50\,mV$, diffusion of the reducing constituents to the electrode was the rate-limiting process; this can be seen from the current-potential curve shown in Figure VII.7. To detect the influence of OH radicals, and to minimize the back-reaction $H + OH \rightarrow H_2O$ in the solution and at the electrode surface, Zalkind and Veselovsky added oxalic acid to the solution.

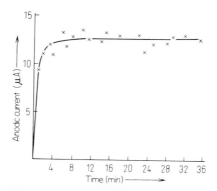

Figure VII.6. Dependence of the anodic current at a γ-irradiated platinum electrode in H_2SO_4 on the duration of irradiation (after Zalkind and Veselovsky[15]).

This reacts with hydroxyl radicals:

$$H_2C_2O_4 + 2OH \rightarrow 2CO_2 + 2H_2O$$

Under these conditions there was a five-fold increase of anodic current.

According to available information[16,17], it is intended in the U.S.A. to study application of the principle proposed by Veselovsky[14] in more detail, using a homogeneous reactor. Platinum and gold, in acid uranyl sulphate solution, show a rest voltage between 0.7 and 0.9 V.

These 'radiogalvanic cells' naturally introduce a series of new problems. The physical form and materials of the electrodes used must not appreciably

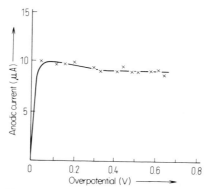

Figure VII.7. Anodic limiting current of the γ-irradiated platinum electrode (after Zalkind and Veselovsky[15]).

hinder the nuclear fission process. Too rapid damage to the catalyst carrier—for which only graphite can be considered—by neutron absorption must be avoided. The catalysts must not rapidly undergo transmutation by neutron capture.

VII.5.2.2   *Addition of Radical Traps; the Radiolytically Regenerated H_2–Fe^{3+} Cell*

Ferrous ions in oxy-acid solutions of pH < 3 react not only with OH radicals and hydrogen peroxide, but also with hydrogen molecule-ions formed by $H + H^+ \rightarrow H_2^+$:

$$Fe^{2+} + H_2^+ \rightarrow Fe^{3+} + H_2$$

According to Hochanadel and Lind[18], for pure H_2SO_4 solution with added Fe^{2+} ions, yields per 100 eV of ^{60}Co γ-irradiation are:

$$4.55H_2O \rightsquigarrow 3.65H + 2.95OH + 0.45H_2 + 0.80H_2O_2$$

$$
\begin{array}{lccccc}
 & \downarrow & \downarrow & & \downarrow & \\
Fe^{3+}: & 3.65 & + 2.95 & \Big\downarrow & 1.60 & = 8.20 \\
H_2: & 3.65 & & 0.45 & & = 4.10
\end{array}
$$

Thus, in the overall radiochemical reaction $2Fe^{2+} + 2H^+ \rightarrow 2Fe^{3+} + H_2$, 8.20 Fe^{3+} ions and 4.10 H_2 molecules are formed as the result of absorption of 100 eV of radiant energy by the electrolyte. The initially 'disorganized' and, in part, unstable chemical energy of the mixture (H, OH, H_2, H_2O_2) is transferred to stable components. These are, in addition, of a kind which tend spontaneously to increase their degree of 'order'; the hydrogen is evolved from the solution, which retains the ferric ions. In the galvanic cell the original substances are regenerated with liberation of electrical energy:

$$2Fe^{3+} + 2e^- \rightarrow 2Fe^{2+} \text{ at the cathode, and}$$

$$H_2 \rightarrow 2H^+ + 2e^- \text{ at the anode}$$

The Fe^{3+}, Fe^{2+} redox potential is about 0.77 V positive to that of the hydrogen electrode. The conversion process (H, OH, H_2, H_2O_2) $\rightarrow$ (H_2, Fe^{3+}) must be paid for by a loss in Gibbs free energy of the system, but this loss is outweighed by the greatly decreased effect of the back-reactions already mentioned.

Two possibilities for the design of these electrochemical nuclear energy converters are indicated in Fig. VII.8. In type A the half-cell containing the Fe^{3+} reduction electrode is situated in the radiation field. This type of construction is compact, but there is some difficulty in confining the irradiation to the region of the catholyte containing the Fe^{3+}, Fe^{2+} redox system. There may also be radiation damage to the electrodes and the diaphragm.

In type B the reactor and the electrochemical cell are joined by electrolyte and gas conduits, the cell lying outside the irradiation zone. This requires

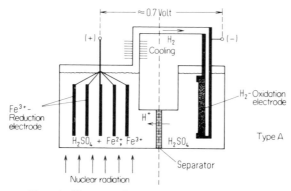

Type A: Close-coupling of the reactor and the cell.

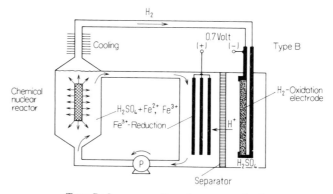

Type B: Loose-coupling of reactor and cell.

Figure VII.8. Constructional diagrams of radiolytically regenerated H_2–Fe^{3+} cells.

circulation of the electrolyte, which involves an additional energy loss; on the other hand flow of electrolyte reduces concentration overpotential. For optimum yield of Fe^{3+} and H_2, the Fe^{3+} concentration must not be allowed to become too high, otherwise there is loss by back-reaction with primarily-formed H atoms.

The first experiments with the system of type A were carried out by Dolin and Duzhenkov[19].

The catholyte, of composition 3N H_2SO_4 + 0.9M Fe^{2+} + 0.1M Fe^{3+}, was separated by a diaphragm from the anode compartment, which contained 3N H_2SO_4. X-rays were used for the radiolysis, the radiation dosage amounting to 1.5×10^{16} eV cm^{-3} sec^{-1}. With 10 cm^3 irradiated volume, and a theoretical $g(Fe^{3+})$ value of 8, the theoretically possible current is $g(Fe^{3+}) \times$ dosage $\times$ volume $\times$ elementary charge $= \dfrac{8 . 1.5 . 10^{16} . 10 . 1.6 . 10^{-19}}{100} = 2$ mA. The authors obtained 0.7 mA at a terminal voltage of 0.6 V. With 0.42 mW load, 27 % of the optimum output was obtained, with reference to the standard emf of 0.77 V.

The theoretical efficiency of this type of converter is

$$\eta_{NE} = \eta_{NC} \cdot \eta_{CE} = \frac{8.2 \cdot 0.77}{100} \cdot \frac{0.75}{0.77} = 6.2\%$$

The value obtained in practice was

$$\eta_{NE} = \frac{8.2 \cdot 0.77}{100} \cdot \frac{0.7}{2.0} \cdot \frac{0.60}{0.77} = 1.7\%$$

The authors attribute the loss of output partly to the back-reaction of Fe^{3+} ions with H radicals, catalysed by the platinum electrodes. They also recognized a second difficulty—that of keeping iron out of the anolyte for an adequate period of time. They used cellophane separators, and their cells 'collapsed' after two or three days.

Yeager, Bennett and Allenson[20] have made considerable technical improvements in the H_2–Fe^{3+} cell. At this time good separators of high conductance were already available. The authors used a microporous polyvinyl chloride separator called 'Porvic', but, because this material is sensitive to both radiation and OH radicals, they confined themselves to the second type of converter, B (Figure VII.8). To meet the problem of the back-reaction, catalysed by platinum, encountered by Dolin and Duzhenkov[19], they used porous graphite as the cathode material, since this is inert to both hydrogen atoms

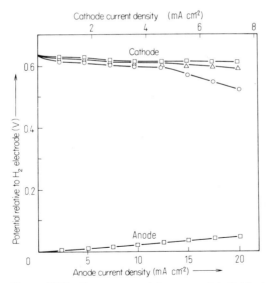

Figure VII.9. Current–potential curves of individual electrodes at 25°C. ○ 1M H_2SO_4; 0.025M $Fe_2(SO_4)_3$; 0.25M $FeSO_4$. △ 1M H_2SO_4; 0.037M $Fe_2(SO_4)_3$; 0.37M $FeSO_4$. □ 1M H_2SO_4; 0.05M $Fe_2(SO_4)_3$; 0.5M $FeSO_4$.

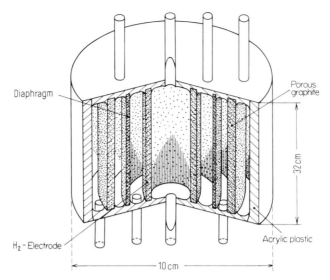

Figure VII.10. Fe^{3+}-hydrogen battery of the 5 watt prototype of Yeager and co-workers[20].

and molecules. The hydrogen-diffusion electrodes were platinized and hydrophobic.

The current–potential curves of the individual electrodes at a working temperature of 25°C are shown in Figure VII.9. At the necessarily small concentration of Fe^{3+} ion, the cell voltage was a little over 0.6 V; for equal initial concentrations (0.04M in Fe^{3+} and Fe^{2+}), the rest voltage was 0.68 V.

Figure VII.10 is a sketch of a cell forming part of a 5-watt prototype battery consisting of six cells in series, providing a working voltage of 3 V. The current densities are $10 \, mA \, cm^{-2}$ (H_2 electrode) and $2.1 \, mA \, cm^{-2}$ (Fe^{3+}, Fe^{2+} electrode). The irradiation of the electrolyte from ^{60}Co takes place in a cylindrical reaction vessel of 1.50 m diameter and 1.80 m height, in which 70 % of the radiation is absorbed. The electrolyte is circulated at $400 \, cm^3 \, min^{-1}$. In use the Fe^{3+} concentration in the battery decreased from 0.025 to 0.015M, and simultaneously $70 \, cm^3 \, min^{-1}$ of hydrogen were evolved and transferred to the anode. The battery could be operated for more than four months at $10 \, mA \, cm^{-2}$ (anode) without interruption. The efficiency obtained was $\eta_{NE} = 3\%$.

VII.5.3   Utilization of the Radiolysis Products of Gases

For the transformation of atomic energy into electrical energy by means of a closed electrochemical system, it is a fundamental necessity to use only working materials that, after release of electrical energy (and perhaps some heat), quantitatively regenerate the initial reactants.

Table VII.5. Some standard redox potentials.

H_2/H^+	0.00 volt	H/H^+	$- 2.106$ volt
O_2/H_2O	1.23 volt	O/H_2O	$+ 2.422$ volt
		O_3/O_2	$+ 2.07$ volt
F_2/F^-	2.87 volt	F/F^-	
Cl_2/Cl^-	1.34 volt	Cl/Cl^-	

From previous experience it seems that of substances liquid under normal conditions, water alone is suitable as working material for a conversion process. Of chemically oxidizable or reducible gases, those should be suitable that form stable diatomic molecules and also give rise to chemically reactive components, e.g., H_2, O_2, Cl_2 and F_2.

The action of high-energy radiation on these molecules produces atoms which can be regarded as free radicals. The electrochemical redox potentials of these free radicals deviate considerably from those of the corresponding stable molecules, as can be seen from Table VII.5.

The principle of the galvanic fuel cell (used in the case of water) is not, however, applicable to the conversion of chemical energy produced by radiolysis in this way. Instead, the following kind of cell is required.

One of the electrodes of the cell must function as a gas-evolution electrode of very low overpotential. At this electrode, that quantity of gas is recovered in molecular form that underwent reaction in radical form at the other electrode, which is a dissolution electrode. The dissolution electrode must have a very high overpotential for reaction of the molecular gas; its potential must be determined by the radicals produced by the radiolysis.

The problem is to find dissolution electrodes of adequate capacity, and in recent years it has appeared most likely to be solved for the case of oxygen.

VII.5.3.1   The Radiolytically Regenerated Oxygen–ozone Cell

At the 49th National meeting of the American Institute of Chemical Engineers, information was given by Ju Chin Chu[21] on the work of Eerkens and Reder[22] concerning a photolytically or radiolytically regenerable cell, working at 400°K, with a rest voltage of 0.87 V, and a short-circuit current density of 95 mA cm^{-2}. Such a cell should have an optimum power output of about 20 mW cm^{-2}, and should be capable of development as an additional energy source for space vehicles. A 28 V, 1 kW unit can be estimated to weigh 54 kg. The conversion of the energy of ultraviolet light, plentiful in the exosphere, or of X-rays, γ-, α-, proton- or β-radiation, or of uranium fission products, into chemical energy occurs by the photolysis or radiolysis of oxygen.

Since the report[22] on this cell was not available, only a conjecture could be made as to its construction and mode of operation. It may be suggested that

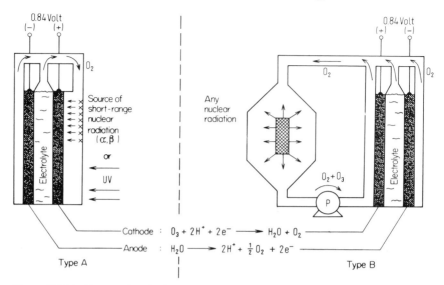

Figure VII.11. Constructional schemes for radiolytically regenerated oxygen–ozone cells.

constructional schemes A and B of Figure VII.11 are practicable. Type A involves a combination of reactor and cell suitable for the conversion of shorter-range nuclear radiation, or of UV, in gaseous oxygen, whereas in Type B the cell is outside the irradiation zone of the reactor, and this would be suitable for mixed, or long-range radiation. If the cell is to derive electrical energy from the overall reaction $O_3 \rightarrow O_2 + \frac{1}{2}O_2$, an electrode is required at which the cathodic reduction

$$O_3 + 2H^+ + 2e^- \rightarrow H_2O + O_2 \tag{VII.6}$$

will occur, but for which the O_2 overpotential is very high. At the other electrode oxygen evolution must take place at very low overpotential,

$$H_2O \rightarrow 2H^+ + \frac{1}{2}O_2 + 2e^- \tag{VII.7}$$

the oxygen being returned for repeated radiolysis. Under standard conditions the difference of the electrode potentials corresponding to reactions (VII.6) and (VII.7) is 0.84 V.

Since the life of the photolysis product, O_3, is about 154 hr at 127°C, oxygen–ozone mixtures can be circulated without loss.

VII.5.3.2   Estimation of Optimum Conversion Efficiency

For the formation of ozone by the action of fast electrons on liquid oxygen (90°K), the radiation yield is $g(O_3) = 12$–15 (O_3 molecules per 100 eV of energy absorbed)[23,24]. The energy stored in O_3, in excess of that in O_2, is, from thermodynamic data, $2 \cdot 0.84\,\mathrm{eV}$. This gives an optimum efficiency of radiochemical conversion of

$$\eta_{\mathrm{NC}} = \frac{12\,(\text{or }15) \cdot 2 \cdot 0.84}{100} \sim 20 \text{ or } 25\%$$

Ju Chin Chu reports $\eta = 10\text{–}15\%$.

VII.5.3.3   *The Radiolysis of Oxygen*

Oxygen is of special interest because of its high radiochemical conversion efficiency of 20 to 25%. This high efficiency, however, relates only to pure, liquid oxygen at its boiling point of 90°K. $g(O_3)$ values of 12 to 15 were found. Decreased efficiencies are found for liquid N_2–O_2 mixtures under γ-irradiation, as shown in Table VII.6.

Table VII.6. $g(O_3)$ Values for liquid N_2–O_2 mixtures.

$g(O_3)$	$T(°K)$	$O_2(\%)$	$N_2(\%)$
12.5	77	100.00	0.00
8.1	77	10.2	89.8
5.8	77	1.24	98.76
4.5	77	0.11	99.89
4.0	77	0.05	99.95
13.0	90	100.00	0.00

    Ozone is formed by the irradiation of oxygen by UV light of wavelengths 1751 to 1900, 2070 to 2200 and 2530 Å. The quantum energies corresponding are from 7.6 to 5 eV. Comparison of these quantities with the energy levels of excited states of the oxygen molecule leads to the conclusion that for the second and third wavelength ranges quoted above, ozone formation apparently proceeds via excited states (e.g., $3\Sigma_u^+$, excitation energy 4.9 eV; $^3\Sigma_u^-$, excitation energy 6.09 eV), probably according to

$$O_2 + O_2 \rightarrow O_3 + O \tag{VII.8}$$

$$O + O_2 + M \rightarrow O_3 + M \tag{VII.9}$$

    Irradiation by UV of 1751 Å dissociates the oxygen molecule, forming atoms in ^3P and ^1D states. Under these conditions, ozone formation occurs predominantly by reaction (VII.9).

VII.6   A COMPARISON OF TYPES OF CONVERTER; CONCLUDING CONSIDERATIONS

The electrochemical methods of nuclear energy conversion discussed in the preceding sections are collected in Table VII.7, together with the physical methods, scientific and—to some extent—technical features of which are well established. A comparison indicates that the electrochemical methods do not compete with the physical methods but, rather usefully, extend them. They are about equally matched in the efficiencies yet attained or attainable. Since the electrochemical converter employs materials in three states of aggregation, its

Table VII.7. Comparison

Converter		Nuclear energy or form of radiation	Method
A. Physical			
1. High-voltage nuclear battery (direct charging converter)	Vacuum Dielectric	α- or β-rays. Monoenergetic: β-radiation	Monoenergetic nuclear particles as charge carriers
2. Gas ionization nuclear battery (contact potential converter)		α- or β-rays	Ionization of a gas by nuclear radiation, in the field of a contact potential difference.
3. Semiconductor nuclear battery (semiconductor junction battery)		Soft, pure β^--radiation	Electrons and positive holes, separated at a semiconductor junction by monoenergetic β^--radiation
4. Photoelectric converter (Scintillation converter)		Soft, pure β^--radiation	Photoelectric use of light produced by monoenergetic β^--radiation
5. Neutron converter		Slow neutrons $\rightarrow \alpha$- or β^--rays	Neutrons of certain kinetic energy transmute certain isotopes into short-lived α- or β-emitters, used as in method 1.
B. Electrochemical			
1. Radiolytically regenerated . . . H_2–O_2 fuel cell		All radiations that will radiolyse water; mixed radiation	Electrochemical use of . . . separate molecular products of water radiolysis: H_2 and O_2 (or H_2O_2).
2. . . . oxyhydrogen cell		As above	. . . Molecular water-radiolysis products, $(H_2 + \frac{1}{2}O_2)$ as stoichiometric mixture
3. . . . H, OH cell		UV, X, γ- or β^--radiation	. . . chiefly the radical water-radiolysis products (H, OH, HO_2)
4. . . . H_2–Fe^{3+} cell	Type A. Type B.	UV, X, α- or β-radiation UV, X, γ-, α-, β-or mixed radiation	. . . both radical and molecular water radiolysis products, in the form of the separate reactants, H_2 and Fe^{3+}

of types of converter.

Efficiency η obtained (%)	η optimal (%)	Maximum attainable voltage of a single cell	Preferred application	Limitation on cell life by effect of radiation	Figure no. in this book	Literature reference
15–20		365,000	Superior in temp. range −50 to +50°C	Vacuum must be continuously maintained. Radiation damage to solid dielectrics	VII.2	24, 25
< 1	2–3	A few thousand				26, 27
1.0–1.5	2–3	0.5–2.0	Small portable batteries	Highly electronegative gas (e.g. O_2) liberated by action of radiation on electrodes must be removed from working gas		
0.1–2	5–10	< 0.25	Combination with transistors	Radiation damage to semiconductors by use of too high energy radiation		26
0.1–2		< 0.25	Space temperatures	Radiation damage to semiconductors by use of too high energy radiation		26
< 0.7	1.5–2.0	cf. method 1	Neutron flux measuring apparatus	Undesired nuclear reactions; otherwise as 1		28
< 1	2.1 For fission products ?	0.95	Use of radioactive waste on the large scale	None	VII.4	11
	As above	0.95	As above	None	VII.5	13
1.3(?)		0.7–0.9	Electrochemical counterpart of photocell; only of scientific interest	Decay of catalytically selective electrode properties		14
1.7–2	55–6	0.77	Galvanic nuclear battery of medium size: portable units Larger units: possible combination with nuclear reactors	Radiation damage to separator None	VII.8A VII.8B	15

Table VII.7. Comparison of

Converter		Nuclear energy or form of radiation	Method
5. ... O_2–O_3 cell	Type A.	UV, X, α- or β- radiation	... ozone produced by irradiation of oxygen
	Type B.	UV, X, γ-, α-, β- or mixed radiation	
6. Electrochemical homogeneous reactor		All radiations that will radiolyse water; mixed radiations from homogeneous reactor, fission products	Combination of 3 with 1 or 2

construction and operation are less simple. It therefore cannot be adapted to a number of special purposes requiring 'miniature' equipment, to which several physical methods lend themselves very well. The temperature range accessible to electrochemical methods is limited by the use of liquid electrolytes. The strength of these methods lies in the comparatively small variations in construction by which they can be adapted to a variety of radiations, or at least partially, to non-uniform radiation mixtures. In addition, their field of application includes, for example, the hydrogen–oxygen mixture coming from atomic waste stored in water, or the (partial) direct use of nuclear-fission energy, for which other direct methods are unsuited.

REFERENCES

1. S. C. Lind, *Radiation Chemistry of Gases,* Reinhold Publishing Corporation, New York, 1961.
2. I. L. Magee, *Ann. Rev. nuclear Sci.* **3**, 171 (1953).
3. P. Y. Feng, AFOSR Document No. AFOSR–241 (21.2.1961) Contract No. AF 18 (603)–121.
4. R. L. Platzmann, 'The Physical and Chemical Basis of Mechanisms in Radiation Biology', in *Radiation Biology and Medicine,* pp. 15–72 and back references, editor W. D. Claus, Addison-Wesley, New York, 1956.
5. H. A. Dewhurst, A. H. Samuel and J. L. Magee, *Radiat. Res.* **1**, 62 (1954).
6. L. H. Gray, *Brit. med. Bull.* **4**, 11 (1946).
7. L. H. Gray, *Brit. J. Radiol.* **26**, 609 (1953).
8. A. O. Allen, *Discuss. Faraday Soc.* **12**, 79 (1952).
9. J. Weiss, *Ann. Rev. physic. Chem.* **4**, 143 (1953).
10. E. J. Hart and R. L. Platzmann, *Mechanisms in Radiobiology,* Vol. 1, ch. 2, editors M. Errera, A. Forssberg, Academic Press, New York, 1961.
10a. F. Cap, *Physik und Technik des Atomreaktoren,* Springer, Wien, 1957, p. 370.
11. L. Rosenblum and R. English, *Advanced Energy Sources and Conversion Techniques,* Vol. I, S. 243/53, Department of Commerce OTS No. PB 151461.
12. E. Justi, DP-Anmeld. J. 11654 Ia/14 h v. 7.5.56 (unobtainable) quoted from, E. Justi and A. Winsel, *Kalte Verbrennung,* Steiner, Wiesbaden, 1962.

types of converter—continued.

Efficiency η obtained (%)	η optimal (%)	Maximum attainable voltage of a single cell	Preferred application	Limitation on cell life by effect of radiation	Figure no. in this book	Litera- ture refer- ence
(?)	20–25	0.84	Universally applicable in the range 15–200°C	Not known	VII.11A	21,22
12–15			Very small or large units	None	VII.11B	
	5.5–6	0.95	In combination with homogeneous reactor	1. Decay of catalyst selective electrode properties. 2. Transmutation of electrode materials		16,17

13. G. Grüneberg, W. Wicke and E. Justi, French Pat. 1321373.

13a. G. Goebel, B. D. Struck and W. Vielstich, unpublished.

14. Ts. I. Zalkind, V. I. Veselovsky and G. Z. Gochaliev, All-union Conference on the Application of Radioactive and Stable Isotopes and Radiation in the National Economy and Science, Session on Radiation Chemistry, Moscow, from 25.3–2.4.1957.

15. V. I. Veselovsky, Proc. Intern. Conf. Peaceful Uses Atomic Energy, Geneva 7, 599 (1956). I. Zalkind and V. I. Veselovsky, *Sbornik Rabot Radiatsionnoi Khim. Akad. Nauk SSSR,* 1955, S. 66.

16. *Dtsch. Forschungsdienst* **8**, No. 9(1963).

17. *Frankfurter Allg. Zeitung* of 12.3.1963.

18. C. J. Hochanadel and S. C. Lind, *Radiation Chemistry*, Ann. Rev. Physics. Chem. Vol. 7, 83 (1956).

19. P. I. Dolin and V. I. Duzhenkov, *Trudy Pervogo Vsesoyuz. Soveshchaniy. Po Radiatsion. Khim., Akad. Nauk SSSR,* Otdel Khim. Nauk. Moscow 1957, 135–38 (Publ. 1958).

20. J. F. Yeager, R. J. Bennett and D. R. Allenson, Nuclear Regenerative Fuel Cell, Res. Lab. UCC, Parma 30, Ohio (presumably 1961).

21. *Chemie-Ing.-Techn.* **35**, 541 (1963), Notice from a lecture by Ju Chin Chu at the 49. Nat. Meeting of the American Inst. Chem. Engrs. from 10.–14.3.1963 in New Orleans.

22. J. W. Eerkens and M. C. Reder (Aerojet-General Nucleonics, San Ramon, Calif.) NSA 15–2384, Paper No. 1306–60, presented at the ARS Space Power Systems Conference, Santa Monica, California, Sept. 1960, American Rocket Society, New York, 1960.

23. S. J. Pscheshezki, I. A. Mjasnikov and N. A. Bunejev, 'Die Bildung von Ozon in flüssigem Sauerstoff bei Einwirkung schneller Elektronen,' in N. A. Bach, *Arbeiten über Strahlenchemie,* Akademie-Verlag, Berlin, 1960.

24. J. F. Riley, 'Radiolysis of Liquid Oxygen and Oxygen–Nitrogen Mixtures,' in E. H. Taylor and M. A. Bredig, Oak Ridge National Lab., Tennessee, *Chemistry Division Annual Progress Report, June 1962,* ORNL–3320, UC–4 Chemistry.

25. A. Thomas, *Nucleonics* **13**, 129 (1955).

26. K. Meyer, *Energietechnik* **12**, 496 (1962).

27. K. H. Beckurts and K. Schretzmann, *Atomwirtschaft* **5**, 69 (1962).

28. Sh. S. L. Chang, *Energy Conversion,* Prentice Hall, Englewood Cliffs, N.J. (1963).

CHAPTER VIII

Electrochemical Storage of Energy

The development of fuel cells has been accompanied by renewed interest in alternative electrochemical methods of storing energy. This chapter deals first with the use of the electrolysis of water in conjunction with the hydrogen–oxygen fuel cell for this purpose, and secondly with the application of redox systems, such as Fe,Fe^{2+}, Fe^{2+},Fe^{3+} and Ti^{3+},Ti^{4+}, to energy storage.

The adaptation of such new methods for use in satellites is receiving much attention—to make the electricity generated by solar batteries available during periods of darkness. Another conceivable application is to meet the peak load demands of power stations[1a], although at present hydroelectric storage is used where conditions are favourable, water being pumped to a high-level reservoir during off-peak periods. For this, however, geographical requirements must be met, and adequate water supplies must be available. For 100 m fall, about $4m^3$ of water per kW hr are needed. At the present time, however, the capital and running costs of the electrochemical analogue of hydroelectric energy storage are difficult to estimate, and the working life of the best fuel cell yet developed would hardly be adequate.

VIII.1 THE COMBINATION OF WATER ELECTROLYSIS WITH THE HYDROGEN–OXYGEN FUEL CELL

Limitations of available secondary batteries, primarily in relation to energy density and cycle life, have provided the incentives for the development of an electrolytically regenerative hydrogen–oxygen fuel cell.

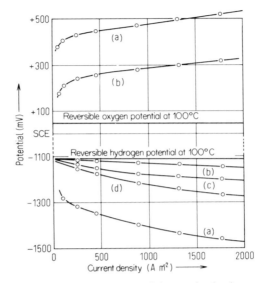

Figure VIII.1. Current–potential curves for the electrolysis of water at DSK-electrodes in 5.5N KOH (after Justi, Winsel, Grüneberg, Spengler and Vielstich[1,2]). Potentials relative to SCE. (a) smooth nickel, 100°C; (b) Raney-nickel, 100°C; (c) Raney-nickel, 60°C; (d) Raney-nickel, 40°C.

The combination of a unit for decomposing water with a hydrogen–oxygen fuel cell can be regarded as a *'gas accumulator'*. The hydrogen and oxygen can be stored under pressure and used in the galvanic cell on demand (cf. Figure I.10).

The terminal voltage required to electrolyse water or an aqueous solution is usually considerably greater than that calculated thermodynamically. This is mainly due to overpotential at the oxygen electrode (cf. Figure VIII.1). The additional voltage required increases with rising current density. The unavoidable voltage loss in the electrolyte (including that in a diaphragm) is minimized as far as possible by reducing the distance between the electrodes and by using an electrolyte of high conductance (perhaps 6N KOH). The other contributions to overvoltage are concentration overpotential and charge-transfer overpotential, but the first of these is reduced to negligible proportions by the use of an alkaline electrolyte (cf. equations III.2 and III.37) and by the effective stirring due to gas evolution[1].

It is therefore the charge-transfer overpotential that essentially determines the total overvoltage. This, in turn, mainly depends on the catalytic activity of the electrode materials (cf. Figure III.2) and on the degree of roughness of the electrode surfaces. Platinum and palladium are excellent catalysts for the electrolysis of acidic aqueous solutions but are far too costly for use on a large technical scale (as in power stations). Technical electrolysis of water is carried out with nickel or iron electrodes in alkaline electrolytes.

In the course of development of hydrogen-diffusion electrodes, it was found that cathodic hydrogen evolution took place at skeleton-nickel* (particularly at DSK- and MSK- electrodes, cf. section IV.1.1.1), at an electrode potential close to the thermodynamic value, even at 'technical' current densities[2].

The properties of DSK electrodes in relation to the electrolysis of water have been studied in detail by Justi and co-workers[2]. Figure VIII.1 shows current–potential curves for oxygen and hydrogen evolution at DSK-electrodes in comparison with those for smooth nickel: potentials relative to the saturated calomel electrode (SCE) are plotted against current density. Hydrogen evolution begins at the thermodynamic potential and the overpotential only amounts to 30 mV at 1500 A m^{-2} at 100°C.

Under these conditions of electrolysis the charge-transfer overpotential at the hydrogen electrode is obviously negligible. Only at 40–60°C does kinetic hindrance become appreciable compared with voltage drop in the electrolyte; the overpotential is increased, and the current–potential plots are more curved (curves e and d). For the oxygen electrode, the main reduction of overpotential, as compared with smooth nickel, is to be attributed to the greatly roughened surface of the DSK-electrode.

The DSK material used in this work was prepared in the following way[1,2]. Raney-nickel (60% Al, 40% Ni) was reduced in a ball-mill to a powder of 20–60 μm particle size. This was mixed with carbonyl nickel powder (5–15 μm in a volume ratio of 1:2 and the mixture was spread in a thin layer (approx 0.3 mm†) on a nickel screen as carrier ('economy electrode') pressed at 5 ton cm^{-2}, and sintered for 3 minutes in a reducing atmosphere at about 700°C. The aluminium was then dissolved out in concentrated alkali at 80–100°C.

In Figure VIII.2 the characteristic curve of a laboratory cell with DSK-electrodes (each 12 cm^2 in area, 7 mm apart, with a plastic screen for separating the gases; 5.5N KOH, 100°C) is compared with the current–voltage curve of a corresponding cell with smooth nickel foil electrodes. Curves for technical water-decomposition units are shown in the same diagram. It is clear that the use of skeleton-nickel electrodes for the electrolysis of water can effect an economy of energy of about 20%.

Experiments prolonged for more than a year have shown that the activity of the electrodes hardly deteriorates at all; at a constant current density of 1500 A m^{-2} an increase of cell voltage of no more than 40 mV was observed. Hydrogen evolution appears to be responsible for a continuous reactivation of the electrode. An additional stability is also conferred on Raney-nickel by addition of small quantities of platinum (cf. section IV.1.1.1. and reference 29 of chapter IV).

Figure VIII.3 shows efficiencies for a few examples of electrolyses and of hydrogen–oxygen cells as functions of current density, based on the thermo-

* See also L. Kandler, Deutsche Auslegeschrift (DAS), 1183892 (14.6.1961).
† A greater thickness than this leads to a marked reduction in electrode activity (cf. W. Ermrich, Dipl-Arbeit, Braunschweig, 1957, as well as E. Justi and A. Winsel[2]).

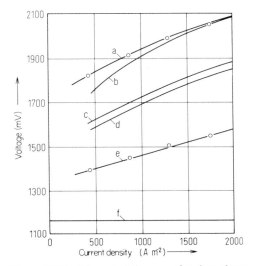

Figure VIII.2. Characteristic curves for electrolyses. a, Laboratory cell with smooth nickel electrodes, 5.5N KOH at 100°C (after Vielstich[1]); b, 'Bamag-Zersetzer' (after Henglein[4]); c, 'Demag-Zersetzer' (after v. Pichler[5]); d, Lurgi high pressure decomposer, 5.5N KOH, 100°C, 30 atm[6]; e, Laboratory cell with DSK electrodes (cf. Figure VIII.1). f, Thermodynamic decomposition potential at 100°C.

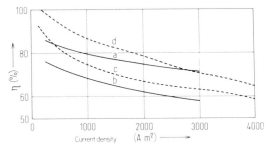

Figure VIII.3. Efficiencies, η, for water electrolysis and hydrogen–oxygen fuel cell in dependence on load. a, Electrolysis with Raney-nickel-DSK electrodes (cf. curve e, Figure VIII.2); b, electrolysis with 'Demag-Zersetzer' (cf. curve c, Figure VIII.2); c, hydrogen–oxygen cell of Allis-Chalmers (cf. section IV.1.2); KOH-impregnated asbestos, 80°C; d, hydrogen–oxygen cell of Pratt and Whitney (cf. section IV.2.3); 85% KOH, 200°C.

dynamic terminal voltage of 1.15 V at 100°C. The efficiency of electrolysis is given by the ratio of the thermodynamic decomposition voltage to the terminal voltage used:

$$\eta_{\text{decomp.}} = \frac{1.15}{E_{\text{term}}(j)}$$

That for the fuel cell is, correspondingly,

$$\eta_{\text{H}_2,\text{O}_2} = \frac{E_{\text{term}}(j)}{1.15}$$

assuming in each case complete reaction of the gases.

In Figure VIII.4, the *overall efficiency* of electrochemical energy storage is shown as a function of fractional load (200 A m^{-2} is assumed as full load), in comparison with a hydroelectric system[3]. Whilst the efficiency of the latter falls steeply at less than one quarter load, the electrochemical system increases in efficiency. If transformation to ac is necessary, the overall efficiency of the electrochemical method is reduced by 10–15%. In practice electrical storage systems are required to deal with fluctuating load and it is therefore the average efficiency which is significant. For the modern hydroelectric unit quoted, this amounts to 68%; for electrochemical storage, for dc use, 60 (curve b) to 70% (curve c). If for the fuel-cell section a maximum load of only 1500 or 1000 A cm^{-2} is provided for, the efficiency is increased by 3–6%.

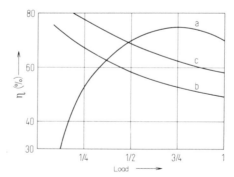

Figure VIII.4. Comparison of efficiencies, η, of electrochemical and hydroelectric energy storage. a, Modern hydroelectric system (after Böhler[3]); b, electrolysis cell with DSK electrodes and the Allis-Chalmers hydrogen–oxygen cell (max. load 2000 A m^{-2}); c, electrolysis cell with DSK electrodes and the Pratt and Whitney hydrogen–oxygen cell (max. load 2000 A m^{-2}).

Ching and Cohen[7] have discussed the application of an electrolysis cell with a modified Bacon hydrogen–oxygen cell (Pratt and Whitney, cf. Figures VIII.3 and 4, as well as section IV.2.3) for use in space vehicles.

For the hydrogen–oxygen cell, $170\ mA\ cm^{-2}$ and a working temperature of 250°C were adopted, and a battery of 28 V terminal voltage and 500 W rating was proposed. Photocells would provide energy during sunlit periods. Besides the 28 V battery, a 2.5 V unit would provide the voltage required to electrolyse the water produced by the hydrogen–oxygen cell. Hydrogen and oxygen would be collected in gas vessels and used to supply the fuel battery during periods of darkness. The battery would, when not in operation, be kept at the working temperature with a 20 W heater. The electrolysis cell would operate at $540\ mA\ cm^{-2}$ at 120°C; because of this high loading the total efficiency under these conditions was rather less than 40%.

Electrochemical storage by electrolysis and hydrogen–oxygen cell can, if necessary, be accomplished by means of a *combined* cell. This does, however, introduce additional technical difficulties with a consequent loss of efficiency, since electrode systems cannot simultaneously be of optimum efficiency for reaction in either direction. For this purpose a hydrogen–oxygen cell with a membrane as electrolyte carrier (cf. section IV.1.1.2) could be used. The first experiments in this direction were described by Bone[8] by using an ion-exchange cell.

As indicated in Figure VIII.5, the membrane separates two gas compartments. Hydrogen and oxygen are evolved at the electrodes in contact with the membrane on application of an external emf; this mainly occurs on the parts of the electrodes dipping into the electrolyte (KOH or H_2SO_4). During discharge of the cell it is mainly the electrode areas not covered by electrolyte that are active. Water is regenerated during discharge.

With such a membrane-electrode assembly ('bonded electrodes'), Bone[8] could charge the storage cell at $30\ mA\ cm^{-2}$ at a terminal voltage of 1.7 V. At a discharge current density of 15–$20\ mA\ cm^{-2}$ the terminal voltage was between 0.6 and 0.7 V.

Exchange membranes with attached electrodes have not, however, proved to be durable; even at current densities of but a few $mA\ cm^{-2}$, deterioration occurs on the oxygen side[9] probably because of attack on the membrane material by active oxygen formed in the electrolysis. For this reason, the

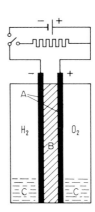

Figure VIII.5. Combination storage cell with ion-exchange membrane as electrolyte carrier (after Bone[8]). A, Ion-exchange membrane (OH^- or H^+ type); B, electrodes; C, H_2O, KOH (H_2SO_4).

further experiments have been conducted with electrodes (made from a mixture of platinum and Teflon) only in mechanical contact with the (alkaline) membrane.

Short-period current–voltage curves for these cells gave the following values:

$$Discharge \quad 20 \, mA \, cm^{-2} \quad 0.8 \, V$$

$$40 \, mA \, cm^{-2} \quad 0.7 \, V$$

$$Charge \quad 20 \, mA \, cm^{-2} \quad 1.6 \, V$$

In these, the membrane dipped into 2–10N KOH. More than 100 charge–discharge cycles were achieved with charging at 4 mA cm^{-2} (1.78 V) and discharge at 15 mA cm^{-2} (0.75 V). Under these conditions the efficiency was about 40%.

A modification of the Bone-type of combination cell has been proposed by Findl and Klein[9a]. The electrodes are similarly placed directly on either side of an asbestos membrane. Free electrolyte is, however, eliminated, and only part of the water absorbed in the porous membrane is decomposed during the 'charging' of the cell (Figure VIII.5a).

During charge, i.e. the electrolysis mode of operation, 0.336 g of water is decomposed per A hr. The extent of charging is limited, not only by the increase of resistance due to loss of ionic conductivity, but also by the appreciable cross-leakage of hydrogen and oxygen through the asbestos matrix that takes place if charging is unduly prolonged. The gases evolved are led by appropriate

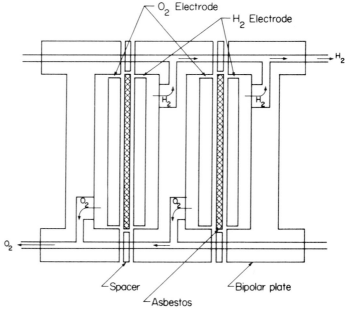

Figure VIII.5a. Schematic diagram of a regenerative hydrogen–oxygen fuel cell (after Findl and Klein[9a]).

manifolds to storage tanks and, during discharge of the cell, are recombined at the electrodes to form water which is reabsorbed by the asbestos matrix. The same reacting surfaces are used for charge and discharge modes of operation.

A ring spacer is used to provide the desired compression of the asbestos. Typically, fuel-cell grade asbestos of dry thickness 0.060 in is used, but is compressed between the electrodes during cell assembly to 0.040 in; this substantially reduces cross-leakage of the gases and causes some penetration of the electrolyte into the electrodes. The electrodes, 0.5 mm thick, are treated with 20 mg cm^{-2} of platinum catalyst and are made of sintered-carbonyl nickel, carried by nickel gauze to facilitate conduction.

Most of the experiments with this cell have been carried out at high rates of charge and discharge—typically for 65 and 35 minutes respectively. This 100-minute cycle is representative of the battery load characteristics of a 300 nautical mile orbiting satellite.

The performance of a single cell with 6 in diameter electrode over a test cycle is shown in Figure VIII.5b; the linear dependence of gas pressure on time in charge and discharge is to be noted. These data indicate current, voltage and energy storage efficiencies of 97, 55 and 53% respectively. Figure VIII.5c shows the results of long-term cycling of a multicell unit nominally rated at 75 W. It is seen that there is a progressive increase in charge and decrease in discharge voltage. The deterioration is attributed to the oxygen electrode occurring as a result of oxidation of the porous nickel substrate (cf. the Bone cell, above). The largest unit so far built is of 34 cells and is rated at 500 W; typical data are: charge, 10 A, 50 V, 65 min; discharge, 18.2 A, 30 V, 35 min.

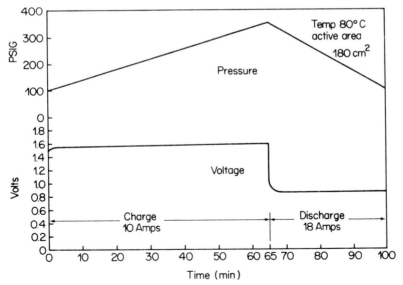

Figure VIII.5b. Typical cycle of a single combination storage cell with asbestos membrane (after Findl and Klein[9a]).

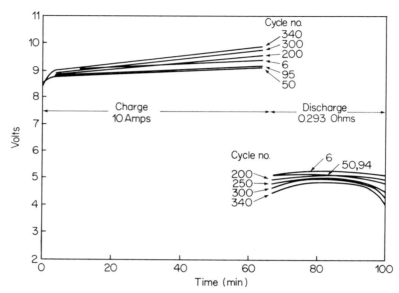

Figure VIII.5c. Voltage performance data during multi-cycle testing of a 75-watt, six-cell, hydrogen–oxygen regenerative battery (after Findl and Klein[9a]).

The gas storage compartments are of 2:1 volume ratio, to store the gases produced during charging at equal pressures. Imperfections in volume balance are taken up by 'floating bellows'; pressure differences for the 500 W unit are $\frac{1}{2}$ to 1 psi, but can rise to 10 psi without noticeable effect on performance. The unit, with aluminium tanks, weighs about 40 lb and, for 600 W hr, yields 15 W hr lb^{-1}. The volume is approximately 1800 cubic inches, giving 0.3 W hr in^{-3}.

Another kind of combination cell is based on the so-called 'valve electrodes' of Friese, Justi and Winsel[10,11]*. The principle of these is indicated in Figure VIII.6. The essential feature is a porous double-layer electrode with a coarse-pored, catalytically active, working layer, with a thin, fine-pored surface layer facing the electrolyte. The overpotential for gas evolution at this surface layer must be so large that at the working potential the reaction proceeds almost exclusively at the coarse-pored, active layer behind it. The gas generated in this layer cannot, if pore-widths are suitably chosen, penetrate back through the surface layer into the electrolyte, and is therefore retained in the gas compartment attached to the electrode.

With two such electrodes (for H_2 and O_2), the electrolysis of water can be conducted without a diaphragm, and, since the electrodes will work at a pressure of several atmospheres, an additional compression of the stored gases can be avoided. On disconnexion of the applied voltage used for electrolysis, current can be drawn from the cell immediately, since the electrodes are in a position

* See also R. Sommer and H. Müller, Austrian Pat. 209877 (8.5.1957).

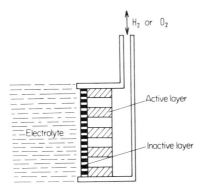

Figure VIII.6. Valve electrode for gas evolution and dissolution (after Friese, Justi and Winsel[10,11]).

to bring the gases into electrochemical reaction, i.e. the cell works in the reverse direction as a hydrogen–oxygen fuel cell.

For a hydrogen *valve-electrode* in alkaline solution, skeleton-nickel is suitable as catalyst material for reaction in either direction. Friese, Justi and Winsel[11] made a double-layer electrode from DSK-Raney-*nickel* and DSK-Raney-*copper*. For the working layer the mixed powder consisted of 1 part by weight of Raney-nickel (1:1; average particle size, 50–75 μ) with 1.5 parts of carbonyl powder for a supporting framework. The material for the inactive layer was composed of 1 part by weight of Raney-copper (1:1, mean particle size <35 μ) and 1.2 parts of copper powder. The two powders, 11 g and 1.4 g, were suitably arranged in layers in a press matrix (diameter 40 mm) and hot-pressed for about 10 min at 4 ton cm^{-2} at 380°C. The disc electrodes so prepared had a working layer 2–2.5 mm thick, and a surface layer about 0.2 mm thick. Activation was carried out in the usual way (cf. section IV.1.1.1).

Current–potential curves for hydrogen evolution from 5N KOH are shown in Figure VIII.7. The excess hydrogen pressure on the gas-side of the electrode

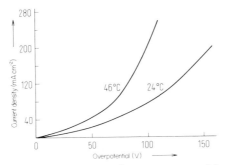

Figure VIII.7. Cathodic current–potential curves for hydrogen evolution at a valve electrode (after Justi and Winsel[11]).

was 1 atm, and at 24°C the yield of hydrogen was 97 % of theoretical at current densities up to 200 mA cm^{-2}. At higher current densities the yield decreased somewhat because, at the potentials concerned, some hydrogen deposition began at the copper layer. With the valve-electrodes described hydrogen could be generated at 3 atm excess pressure; no doubt improvement could be achieved by reduction of the pore width of the inactive layer.

For development of the analogous *oxygen* valve-electrode, Justi and Winsel[11] have used, similarly, Raney-nickel for the active layer, and have tried titanium for the surface layer. Titanium has a sufficiently high overpotential for oxygen evolution because of its surface layer of oxide. With electrodes prepared by the same hot-pressing technique, a particular difficulty was encountered in the activation process. The hydrogen evolved during the dissolution of the aluminium in alkali entered the titanium lattice, producing considerable expansion. This caused deformation and finally exfoliation. This difficulty should, however, be overcome by suitable modification of the activation procedure.

As already discussed in section III.1.7, the pore-free, palladium–silver alloy electrode can be used as a hydrogen valve-electrode; with this, storage pressures of 40 atm and more can be attained at high current densities.

VIII.2   ENERGY STORAGE BY MEANS OF REDOX SYSTEMS

Redox systems conceivably of use for storing electrical energy in solutions have recently been studied by Kangro and Pieper[12]. Any such electrochemical systems, to be of technical interest, must meet the following requirements:

1. Anode and cathode solutions must be reversibly interconvertible with suitable catalyst electrodes, and, in the 'fully charged' condition, must give rise to a sufficiently large terminal voltage.
2. No side reactions must occur at the electrodes, particularly none involving gaseous products.
3. The charged solutions must retain their charged state without change for long periods.
4. If possible, charge and discharge processes should occur in one and the same cell, i.e. they should proceed equally well at the same catalyst electrodes.
5. The active components must be sufficiently soluble in the electrolyte.

Two systems appear to be specially suitable in the light of these criteria. The first is the titanium–iron storage cell, studied quantitatively by Kangro and Pieper[12], and the second is the iron storage cell, described by Dolmetsch and Mylius[13], which contains Fe(II) chloride solution alone—metallic iron is deposited cathodically on charging.

VIII.2.1   The Titanium–iron Storage Cell

The overall reaction in HCl as electrolyte is as follows:

$$H_2Ti^{IV}Cl_6 + Fe^{II}Cl_2 \underset{\text{discharge}}{\overset{\text{charge}}{\rightleftarrows}} Ti^{III}Cl_3 + Fe^{III}Cl_3 + 2HCl$$

On charging, Ti^{IV} is reduced to Ti^{III} at the cathode, and Fe^{II} is oxidized to Fe^{III} at the anode. Anolyte and catholyte must therefore be separated by a diaphragm. Kangro and Pieper[12] used, before charging, a uniform electrolyte solution (1M $TiCl_4$ + 1M $FeCl_2$ + 1.5M HCl); after charging, therefore, the anolyte contained Ti^{4+} and the catholyte Fe^{2+} ions. If care is taken to exclude air, titanium(III) chloride solutions are quite stable, all the more so if the activity of water is reduced by the addition of strongly hydrated salts, such as the chlorides of magnesium, calcium or aluminium. Similar precautions are effective for iron(II) chloride solutions, which, so protected, are oxidized to the extent of only 2–3% in three weeks at room temperature.

Kangro and Pieper[12] used an experimental cell with porous graphite electrodes 13 cm^2 in area, and 15 cm^3 of electrolyte, which was allowed to flow through the cell at 0.4–1.1 cm^3 hr^{-1}. A charging current density of 1 mA cm^{-2} required a terminal voltage of 680 mV, and the current efficiency under these conditions was 76%. Discharge at 1.2 mA cm^{-2} took place at a terminal voltage of 520 mV, yielding 93% of the stored charge. Hence the overall efficiency of the charge–discharge cycle was about 55%.

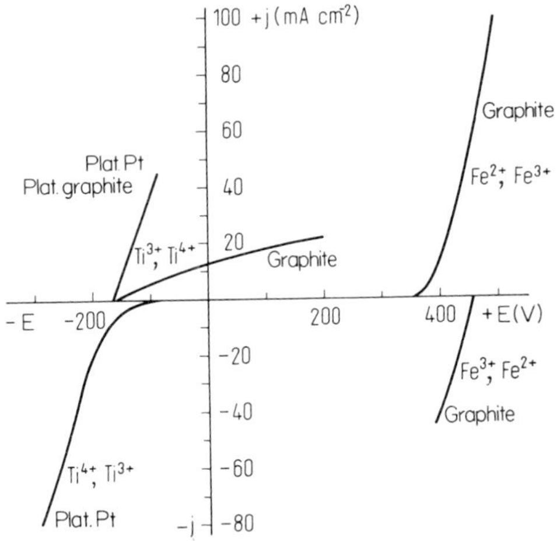

Figure VIII.8. Individual current–potential curves for the redox systems Fe^{II}, Fe^{III} and Ti^{III}, Ti^{IV} at 20°C (after Vielstich and Vogel[14]). Electrolyte 1.5N HCl, unimolar with respect to the salts, magnetically stirred. Potentials relative to SCE.

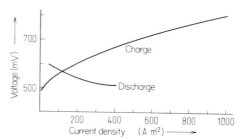

Figure VIII.9. Current–voltage curves of a titanium–iron storage cell for charge and discharge
(after Vielstich and Vogel[14]). Electrodes; graphite
(iron half-cell) and platinized graphite (titanium
half-cell). Ion-exchange membrane diaphragm
(TNO A 60). Electrolyte magnetically stirred on
either side.

Since no gas was evolved at either electrode of this system, Kangro and
Pieper[12] thought that the main loss of charge could be attributed to mixing of
the active components through the relatively coarse diaphragm used (a Perspex
partition with 60 holes of 0.6 mm diameter per cm^2). Complete reaction of the
active solutions could not be attained.

Vielstich and Vogel[14], investigating half-cells with graphite, plantinized
graphite, smooth and platinized platinum electrodes, have shown that the
low current densities observed by Kangro and Pieper[12] were due to the
inadequate catalytic activity of graphite in relation to the Ti^{IV}, Ti^{III} system.
Use of platinized platinum or platinized graphite makes possible current
densities of more than 50 mA cm^{-2} (Figure VIII.8). For the iron half-cell, on
the other hand, it is immaterial which of the above electrodes is used.

Figure VIII.9 shows current–voltage curves for charge and discharge of a
complete cell, with use of an ion-exchange membrane, TNO A 60, and separated
electrolytes (the titanium half-cell contained no iron, and *vice versa*). A charge–
discharge cycle obtained with this arrangement is shown in Figure VIII.10.
The cell was charged at 100 mA cm^{-2} and discharged at 30 mA cm^{-2}. The
energy storage efficiency amounts to 45%.

VIII.2.2   The Iron Storage Cell

In the cell proposed by Dolmetsch and Mylius[13], the electrode compartments
are also separated by a diaphragm, but both initially contain Fe(II) chloride
solution. Metallic iron is deposited at the cathode, and Fe(II) ion is oxidized
to Fe(III) at the anode. To obtain a smooth deposit of iron and avoid the formation of oxides, the Fe(II) chloride solution (600 g l^{-1} of $FeCl_2$) contains the

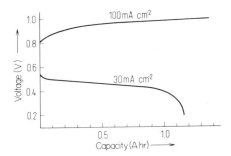

Figure VIII.10. Charge and discharge charac-
teristics of a titanium–iron storage cell (after
Vielstich and Vogel[14]); cf. legend of Figure
VIII.9.

following additional substances: 6 g $TiCl_3$, 0.6 g $AlCl_3$, 6 g H_3BO_3, 0.01 g HCl
and 0.25 mg of β-aminopropionic acid (per litre of electrolyte in each case).

A laboratory cell with graphite electrodes and polyethylene diaphragm was
charged with an applied terminal voltage of 2 V at 80–100 mA cm^{-2}, attaining
a temperature of 50–60°C during the process. The rest voltage of the charged
cell was 1.1 V. The most favourable discharge current density was 20–30 mA
cm^{-2} at 1.0 to 0.8 V; efficiency of discharge $\sim 80\%$.

VIII.2.3　Capacity in Relation to Volume and Weight—Comparison with the Lead Accumulator

Whether these 'liquid-phase storage cells' will be useful in practice can be
studied first by considering the number of watt-hours they can store per unit
volume and weight. Ideally, a cell with 1 litre of a 1M solution of an ionic solute
capable of unit valency change, and with a terminal voltage of 1 V can store
26.8 W hr. For the titanium cell, with 80% current efficiency of discharge
at 0.5 V, 1–3M solutions would provide 10–30 W hr l^{-1}, corresponding to
7–20 W hr kg^{-1}. The higher terminal voltage of the iron cell leads to the better
values of 60 W hr l^{-1} and 50 W hr kg^{-1}. These figures may be compared with
those for the lead accumulator, namely 80 W hr l^{-1} and 20 W hr kg^{-1}, but
other advantages of the conventional accumulator are its higher voltage,
energy efficiency and charging rate, and that it needs no electrolyte circulation.
It is clear that application of the redox systems discussed in this section could
only become feasible if economic factors (costs of installation, maintenance,
working life) could be shown to give them an advantage over existing storage
methods. Their relatively low rating in relation to size and weight is unfavourable to their use in space vehicles.

REFERENCES

1. W. Vielstich, *Chemie-Ing-.Techn.* **33**, 75 (1961).

1a. 'Large-scale storage of energy,' in A. B. Hart and G. J. Womak, *Fuel Cells*, Chapman and Hall, 1967, pp. 265ff.

2. E. Justi, A. Winsel, G. Grüneberg, H. Spengler and W. Vielstich, *Ger. Pat.* 1 065 821 (1.8.1957); E. Justi and A. Winsel, *Fuel Cells—Kalte Verbrennung*, Steiner, Wiesbaden, 1962.

3. K. Böhler, *Elektrizitätswirtschaft* **56**, 341 (1957).

4. F. A. Henglein, *Grundriß der chemischen Technik*, Verlag Chemie, Weinheim/Bergstr., 8. Aufl., 1954, p. 330.

5. A. v. Pichler, *Fette, Seifen, Anstrichmittel* **58**, 371 (1956).

6. Lurgi Co., private communication (1958).

7. A. C. Ching and F. Cohen, *Proc. 16th Power Sources Conference*, Atlantic City, 1962, p. 42; see also W. H. Podolny, *Proc. 14th Power Sources Conference*, 1960, p. 64, and J. M. Lee and R. M. Handlewich, *Proc. 15th Power Sources Conference*, 1961, p. 43.

8. J. S. Bone, *Proc. 14th Power Sources Conference*, Atlantic City, 1960, p. 62.

9. J. S. Bone, S. Gilman, L. W. Niedrach and M. D. Read, *Proc. 15th Power Sources Conference*, Atlantic City, 1961, p. 47.

9a. E. Findl and M. Klein, *Proc. 20th Power Sources Conference*, Atlantic City, 1966, p. 49.

10. K. H. Friese, E. Justi and A. Winsel, *Ger. Pat. Appl.* 1958.

11. E. Justi and A. Winsel, *Fuel Cells—Kalte Verbrennung*, p. 248ff.

12. W. Kangro and H. Pieper, *Electrochim. Acta (London)* **7**, 435 (1962); see also W. Kangro *Ger. Pat.* 914 264 (28.6.1949).

13. H. Dolmetsch, *Ger. Pat.* 954 890, (7.5.1951), H. Mylius, *Ger. Pat.* 1 161 966 (11.7.1962).

14. W. Vielstich and U. Vogel, unpublished.

CHAPTER IX

Electrochemical Enrichment of Deuterium and Tritium— Concentrator Cells for Hydrogen and Oxygen

Heavy water can be obtained as a by-product of the electrochemical production of hydrogen and oxygen. Measurements of deuterium and tritium separation factors are reported as functions of cathodic overpotential.

While at smooth electrodes separation factors change markedly with time because of roughening of the electrode surface, they are reproducible when sintered electrodes are used. Sintered iron gives the highest H, D separation in KOH solutions; the lowest overpotential with good separation is obtained with Raney-nickel electrodes.

A discussion is given of the application of a cell with hydrogen anode and hydrogen cathode to the concentration of deuterium, and also of the recovery of electrical energy and the attainment of higher enrichment by the recombination of electrolysis gases in hydrogen–oxygen fuel cells.

Measurements of H, D separation factors associated with the anodic oxidation of hydrogen ('anodic separation factors') are of interest in this connexion.

An interesting offshoot of the development of hydrogen anodes and oxygen cathodes in fuel cell research is the provision of concentration cells for hydrogen and oxygen. Electrolytically pure hydrogen can be got from a gas mixture containing hydrogen, by means of a hydrogen concentrator; similarly, pure oxygen can be obtained from air by use of an oxygen–oxygen cell. Applications in hospitals or aircraft suggest themselves.

IX.1 DEUTERIUM OXIDE AS A BY-PRODUCT OF WATER ELECTROLYSIS

Direct contributions of fuel-cell research to the technical electrolysis of water have been recorded in the preceding chapter. It is now possible to decompose water on the technical scale with an applied terminal voltage of 1.6 V or less, using electrode materials far less costly than were formerly necessary.

Heavy water (D_2O) can be obtained as a by-product by suitable *cascade connexion* of the electrolysis cells used in large installations for the electrolytic preparation of hydrogen and oxygen[1]. Deuterium is present in natural water to the extent of 0.015%, and, because of the slower reaction of the heavier isotope in hydrogen evolution, deuterium is enriched in the residual electrolyte. In the cascade system of cells, the water enriched in deuterium from one stage of electrolysis is subjected to further electrolysis in the next stage, so that its deuterium content is progressively increased. In the procedures in use at the present time, 99.7% D_2O is obtained after 15 stages.

The concentration shift between solution and gas phases is expressed in terms of the *separation factor*, S, defined as the quotient of the concentration ratios of heavy to light hydrogen in the two phases:

$$S_{(H,D)} = [(D/H)_{liq}]/[(D/H)_{gas}] \qquad (IX.1)$$

It has been found that the separation factor depends on electrode material, temperature and current density, and values between 2 and 15 can be realized[1,2]. Higher separation factors have been expected for metals with lower hydrogen overpotential[7], so that hydrogen cathodes based on skeleton-nickel (see chapter VIII) should provide very good separation.

In the last few years the dependence of hydrogen–deuterium[3] and hydrogen–tritium[5] separation factors on *overpotential* has been more closely investigated for a number of metals, including the nickel DSK electrode (cf. section IV.1). The results most significant for technical application are collected below, and in section IX.3 some practical implications are discussed.

IX.2 THE DEUTERIUM SEPARATION FACTOR AS A FUNCTION OF CATHODE MATERIAL AND OVERPOTENTIAL

Earlier exploratory measurements by various authors[2,6] have not established any unambiguous relationship between separation factor and current density or overpotential.

In technical electrolyses with nickel, iron or steel electrodes, a slight *increase* of separation factor with rise of current density has been observed. Similarly, Horiuti[6] has found that platinum gives separation factors of 6 to 7 at higher overpotentials (0.3 to 0.5 V), but values only of 3 to 4 at 0.1 V; this he attributed to a change in the mechanism of discharge. On the other hand, for mercury in

$0.1_N H_2SO_4$ (1.5–2.5 % D), Rome and Hiskey[16] found a definite *decrease* in separation factor from 3.8 to 3.1 on increase of current density from 0.1 to 30 mA cm^{-2}.

Vielstich, Schuchardt and von Stackelberg[3] and von Buttlar, Vielstich and Barth[5] chose the following electrode metals for a thorough investigation of the relation between separation factor and overpotential:

1. Smooth metal foil electrodes: Fe, Zn, Ni, Pt and Ag,
2. Sintered electrodes: Fe, Fe (90 %) Ni (10 %), Raney-Ni and Raney-Ag, with Ni supporting skeleton,
3. Mercury.

The Raney-metal electrodes of nickel and silver were DSK electrodes of the type used by Justi and co-workers[7] (cf. section IV.1). The measurements were mainly carried out with use of 5_N or 6_N KOH, since this is the electrolyte used technically, but a few separation factor–overpotential curves were taken with $1_N H_2SO_4$ as electrolyte, for purposes of comparison. Electrolyte solutions were prepared from analytical-grade reagents and double-distilled water, and, to improve the accuracy of isotopic analysis, slightly enriched in deuterium (to about six times the natural deuterium concentration).

The experimental cell was initially swept free of air with argon, and then evacuated. Cathode and anode compartments were separated by a sintered-glass partition. An electronic potentiostat was used to keep the potential of the experimental electrode constant with respect to a reference electrode during each measurement*. The hydrogen evolved was dried with a cold trap, purified by passage through a 'palladium valve', stored in glass vessels, and analysed for isotopic composition by a mass spectrometer. Experimental errors of individual measurements amounted to 2.5 %. All measurements were carried out at 20°C.

IX.2.1   Results of Measurements

IX.2.1.1   The Time-dependence of the Separation Factor

Takahashi, Oka and Oikawa[2] have already found, for electrolyses at constant current density, that the separation factors for metals such as Cu, Ni, Pt and Sn are strongly dependent on time for the first 2–5 hours. On the other hand, the values for Fe, Ag, Au and Hg (all measured with $0.2_N H_2SO_4$ as electrolyte) hardly vary at all.

Analogous measurements with a Ag/Ni DSK electrode in 6_N KOH are illustrated in Figure IX.1. A constant potential is not established until after about 100 hours—when the separation factor shows no further change with time.

Corresponding measurements with *foil electrodes* showed that this effect depends upon a change of surface structure. Vigorous gas evolution causes roughening of the initially smooth electrode surface, causing it to assume a grey to deep black appearance. It is obvious that at high overpotentials and high degrees of covering of the surface with adsorbed hydrogen atoms, these atoms penetrate into the surface layers of the metal. Inside, small cavities are formed in which hydrogen pressure builds up until it is sufficient to disrupt the metal surface. This leads to the formation of a porous surface layer which progressively increases in thickness at the expense of the compact metal substrate. The roughened surface is not, however, stable; it 'recovers' on cessation of flow of current.

* For details of the apparatus and procedure, see reference 5.

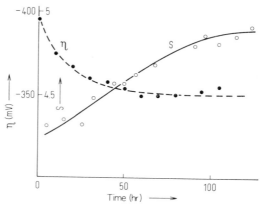

Figure IX.1. Time-dependence of the deuterium separation factor, $S_{(H,D)}$, and the overpotential, η, at an Ag/Ni sinter-electrode in 6N KOH at 100 mA cm^{-2} (after Vielstich, Schuchardt and v. Stackelberg[3]).

    At *sintered electrodes* of iron or nickel, the change of separation factor, or of current density at constant overpotential, was relatively small. These electrodes therefore provided satisfactory reproducibility for the measurements contemplated. In the case of Raney-nickel electrodes of high hydrogen content, care was necessary to ensure attainment of a stationary isotope distribution between metal and solution phases.

IX.2.1.2   *The Potential-dependence of the Separation Factor at Low Overpotentials*

Figure IX.2 shows $S_{H,D}-\eta$ curves for some sintered electrodes in 6N KOH; current densities for a few points on the curves are indicated. The following generalizations can be made from these results.

    At the limit of low current density, the *equilibrium* separation factor for the exchange reaction

$$HDO_{liq} + H_{2\,gas} \rightleftharpoons H_2O_{liq} + HD_{gas} \qquad (IX.2)$$

must be established. It has been theoretically calculated many times, and experimentally confirmed: $S_{equil.} = 3.8$. This value applies for 25°C and the limit of low D concentration. With increasing D/H ratio, the separation factor decreases, ultimately to the value of 3.3, characteristic of the equilibrium $D_2O + HD \rightleftharpoons HDO + D_2$.

    The metal-catalysed *exchange reaction* proceeds by an electrochemical mechanism[5], and is thus characterized by the H_2-H^+ exchange current density. Under the influence of a 'forced' externally supplied current, the separation

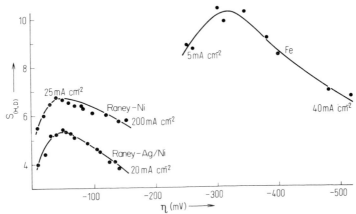

Figure IX.2. Dependence of the deuterium separation factor, $S_{(H,D)}$, on over-potential, η, at Ni-, Ag/Ni- and Fe-sintered electrodes in 6N KOH after stabilization of electrodes by pre-electrolysis for 100 hr at 100 mA cm^{-2} (after Vielstich, Schuchardt and v. Stackelberg[3] and v. Buttlar, Vielstich and Barth[5]).

factor is displaced towards the so-called '*kinetic* separation factor', S_{kin}, to the greater extent the greater the imposed current density. The initial rise seen in the curves of Figure IX.2 indicates that here S_{kin} is greater than S_{equil}.

The S–η curves, however, pass through a *maximum*, at a current density about ten times the exchange current density (determined from the Tafel line, cf. section II.2). At this point it is evident that S must be almost equal to S_{kin}, and it follows that the falling away of S from the maximum must be due to a decrease of S_{kin} with increasing η. This may indicate an alteration of reaction mechanism, but it could also be explained on the basis of small differences in the transfer coefficients, α, for hydrogen and deuterium deposition[5].

The cathodic separation factor has frequently been interpreted in terms of the kinetic model developed by Eyring[19]. According to Horiuti[6], it is determined exclusively by the rate-limiting discharge step. A change of S with overpotential could then be a function of change in mechanism (see Table IX.1). Conway[10] assumes that the extent to which the electrode surface is covered with hydrogen has an influence on the separation factor. By consideration of the desorption step, as well as the possible influence of the tunnel effect, Bockris and Srinavasan[21] have recently obtained a revised series of values for H,D and H,T separation factors (Table IX.1). It is notable that, according to the new theory, the dependence of the separation factor on potential can arise other than by a change of mechanism, and, in principle, the experimental S–η curves can be interpreted with this theory. For confirmation, however, an independent identification of the rate-determining step in the relevant range of potentials is required.

The entire maxima of these curves can be traced out only from measurements conducted with electrodes of large true surface area, because, at the low current densities corresponding to $\eta < 100$ mV, quantities of gas adequate for analysis cannot otherwise be obtained in a practicable time. This is why, for foil electrodes, the curves begin at relatively high overpotentials (Figure IX.3).

The fall from the first maximum is, however, unequivocally identifiable in the curves obtained with smooth electrodes. Then, after passing through a minimum, a second maximum is attained. In this region of high current densities, considerable roughening of the electrode surface is observed

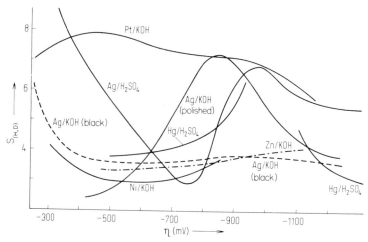

Figure IX.3. H,D separation factors as functions of overpotential at smooth electrodes (after Vielstich, Schuchardt and v. Stackelberg[3] and v. Buttlar, Vielstich and Barth[5]).

(see time-dependence), and the S values vary from experiment to experiment, and as functions of time.

After prolonged electrolysis, *silver foil*, for example, gave the lower, flattened curve denoted Ag (black) in Figure IX.3. If the foil was polished before each measurement, the measurements formed a curve with a pronounced maximum—the upper curve, Ag (polished) in Figure IX.3.

Particularly scattered values were obtained with *platinum*. If however, special care was taken to maintain constant activity of the surface, a reproducible curve (Figure IX.3) was obtained. This involved a pretreatment consisting of polarization by square-wave ac, each positive half-wave taking the electrode to a potential providing a surface oxygen layer.

To summarize:

1. No fixed separation factor can be assigned to any individual metal.
2. The hydrogen–deuterium separation factor is strongly dependent on over-potential or current density, and two maxima can be observed in the separa-tion factor as a function of overpotential*. At low overpotentials, the kinetic separation factor tends towards the equilibrium value (3.8 at 25°C).
3. The surface state of an electrode has an effect on the separation factor, especially for smooth electrodes and high current densities. This leads to a time-dependence of the separation factor.
4. No definite relationship exists between the separation factor and catalytic activity for hydrogen deposition. The factor for iron is higher than that for Raney-nickel. On the other hand, silver shows about as high a separation factor as that of Raney-nickel.

---

* Measurements of the dependence of separation factor on overpotential at Ni, Pt, Ag, Pd, Pb and Hg, with similar results, have been published by Lewis and Ruetschi[10].

Table IX.1. Calculated H,D and H,T separation factors for cathodic hydrogen evolution (after Horiuti[6], Conway[20] and Bockris and Srinavasan[21]).

Hindered reaction step	H,D separation factor according to		Reaction sequence	Separation factors according to Bockris and Srinavasan[21]		Remarks[21]
	Horiuti[6]	Conway[20]		$S_{H,D}$	$S_{H,T}$	
$H^+ + e^- \rightarrow H_{ad}$ (Volmer)	12–13	3.0	Volmer–Tafel Volmer–Horiuti	2.4–3.0 3.8	3.4–4.6 6.2	For large η
$2H_{ad} \rightarrow H_2$ (Tafel)	6–7	9.2	Volmer–Tafel	4.9–6.4	9.4–15.7	For all η (specifically calculated for Ni or Pt)
$H^+ + H_{ad} + e^- \rightarrow H_2$ (Horiuti or Heyrovsky)	3	6.1	Volmer–Horiuti	8.3–10.7 3.4–5.7	19.8–28.7 5.4–7.5	Low and medium η High η (specifically calculated for Ni)

5. Hydrogen pressure over the electrolyte has an effect on the value of the separation factor. Measurements at normal pressure with smooth platinum in KOH solution have shown a steady increase in separation factor from 6 at -200 mV to 8 at -1500 mV in consequence of this[25].

IX.3 THE RELATION BETWEEN HYDROGEN–TRITIUM (H, T) AND HYDROGEN–DEUTERIUM (H, D) SEPARATION FACTORS

The authors of the older work[8] proceeded on the assumption that the ratio $S_{(H,T)}/S_{(H,D)} = 2.1 \pm 0.1$, independently of the electrode material, its surface state, the electrolyte and current density or overpotential. Recently, Bigeleisen[4] has derived the theoretical relation

$$\ln S_{(H,T)}/\ln S_{(H,D)} = 1.38 + 0.06 \ln S_{(H,D)} \qquad \text{(IX.3)}$$

Oestlund and Werner[9] have found, for iron-foil electrodes in KOH, through almost complete decomposition of test samples of water, values in the range $1.35 < \ln S_{(H,T)}/\ln S_{(H,D)} < 1.41$, in good agreement with Bigeleisen's theory.

 To test this relationship further, and to extend the information gained from the deuterium measurements, von Buttlar, Vielstich and Barth[5] have simultaneously followed tritium enrichment in several series of measurements.

 For this purpose tritium-enriched electrolyte was used and measurements of $S_{(H,T)}$ values were carried out by an appropriate counting technique with an error of 3 % or less*.

IX.3.1 Results

A series of tritium measurements in terms of $\ln S_T/\ln S_D$ plotted against overpotential are shown in Figure IX.4. It is clearly seen that this ratio does not depend either on overpotential or current density, or on surface state. The dependence of the tritium separation factor on overpotential is therefore similar to that of the deuterium separation factor throughout (Figure IX.5.). The value of the ratio $\ln S_{(H,T)}/\ln S_{(H,D)}$ is close to 1.4, but varies somewhat from one electrode material to another. Although individual values vary from 1.29 to 1.53, it appears that Bigeleisen's theory is essentially correct.

IX.4 PRACTICAL PROPOSALS

IX.4.1 Suitable Cathodes for Deuterium and Tritium Concentration

The working conditions of large-scale electrolysis plants are adjusted to the economic production of hydrogen and oxygen, but are suitable also for the

---

* For details of the method, see reference 5.

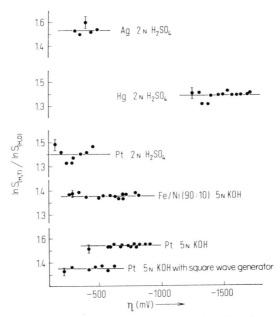

Figure IX.4. Plot of $\ln S_{(H,T)}/\ln S_{(H,D)}$ as a function of over-potential (after von Buttlar, Vielstich and Barth[5]).

extraction of heavy water as a by-product. Conditions found advantageous for the decomposition of water are a current density of 100–150 mA cm^{-2}, 70°C, 6N KOH and atmospheric pressure. The normal commercial electrolysis installations of Norsk Hydro[24] work under these conditions (5–7N KOH or

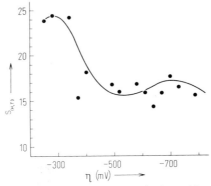

Figure IX.5. Dependence of the tritium separation factor $S_{(H,T)}$ on the overpotential, η, at a sintered iron electrode (after v. Buttlar, Vielstich and Barth[5]); cf. Figure IX.2.

NaOH, steel cathodes), with a hydrogen overpotential of about 300 mV, and a hydrogen–deuterium separation factor between 3 and 4 (in the first step which largely determines the total energy demand*).

Hydrogen–deuterium and hydrogen–tritium separation factors determined at 100 mA cm^{-2} and 20°C by von Buttlar, Vielstich and Barth[5] are collected in Table IX.2; cathodic overpotentials are also included. Entries in brackets are of poor reproducibility, because of the structural changes, mentioned above, of electrode surfaces during electrolysis.

The results show that sintered electrodes of iron or nickel are especially suitable for technical use. Raney-nickel electrodes[7] (or electrodes from skeleton-nickel) show the lowest overpotential whilst retaining a good separation factor. Although iron electrodes show higher separation factors, the overpotentials, and therefore the energy requirements at a given current density, are considerably greater. Raney-nickel electrodes probably provide the best compromise but iron does have the advantages of a high separation factor, even at an overpotential of 1000 mV, and properties which hardly alter under the conditions of electrolysis. At working temperatures of 60–80°C, separation factors 10–20 % lower than at 20°C are to be expected. The effect of gas pressure during electrolysis on the separation factor is also to be taken into account.

Table IX.2. Separation factors and overpotentials at 100 mA cm^{-2} and 20°C (after von Buttlar, Vielstich and Barth[5]).

Electrode	Electrolyte	Overvoltage (mV)	Separation factor	
			$S_{(H,D)}$	$S_{(H,T)}$
Fe (sintered, $\rho = 6.45$)	5N KOH	520	8.2	—
Fe (sintered, $\rho = 7.0$)	5N KOH	800	6.1	11.0
Fe/Ni (90/10, sintered)	5N KOH	650	7.7	16.2
Fe foil (puriss.)	5N KOH	(875)	(6)	(20)
Ni foil	6N KOH	400	(2.9)	—
Raney-Ni (sintered)	6N KOH	75	6.4	—
Ag foil	6N KOH	(520)	(3.7)	—
Ag foil	2N H$_2$SO$_4$	(330)	(8.5)	(24.5)
Raney-Ag/Ni (sintered)	6N KOH	350	4.0	—
Pt foil	5N KOH	390	7.7	17.1
Pt foil	2N H$_2$SO$_4$	(300)	(7.4)	(21.5)
Zn foil	6N KOH	(900)	(3.9)	—
Hg	2N H$_2$SO$_4$	1580	3.3	5.1

* In later enrichment steps, special cathode materials are used which increase the separation factor to about 9.

IX.4.2   Deuterium Enrichment in a Cell with Two Hydrogen Electrodes

In the electrolysis of water, displacement of the H/D distribution between liquid and gas phases occurs only at the cathode; evolution of oxygen at the anode has no effect on the separation process. Now, if the main objective of the electrolytic process is to obtain D_2O, it is suitable to combine the cathodic hydrogen-evolution reaction with a second electrode process that occurs at a potential as close as possible to the hydrogen potential. Thus, a hydrogen-deposition electrode, combined with a hydrogen-dissolution electrode, forms a cell in which deuterium can be enriched with a very low applied terminal voltage, e.g. 100–150 mV[7,11]. Figure IX.6 shows the principle of such an enrichment cell.

In connexion with such cells it is to be noted that the H/D ratio can also be displaced at the anode, provided that not all of the gas taken to the hydrogen-dissolution electrode is converted into water ($\eta_r < 100\%$). The *anodic separation factor*, $S_{(H,D)}^+$ can be defined as the ratio $(D/H)_g$ in the gas taken to the anode divided by the ratio $(D/H)_r$ in the product of the anodic reaction:

$$S_{(H,D)}^+ = (D/H)_g/(D/H)_r \tag{IX.4}$$

Hitherto very little has been known about the behaviour of the heavy isotopes of hydrogen in anodic oxidation, and, in particular, separation factors as such have not been investigated. Schuldiner and Hoare[22], however, have shown that light hydrogen dissolved in palladium is oxidized,

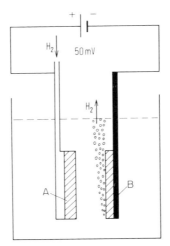

Figure IX.6. Principle of an enrichment cell with hydrogen anode and hydrogen cathode (after Vielstich and Winsel[11]). (A) H_2 diffusion-electrode, (B) H_2 evolution electrode.

at low overpotentials ($0 < \eta < +5$ mV) in 2N H_2SO_4, about twice as fast as deuterium dissolved in palladium (electrolyte 2N D_2SO_4). Rosental, Dolin and Ershler[23] have also found that H atoms adsorbed on platinum are oxidized faster than adsorbed D atoms by a factor of 1.5–2.

These results do not, however, necessarily lead to the conclusion that there is indeed an effective anodic separation factor, since at low overpotentials (<100 mV) isotopic exchange at the metals concerned can compensate the effect of different rates of oxidation, and at high overpotentials the charge-transfer step is seldom rate-determining.

Recently, Lauer and Vielstich[12] have studied this matter by circulation of hydrogen around various gauze anodes (gold, platinum, platinized platinum) during anodic oxidation, following change in the deuterium content of the gas mass-spectrometrically. To ensure good mass-transfer, spinning electrodes (3000–6000 rpm) were used. Over the whole potential range studied, up to overpotentials of 1 V, the expected anodic separation effect could not be established. The measured values of $S^+_{(H,D)}$ lay between 0.95 and 1.1.

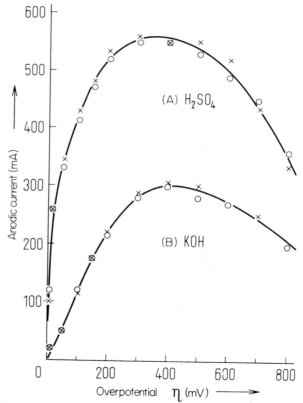

Figure IX.7. Stationary current–potential curves with H_2 ($\bigcirc$) and D_2 ($\times$) bubbling at a rotating Pt gauze anode (geometrical area 25 cm^2) (after Lauer and Vielstich[12]). (A) In 1N H_2SO_4 of natural D content; (B) in 0.1N KOH + 0.5M K_2SO_4, of natural D content.

The current–potential curves obtained with H_2 and D_2 (Figure IX.7) suggest the following explanation for the lack of effective separation. The form of the curves (cf. section III.1) leads to the conclusion that, over the whole potential range, despite strong stirring, it is not the charge-transfer step that is rate-determining (at least not alone)—as is the case with gas-diffusion electrodes. Adsorption and dissociation of molecular hydrogen must be hindered.

This conclusion has been confirmed[12] by experiments with electrodes of palladium–silver or palladium–platinum alloy made in tubular form, through which gas was circulated. Hydrogen dissolved in the alloy and diffused through it to the metal–electrolyte interface, which it reached already in atomic form. In this way, the processes of adsorption and dissociation were spatially separated from the electrochemical reaction proper, and could be selectively catalysed by palladization of the internal surface. The current–potential curves obtained with such an electrode were indeed exponential in the range of potentials between $+100$ and $+700$ mV, and the circulated hydrogen became enriched in deuterium to an extent corresponding to an anodic separation factor of 1.7–1.3 (Figure IX.8)*.

Since the equilibrium $(D/H)_{gas} : (D/H)_{metal} = S_{g,me}$ is established very rapidly, the equilibrium isotopic ratio $(D/H)_{me}$ is attained at the metal–electrolyte interface. The value $S_{g,me} = 1.8$ at 25°C has been found for the extensively studied system Pd–Ag/NaOH. The deuterium content of the metal phase is accordingly poorer by a factor of 1.8 than that of the gas with which it is supplied.

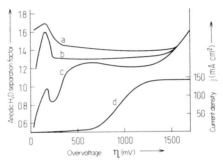

Figure IX.8. Anodic H,D separation factors in dependence on overpotential (a, b, c) and a current–potential curve (d) for a tubular electrode of palladium–silver alloy $(Pd:Ag = 75:25)$ in 4N NaOH at 20°C (after Lauer and Vielstich[12]). Gas-side of electrode palladized, electrolyte side smooth. (a) D content of hydrogen (150 ppm) less than that corresponding to equilibrium with the electrolyte (D content 2600 ppm). (b) D content of hydrogen (660 ppm) in equilibrium with that of the electrolyte (2650 ppm). (c) Hydrogen enriched in deuterium (2040 ppm) compared with the electrolyte (150 ppm).

* E. Wicke and H. Gibmeier (Z. Physikal Chem., N.F., **42**, 112 (1964)) have also studied the isotope effect, using gas highly enriched with deuterium (H:D = 1:1), 0.7N K_2CO_3 as electrolyte, and also a Pd/Ag membrane electrode activated on the gas side by palladization. With increasing current density, the anodic separation factor decreased from 1.9 to 1.5.

It follows from the lower separation factors that are found in practice (1.7–1.3) that the heavy isotope reacts preferentially in the anodic oxidation of hydrogen from Pd–Ag alloy ($H_{me} \rightarrow H^+$), in contrast to the cathodic evolution of hydrogen ($2H^+ \rightarrow H_2$).

The usual technique, involving the porous gas-diffusion electrode can, according to these results, lead to no effective difference in the rates of oxidation of H_2, HD or D_2. The separation factor of the hydrogen–hydrogen cell must very nearly correspond to the separation factor of the cathode.

It is a disadvantage of the hydrogen–hydrogen cell for application to deuterium enrichment that it cannot alone be made the basis of a cascade system that could lead to production of 99 % D_2O. For this, water-decomposition cells must be incorporated in a suitable way[13]. It therefore seems that this method, despite its low energy requirement compared with electrolysis, is inferior to others at present in use for isotope concentration (e.g. H_2S process or distillation of water[14]).

IX.4.3   Recombination of Electrolysis Gases in Fuel Cells

The energy required for the electrolytic preparation of deuterium, or for the electrolysis of water with D_2O as a by-product, can, as discussed above, be decreased by choice of suitable cathodes—in relation to separation factor and overpotential. A low hydrogen overpotential considerably decreases the costs of the production of hydrogen and oxygen.

Additional energy can be saved if the electrolytic gases from the later stages of a cascade electrolysis system are used in a hydrogen–oxygen fuel cell to recover part of the energy required for electrolysis[15]. Only the hydrogen poor in deuterium from the earlier stages can be put to other technical use (e.g. ammonia synthesis). The deuterium-rich hydrogen from a later stage is normally combined catalytically with the oxygen simultaneously evolved, and the water formed returned to an appropriate earlier stage. The later stages, of course, only consume a small fraction of the total energy required, and whether the suggested introduction of fuel cells would be advantageous could only be estimated by a comparative analysis of costs.

IX.5   ELECTROLYTIC HYDROGEN CONCENTRATOR

An electrolytic cell with hydrogen anode and hydrogen cathode, as described in section IX.4.2, can be applied to the electrolytic purification of hydrogen[13,17]. The impure hydrogen, containing perhaps CH_4, N_2, CO is taken to the anode; the cathode provides electrolytically pure hydrogen. This type of purification cell has the advantage of requiring a relatively low working voltage and therefore consumes little energy.

For this application of the hydrogen–hydrogen cell, effective separation of anode and cathode compartments is of course indispensable. A cell with asbestos diaphragm and closely-spaced electrodes, such as described in section IV.1.2.2, is suitable. If the anode gas contains an electrode poison (e.g. CO), the electrode material must be chosen to minimize the poisoning effect, or provision must be made for periodical reactivation of the electrode.

Langer and Haldeman[17] have provided data for a purification cell working at 25°C; cf. Table IX.3. The authors used a membrane cell with electrodes of 5 cm^2 effective area; it provided 15 cm^3 of hydrogen per minute at 2 A. The gas supplied to the cell is saturated with water vapour—as, of course, is the evolved hydrogen, which is of better than 98% purity.

Table IX.3. Electrolysis cell with hydrogen anode and cathode at 25°C (after Langer and Haldeman[17]).

Electrodes	Electrolyte	Diaphragm	Current density (mA cm^{-2})	Terminal voltage (mV)	Anode gas
Pd on stainless steel gauze 11.2 mg cm^{-2}	6N H_2SO_4	Filter paper	60	90	H_2
Pd on stainless steel gauze 11.2 mg cm^{-2}	6N H_2SO_4	Filter paper	200	290	H_2
Pd on stainless steel gauze 11.2 mg cm^{-2}	6N H_2SO_4	Filter paper	200	300	H_2 (40%) N_2 (60%)
Pt on stainless steel gauze 11.2 mg cm^{-2}	6N H_2SO_4	Filter paper	350	410	H_2
Pt on stainless steel gauze 11.2 mg cm^{-2}	6N H_2SO_4	Filter paper	60	990	Town gas: CH_4 (24%) C_2H_6 (3%) CO (18%) H_2 (55%)
Pt on tantalum gauze 9 mg cm^{-2}	2N $HClO_4$	Filter paper	430	260	H_2
Pt on stainless steel gauze 11.2 mg cm^{-2}	5N KOH	Filter paper	40	150	H_2
Pt on stainless steel gauze 11.2 mg cm^{-2}	5N KOH	Filter paper	40	190	H_2 (40%) N_2 (60%)

IX.6 ELECTROLYTIC OXYGEN CONCENTRATOR

There is a very similar application of the analogous cell with two oxygen electrodes for the electrolytic preparation of pure oxygen[13,18]. Pure oxygen can be obtained as anodic product from air supplied to the cathode with a cell terminal voltage of less than 1 V. With suitable choice of working conditions and electrode materials, this cell could also be used for the preparation of hydrogen peroxide.

The principle of the oxygen concentration is as follows. The cathode, supplied with air, acts as an oxygen reduction electrode, as in a primary or fuel cell, according to

$$O_2 (\text{from air}) + 2H_2O + 4e^- \rightarrow 4OH^-$$

The anode works as an oxygen evolution electrode

$$4OH^- \rightarrow O_2 + 2H_2O + 4e^-$$

so that the overall reaction is

$$O_2 \ (\text{from air}) \rightarrow O_2 \ (\text{pure})$$

Accordingly, 4 faradays are required for the preparation of a mole of oxygen, i.e., 3350 A hr provide 1 kg of oxygen, assuming 100% current efficiency. The energy consumption at a given total current is a linear function of the voltage

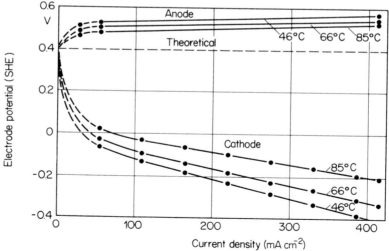

Figure IX.9. Current–density overpotential curves for individual electrodes of an oxygen–oxygen cell (after Tomter[26]); area 58 cm^2; air-flow, twice stoichiometric; air pressure, 0.7 atm above atmospheric; oxygen pressure, atmospheric; electrolyte, 25 per cent KOH.

required to maintain that current density. For a partial pressure of oxygen in air of 0.21 atm (P_1) and a delivered pressure of pure oxygen of 1 atm (P_2), the cell voltage at 25°C can be calculated as

$$E = -\frac{RT}{4F} \ln \frac{P_2}{P_1} \sim 0.01 \text{ V}$$

In practical operation a higher terminal voltage is required because of the incidence of transfer overpotential and unavoidable ohmic potential drop in the electrolyte. This is apparent from the data presented in Figure IX.9.

A unit working on this principle has been built by the Allis-Chalmers Manufacturing Company, Milwaukee, Wis., U.S.A.[26], and is illustrated in Figure IX.10. A hydrophobic cathode and porous anode are separated by a KOH-soaked asbestos membrane which acts as electrolyte. The cathode compartment is supplied with a flow of air; the anode compartment is filled with KOH solution which circulates from a storage tank, serving to remove heat and to regulate the concentration of KOH. The whole unit consists of 15 cells, each of about

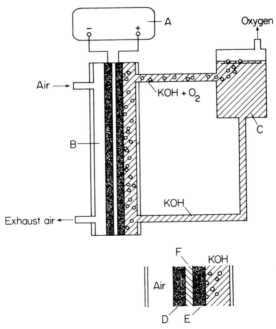

Figure IX.10. Schematic diagram of the oxygen–oxygen cell for preparation of pure oxygen (after Tomter[26]). A, dc source; B, gas space; C, KOH storage vessel; D, hydrophobic porous cathode; E, porous anode; F, asbestos matrix with KOH electrolyte.

100 cm^2 electrode area, working at 220 mA cm^{-2}, 0.7 V and 70°C. The oxygen produced bubbles through the KOH solution and is collected in a 27 l vessel and stored at 2.1 atm. Current and rate of flow of air are regulated to suit the operating conditions, and the whole unit works automatically. It has a capacity of 1 l min^{-1} of pure oxygen and an operating life of more than 1000 hr. About 450 g of oxygen are obtained per kW hr; this is only 30 % of the energy requirement of an equivalent water electrolysis unit—but, of course, the latter provides hydrogen as well. A simplified portable unit of about half this capacity has also been built. A very similar electrolytic oxygen generator has recently been described by Wynveen and Montgomery[27].

Possible applications of such generators seem to be limited to those calling for small supplies of the purest oxygen. Cathode exhaust gas, freed from residual oxygen might be of use as nitrogen (plus noble gases).

REFERENCES

1. M. Benedict, Geneva Conference on Atomic Energy 1955, 8 P 819.
2. A. Eucken and K. Bratzler, Z. physik. Chem., Abt. A **174**, 273 (1935); G. S. Subramanian, Bull. Centr. Electrochem. Res. Inst. **2**, 66 (1955); Y. Takahashi, S. Oka and M. Oikawa, Bull. Chem. Soc. Japan **31**, 220 (1958).
3. W. Vielstich, H. Schuchardt and M. v. Stackelberg, Z. Elektrochem., Ber. Bunsenges. physik. Chem. **63**, 1014 (1959); **67**, 645 (1963).
4. J. Bigeleisen, Report STI Publ. 39, Intern. Atomic Energy Agency, Vol. **1**, 161 (1962).
5. H. v. Buttlar, W. Vielstich and H. Barth, Ber. Bunsenges. Physik. Chem., **67**, 650 (1963).
6. J. Horiuti, Z. Physik. Chem. N.F. **15**, 162 (1958).
7. E. Justi, M. Pilkuhn and A. Winsel, Hochbelastbare Wasserstoff-Diffusionselektroden für Betrieb bei Umgebungstemperatur und Niederdruck, Abh. Mainzer Akad. Nr. 9 (1959), Steiner, Wiesbaden.
8. S. Kaufman and W. F. Libby, Physic. Rev. **93**, 1337 (1954).
9. H. G. Oestlund and E. Werner, Report STI Publ. 39, Intern. Atomic Energy Agency, Vol. **1**, 95 (1962).
10. G. P. Lewis and P. Ruetschi, J. Physic. Chem. **66**, 1487 (1962).
11. W. Vielstich and A. Winsel, Ger. Pat. 1023017 (1956).
12. H. Lauer and W. Vielstich, Ber. Bunsenges. Physik. Chem., **69**, 538 (1965).
13. W. Vielstich, unpublished.
14. E. W. Becker, Angew. Chem. **68**, 6 (1956).
15. E. Justi, Ger. Pat. Appl. 1051820 (1955).
16. M. Rome and C. F. Hiskey, J. Amer. Chem. Soc. **76**, 5207 (1954).
17. S. H. Langer and R. G. Haldeman, Science (Washington) **142**, 225 (1963).
18. S. H. Langer and R. G. Haldeman, J. Physic. Chem., **68**, 962 (1964).
19. B. Topley and H. Eyring, J. Chem. Physics, **2**, 217 (1934).
20. B. E. Conway, Proc. Roy. Soc. London, Ser. A **256**, 128 (1960).
21. J. O'M. Bockris and S. Srinivasan, J. Electrochem. Soc. **111**, 844, 853 u. 858 (1964).
22. S. Schuldiner and J. Hoare, J. Electrochem. Soc. **105**, 278 (1958).
23. K. Rozental, P. Dolin and B. Ershler, Acta physicochim. URSS **22**, 213 (1946).
24. A. K. Larsen, Lecture at GDCh-Fachgruppe, Angewandte Elektrochemie on 26.10.1962 in Frankfurt/M.

25. H. Bahe, Diplomarbeit, Bonn, 1967.
26. S. S. Tomter, *Chem. Eng. Progr.* 62, No. 5, 66 (1966); 58th Ann. Meeting, Amer. Inst. Chem. Eng., Philadelphia, 1965.
27. R. A. Wynveen and K. M. Montgomery, *J. Electrochem. Soc.,* **114**, 589 (1967).

CHAPTER X

The 1967 State of Development;
Possibilities of Future Applications

The state of technical development in the fuel cell field has been frequently reviewed from various points of view in the last several years[1-6]. In this closing chapter, the developments that have been taken to the stage of prototype battery construction are compared with each other*, in the light of the most recent data, presented in tabular form. The comparison also includes conventional sources of electrical energy (accumulators, primary cells, combustion engines and generators).

Following this, possible applications of the new electrical energy sources are discussed. Of the fuel cells that have been studied, only three types have as yet been brought to a stage of development adequate for practical application. These types are:

1. *Hydrogen–oxygen (or air)*
2. *Hydrazine–oxygen (or air or hydrogen peroxide)*
3. *Methanol (or formate)–oxygen (or air)*

For application in space travel, the hydrogen–oxygen cell at present has a clear advantage because of its low weight and volume, including the liquefied reactants, hydrogen and oxygen. It has the further advantage of producing potable water.

In terrestrial applications, fuel cells meet considerable difficulties, technically and economically, in infiltrating the territory of the established energy sources. It is, however, to be expected that in the near future fuel

---

* The construction and mode of action of the cells discussed in chapters IV and V correspond with the stage of development in 1967.

cells will not only serve various military purposes, but also that limited application will be found for them in signalling units, radio stations, emergency power sources, transistorized equipment and eventually also in trucks, fork-lifts, buses and boats, as well as reserves for peak loads at small power stations (cf. chapter VIII).

The topical problem of electric traction is treated in some detail. Experience to date suggests that smaller electric vehicles are best powered by improved lead accumulators, but newly developed metal–air cells (e.g. zinc–air) are discussed in this connexion. Fuel cells play a part in a second stage of evolution, namely the 'Hybrid System' which has already been explored. In this, secondary batteries (e.g. lead accumulators) are coupled with fuel cells (e.g. hydrogen–oxygen cells, with associated converters for cracking ammonia, methanol or hydrocarbons); in periods of light or zero load, the fuel cells charge the accumulators, which provide the excess energy required for starting and acceleration.

X.1   GENERAL COMMENT ON THE STATE OF TECHNOLOGICAL DEVELOPMENT IN THE U.S.A. AND IN EUROPE

Developments in the U.S.A. have been promoted and substantially influenced by the demand for direct application of fuel cells for use in space vehicles and for military purposes. In Europe, on the other hand, the direction of research has been towards commercial applications. Of the numerous cells that have been investigated and discussed, it seems that only three types have at present reached a stage of development adequate to be of practical interest. They are:

1. Hydrogen–oxygen (or air) cells
2. Hydrazine–oxygen (or air, or hydrogen peroxide) cells
3. Methanol (or formate)–oxygen (or air) cells

The hydrogen–oxygen cell—aside from use in space travel—usually requires indirect provision of hydrogen as fuel gas by the decomposition of ammonia or the re-forming of methanol or hydrocarbons. Hydrogen production by such means can be carried out either in a separate reactor, or directly in the anode compartment of the cell. In the latter case, the gas side of the hydrogen electrode must be provided with the catalyst required by the re-forming reaction.

X.1.1   Developments in the U.S.A.

Developments in the U.S.A. have been characterized by an extraordinarily strong interest in new kinds of sources of electrical power, arising from the demands of *space travel* and of *military requirements*. This has meant that factors such as safety in operation and low weight and volume per kW hr have taken precedence of economy in costs in the choice of battery type and construction. It is therefore no accident that the centre of furthest development has been the *hydrogen–oxygen cell*, with uniform use of noble-metal catalysts (10–40 mg cm^{-2} of platinum and palladium). The lead in development has been taken in this field by the General Electric Company (2 kW batteries with ion-exchange membranes for the Gemini space-craft, weight per kW hr ca. 1.5 kg), the Allis-Chalmers Company (batteries up to 5 kW with asbestos membranes, also for space projects, but successfully tested for surface vehicles), the Pratt and Whitney Company (2 kW batteries of the Bacon type for Apollo moon-craft), and the Union Carbide Corporation (units up to 30 kW with Kordesch carbon electrodes, used, for example, in experimental operation of a 'minibus', equipped with a 32 kW battery of 10 kg kW^{-1}, without accessories).

On the other hand, work on liquid fuels has been undertaken on a large scale since about 1960. Hydrazine batteries of 200–5000 W have been built of weight and volume per kW between 9 and 50 kg kW^{-1} and 6 and 100 l kW^{-1} respectively. The leading companies in this field are Allis-Chalmers (golf carts, fork-lifts, miniature submarine); Union Carbide (motor cycle) and Monsanto (jeep with 5 kW modules). Alcohol cells (with acid electrolytes) have been intensively studied only by the Esso Company. The advantages of such cells operating with liquid fuels (ease of handling, low cost, high energy density, storage facility, etc.) make them of interest for civil use. As a consequence of intensive research in the U.S.A., industry is increasingly seeking possibilities of application of these cells for normal, commercial purposes—the hydrogen–oxygen cells developed for the Gemini battery have, for instance, been tried out for use in buoys and signal lights (units of 5 W continuous rating). The decisive criterion for future development in the U.S.A. will be whether economic production methods can be established for batteries mainly manufactured up to now without regard to either cost or commercial application.

X.1.2   Developments in Europe

Research in Europe, in contrast to that in the U.S.A., has from the start been directed to commercial applications. Although there has been some not un-generous support from public funds (e.g. in France, the U.K. and Sweden), expenditure on the development of fuel cells has not been commensurate with their future possibilities or with the resources at the disposal of the American research groups. This in no way indicates, however, that research in Europe has been less promising, or has been confined to side issues, but only that the main effort has been differently directed. This can be seen in the more extensive development of cheaper electrodes not incorporating noble metals (Raney-nickel and nickel boride for fuel electrodes, active carbon and silver for oxygen electrodes), and in the proportionally greater attention given to economically working cells using liquid fuels. *Hydrogen–oxygen batteries* have been made in

Europe predominantly in France (Comp. Générale d'Électricité, Comp. Française Thomson Houston); England (Shell, Electric Power Storage Ltd.); Germany (Varta, Siemens); the U.S.S.R. (Moscow Power Institute) and Switzerland (Brown Boveri u. Cie).

One of the technically most advanced projects is the 200 kW battery built by ASEA in Sweden for submarine propulsion; 20 kW units have been successfully tested for more than 1000 hr. The electrodes work without noble metals; nickel boride is used for fuel electrodes and silver for the oxygen electrodes. Hydrogen is provided by thermal decomposition of ammonia.

In general, it appears that at the present time the indirect combustion of ammonia or of hydrocarbons by preliminary cracking or reforming is the only practicable way of using these cheap fuels in fuel cells. As yet, the main difficulty with hydrocarbons used in this way is that the purification of the gas mixture from the reformer requires sensitive and expensive palladium–silver membranes.

The potentialities of *hydrazine cells*, as portable and cheap energy sources of high storage capability, have long been realized in Germany (Ruhrchemie, Varta, Siemens). Recently the particular type, hydrazine–hydrogen peroxide, has been studied by the Société Alsthom, Paris, and batteries of up to 2 kW have been built without use of noble metals, with weight and volume per kW of $6–8 \text{ kg kW}^{-1}$ and $2–3 \text{ l kW}^{-1}$. Such batteries have civil (reserve power supplies) as well as military applications.

Finally, *methanol– or formate–air cells* have a future in applications specially calling for prolonged operation without attention (navigation-buoys, television relay stations, equipment of low power requirements). It is a disadvantage that in the cells so far available, the electrolyte (KOH) is consumed in use, and also that, as yet, noble metal (at least palladium at about 1 mg cm^{-2}) is needed for the fuel electrode. In Europe alkaline methanol cells have been built by Bosch AG, Battelle Institut, Frankfurt, Brown Boveri u. Cie, and by Varta AG. In addition, other development work is of interest, aimed at using methanol as fuel, without noble metals, in *acid* solutions (Shell, Bosch, AEG).

Cells for medium and high working temperatures with H_2SO_4, H_3PO_4 or carbonate–bicarbonate equilibrium electrolytes (100–200°C), or molten salt (500–700°C) or solid (> 1000°C) electrolytes, are all intended for the direct use of hydrocarbons (CH_4, C_3H_8, etc) and, at the present time, are still at the laboratory stage of development.

X.2 THE CHARACTERISTIC DATA FOR FUEL CELLS IN COMPARISON WITH TRADITIONAL SOURCES OF ENERGY

As a basis for statements about the cells, data for the more important fuels, oxidants and electrolytes[7-9] are assembled in Table X.1. The cell voltages quoted are in each case those for optimum working conditions. These values,

Table X.1. Data for the more important fuels and oxidants.

Fuel or oxidant	Molecular weight	Density (g cm^{-3})	No. of electrons transferred	Capacity (A hr kg^{-1})	(A hr l^{-1})	Cost per kg DM	$	Cell voltage (V)	Cost per kW hr DM	$	Comments
Hydrogen electrolytic, 250 atm	2	0.021	2	26800	562	11.0 to 15.0	3.0 to 4.0	0.9	0.46 to 0.63	0.12 to 0.17	
Hydrogen, liquid	2	0.071	2	26800	1900	170.0		0.9	7.00		Current efficiency >95%
Hydrogen from methanol	32	0.79	6	5025	3970	0.18 to 0.20		0.9	0.04 to 0.05		
Hydrogen from hydrocarbons (benzene)						0.08 to 0.11		0.9	ca. 0.05		
Hydrogen from NH$_3$ cracking (10 atm)	17	0.68	3	4730	3220	0.18		0.9	0.042		
Hydrazine hydrate 80%	50	1.03	4	1715	1770	6.00	1.50	0.7	5.00	1.25	Current efficiency 70–90%
Hydrazine hydrate 100%	50	1.03	4	2145	2205			0.7			
Ammonia 10 atm at 20°C	17	0.68	3	4730	3220	0.18 to 0.20	0.095 to 0.10	0.5	0.08	0.04 to 0.06	
Methanol	32	0.79	6	5025	3970	0.18 to 0.20	0.10 to 0.16	0.5	0.07 to 0.08		
Potassium formate	84.1	1.91	2	636	1205			0.6			Current efficiency 85–98%
Sodium formate	68	1.92	2	787	1490	0.60		0.6	1.28		
Formic acid, techn. 85%	46	1.22	2	1165	1420	0.95		0.6	1.35		
Glycol	62	1.11	8	3460	3840	1.00		0.7	0.42		
Methane, 250 atm	16		8	13400		0.50		0.4	0.093		

Propane, liq.	44.1	0.50	20	12150	6075	1.00	0.04 to 0.10	0.4	0.21	0.01 to 0.02	
Carbon monoxide 99 %, 250 atm	28		2	1915		0.15		0.5	0.16		
Aluminium	27	2.70	3	2980	8050	2.25		1.4	0.56		
Zinc	65.4	7.14	2	820	5850	1.90 to 2.00	0.29 to 0.44	1.25	1.85 to 1.95	0.28 to 0.43	Current efficiency 80–85 %
Magnesium	24.3	1.74	2	2205	3840			1.0			
Sodium	23	0.97	1	1165	1130		0.38 to 0.66	1.4		0.23 to 0.41	Operating voltage of an amalgam cell = 1.4 V
Lead	207.2	11.34	2	259	2940	0.40 to 0.50		2.0	0.13 to 0.17		
Oxygen 99.5 %, 150 atm	32	0.333	4	3345		0.80		0.9			
Oxygen, electrolytic, 250 atm	32		4	3345	1115		0.09 to 0.18	0.9	0.27	0.03 to 0.06	Current efficiency >95 %
Oxygen, liq.	32	1.14	4	3345	3820	2.20		0.9	0.73		Current efficiency >95 %
Hydrogen peroxide 40 %	34	1.15	2	632	728	1.25		0.7	2.83		} Current efficiency
Hydrogen peroxide 70 %	34	1.25	2	1108	1385	2.30		0.7	2.97		35–70 %
KOH	56		1	960		0.75 to 0.91		0.6	1.30 to 1.60		
NaOH	40		1	1340		0.37 to 0.44		0.6	0.46 to 0.55		

with the theoretical capacities in A hr kg^{-1} and cost per kg, form the basis of calculation of costs per kW hr, assuming 100% utilization of the working materials. This assumption is not justified in practice, especially for aluminium, hydrogen peroxide and also hydrazine ($\eta_r \sim 80\%$).

The cheaper fuels, such as methane and propane, are less active electrochemically, and there are no technically developed batteries designed for using them directly. Fuels such as ammonia, methanol and certain petroleum fractions (e.g. benzine) come into consideration for indirect utilization in hydrogen–oxygen cells by introduction of a reforming treatment.

Electrolyte costs become significant when the electrolyte is itself consumed in the electrochemical process (the reaction of carbon-containing fuels, such as methanol or glycol, in alkaline electrolytes). Thus, although caustic soda is an economically favourable choice, deterioration of properties and limitation of working life are unavoidable.

In Tables X.2(a) and (b), characteristic data for prototypes at present known are collected, operating costs being calculated from the information provided by Table X.1. Large-scale production costs for batteries and their auxiliary equipment are hard to estimate, because no considerable numbers of them have yet been manufactured.

Data for various types of hydrogen–oxygen cells are compared with each other in Table X.2(a); data for hydrogen–oxygen batteries for space vehicle application are given separately in Tables X.5 and X.6. The performance of cells with skeleton-nickel as hydrogen catalyst is considerably inferior to that of cells with noble metals but is adequate for most of the practicable applications. The overall weight and volume per kW hr depend very much on the form of the reactants—gases in steel cylinders, liquid hydrogen and liquid oxygen, or liquid hydrogen-containing fuel, such as ammonia or methanol with air as oxidant. The net weight per kW hr of hydrogen and oxygen without containers or battery, amounts to 0.37 kg kW^{-1} hr^{-1} at a terminal voltage of 0.9 V.

Table X.2(b) sets out batteries which use liquid fuels directly. Batteries of higher rating per unit weight operate either with hydrazine or methanol, acid electrolytes, and electrodes of high noble-metal content. On the other hand, those using methanol or formate and alkaline electrolyte are best suited for low-rating densities and prolonged unattended operation and their noble-metal requirements can be kept relatively low. The performance of the hydrazine cell depends very little on whether skeleton-nickel or noble metal is used as catalyst.

The weight and volume per kW hr required by fuel cells is very much more dependent on duration of discharge than is the case for accumulators (Figure X.1). For short periods of discharge (a few hours), it is the weight of the battery that is the main consideration, but for long periods the weight of the operating materials and their containers becomes decisive. These factors are of special significance in space vehicle applications (see below). Very favourable energy per unit weight and volume data are characteristic of methanol/KOH/air cells

Table X.2. Characteristic data for fuel cell prototypes; (a) hydrogen–oxygen batteries*.

Type	$H_2|KOH|O_2$ (air)	$(NH_3)H_2|KOH|O_2$	(Methanol) $H_2|KOH|$air	(Hydrocarbon) $H_2|KOH|$air
Company	Union Carbide Corporation, U.S.A.	ASEA (Sweden)	Shell Research Limited, Thornton Research Centre, U.K.	Allis-Chalmers Co. (FC unit), Engelhard Ind. Inc. (Reformer)
General description	Parallel plate electrodes, 1 mm apart. Free electrolyte, pump-circulated. Gas circulation by Venturi tube device, gas pressure 0.05 atm above atmospheric at the electrodes. Module size, 1 kW. Temperature control by air stream. Water removal over gas-circulation loop	20 kW batteries of 250 W modules each 8 cells in parallel. Bacon double layer electrodes and free electrolyte. Hydrogen normally from associated NH_3 cracking unit	Steam–methanol reformer, H_2 separation by Pd–Ag. Start-up power from Pb–acid batteries. 2 modules of 62 cells in series (bipolar construction). Gas pressures 0.15–0.25 kg cm^{-2}. Over-all efficiency, 20–25 %	Steam-hydrocarbonreformer,H_2 separation by Pd–Ag. 4 parallel modules, 68 cells each in 34 parallel pairs. Control of temperature and water balance by circulation of 35 % KOH, with use of water transport membrane. Air (CO_2-free), 3–5 psi; H_2, 1–3 psi. 28 V dc to 120 V ac (60 cycles)
Electrolyte	9–12N KOH	30 % KOH	KOH	35 % KOH
Temperature (°C)	65	70–80	65	80
Anode	0.9 mm composite metal electrode with sintered nickel as support for several carbon–Teflon layers, 1–3 mg Pt cm^{-2}	Sintered nickel with nickel boride	Porous plastic substrate (PVC, 0.76 mm; pore size, 5 μ), silver metallized surface, activated with platinum metals <1 mg cm^{-2}	Sintered nickel with Pt and Pd
Cathode	Similar to anode. Spinel, no noble metal (or 1 mg Pt cm^{-2}, 50 mV difference)	Sintered nickel with Ag	As anode	Pt, Pd black plus Teflon on Ni-screen
Electrode size (cm^2)	97 and 125	about 300	1170	500
Current density (mA cm^{-2})	50–100 (150)	150	60	135
Terminal voltage per cell (V)	0.87–0.82	0.7	0.6–0.7	0.83
Rating (W)	3000	20 000	5000	7300
Parasitic power (W)			900	2368
Weight (kg)	19.1 for 78 cells (1 kW at 50 mA cm^2)		600	500

* See Tables X.5 and X.6 for hydrogen–oxygen cells developed for application in space.

Table X.2(a)—*continued*

Type	H_2|KOH|O_2 (air)	$(NH_3)H_2$|KOH|O_2	(Methanol) H_2|KOH|air	(Hydrocarbon) H_2|KOH|air
Volume (l)	15.9			1000
Battery (kg kW⁻¹)	17	ca. 20		68.5
Battery (l kW⁻¹)				137
Stack (kg kW⁻¹)	10			19
Stack (l kW⁻¹)	14			11
⎰kg per kW hr for 5000 hr ⎱operation, including working materials and containers			1.07	
Operation cost per kW hr			(fuel) 3d–6d	
Working life	> 2000 hr	> 1000 hr	Electrodes and catalyst 1000 hr, unit 10 000 hr	Tested for 200 hr (goal 2000 hr)
Remarks	Use of air requires separate CO_2 purifier, otherwise 50 g KOH per kW hr are consumed		Water/methanol ratio of 1.2:1 completed 1964, based on 1962 technology. A unit built today would be much more compact and more efficient	Efficiencies: reformer 68%, fuel cell 56%, inverter 80%, overall 25%. Start-up time 45 min

Table X.2. Characteristic data for fuel cell prototypes; (b) liquid fuel batteries.

Type	N_2H_4|KOH|air	N_2H_4|KOH|air	N_2H_4|KOH|O_2	N_2H_4|KOH|H_2O_2	CH_3OH|KOH|O_2	CH_3OH|H_2SO_4|air	Formate|KOH|air
Company	Union Carbide Corpn.	Monsanto Research (U.S.A.)	Arbeitsgruppe Aachen	Arbeitsgruppe Aachen	Robert Bosch GmbH, Stuttgart	Esso Research and Engineering Co. (U.S.A.)	Arbeitsgruppe Aachen
General description	78 cells in parallel pairs. Free electrolyte between electrodes. Anolyte pump-circulated. Automatic feed of 100% $N_2H_4.H_2O$. 5 W for air-cooling of electrolyte tank, 5 W air compressor for supply of air to cathode	KOH/N_2H_4 and air supplied from reverse sides of electrodes separated by asbestos (0.4 mm). Air (blower) 4–10 times stoichiometric, air-cooled heat exchanger. N_2H_4 control by voltage difference between N_2H_4 and H_2O electrolysis	Anode and cathode firmly pressed to asbestos membrane. Waffle-like (indented) cell walls pressed from fibre-glass reinforced epoxy-resin. Cell immersed in anolyte. Circulation regulated by gas and heat evolution. Excess p_{O_2} = 0.5–0.6 atm. 2 × 10 cells per module, 6 mm per cell. Fuel fed by common peristaltic pump electronically controlled by 4 probe electrodes	Anode and cathode firmly pressed to asbestos membrane. Cell walls serve for heat exchange; cell parts pressed from epoxy-resin. 13 cells per module. Electrolyte supply by synchronous gear-pump (1 l min⁻¹ per module). Electronic control of fuel feed by anode or cathode potentials	2 modules of 24 double cells (O_2 electrodes outermost) in parallel. Voltage regulated to 12 V, impulse control. Automatic fuel feed and regulation to optimal working conditions. Heating-up time 20–30 min	20 cells in series, voltage regulated to 6 V. Cell casings of polypropylene. Electrodes separated by 2 asbestos and 1 ion-exchange membrane (Permion 1010). Air automatically brought to humidity appropriate to working temperature in water economy unit. Anolyte, pump circulated. Automatic control of H_2SO_4 and CH_3OH concentrations	Electrolyte container with renewable pairs of (cylindrical) electrodes. Air supply by diffusion at atmospheric pressure. Renewal of fuel-KOH mixture by exchange of container. dc transformer (0.6 to 6 V) in lid of cell
Electrolyte	5–10N KOH + 0.3–0.7% $N_2H_4.H_2O$	5N KOH + 0.5–1.5M $N_2H_4.H_2O$	6–8N KOH + 0.5M $N_2H_4.H_2O$	8–4N KOH + 0.5M $N_2H_4.H_2O$ + 0.5–1M H_2O_2	6N KOH + 3M CH_3OH	3.7M H_2SO_4 + 0.75M CH_3OH	6.0N KOH + 7.0M formate
Temperature (°C)	55–65	60–75	80–90	40	65–70	60–80	−30 to +30
Anode	Sintered Ni + ~1 mg cm⁻² Pd-Catalyst	Sintered nickel with Pd	Raney-Ni, flame-sprayed on Ni gauze (50 mg Ra-Ni cm⁻²)	Raney-Ni, flame-sprayed on Ni gauze (80 mg Ra-Ni cm⁻²)	Nickel, with platinum metal alloy on Raney basis	20 mg platinum metal cm⁻² on tantalum screen	Sintered nickel foil, electroplated with Pd/Pt (9/1), 2–4.5 mg cm⁻²
Cathode	Carbon on porous nickel	Teflon-bonded carbon with Pt addition	Silvered sintered nickel	Silvered Ni gauze 30–35 mg Ag cm⁻²	Nickel with silver, double layer construction	American Cyanamide AA1 (Teflon mixed with Pt. 9 mg Pt cm⁻²)	Active carbon, with 10–20% by wt polyethylene
Electrode size (cm²)	137 (11 × 12.5)		250	400	50	325 (22.5 × 14.5)	Anode, 150; cathode 115
Current density (mA cm⁻²)	60–65	65–100	200	200 (max 550)	40	40	3–20
Terminal voltage per cell (V)	0.72	0.78	0.6	0.86	0.5	0.4	0.6–0.7

Table X.2(b)—continued

Type	N_2H_4|KOH|air	N_2H_4|KOH|air			N_2H_4|KOH|O_2	N_2H_4|KOH|H_2O_2	CH_3OH|KOH|O_2	CH_3OH|H_2SO_4|air	Formate|KOH|air
Rating (W)	300 (28 ± 1.5 V dc)	60	300	20 000	1200 (4 modules, each 12 V, 50 A)	max 6400 (at 7.3 V per module) 3600 (4 modules at 12 ± 1 V)	100	80	0.36 (600 mA, 0.6 V)
Parasitic power (W)				600			9	15	Nil
Weight (kg)	15 (with 4.5 kg fuel for 12 hr operation)	6.6		240	80	800	160	43	2.35 (including 1.65 kg fuel-electrolyte mixture)
Volume (l)	25	11		12				100	1.85 (14.5 × 9.5 × 13.5 cm^3)
Battery (kg kW^{-1})	50	110	50						1 charge, 8.5; 10 charges, 6.2
Battery (l kW^{-1})	83	183	95						1 charge, 6.6; 10 charges, 4.2
Stack (kg kW^{-1})	13.5	8.6			21	14 (without electrolyte)			
Stack (l kW^{-1})		6.3			7.5	10.3			
{ kg per kW hr for 5000 hr operation, including working materials and containers					5 (O_2 gas), 1.6 (O_2 liq)	ca. 3.0			6.25
Operation cost per kW hr	$1.90				ca. DM 8	ca. DM 10	DM 1.20		DM 1.80
Working life	Electrolyte change after 500 hr for removal of carbonate	450 hr			not yet established	not yet established	> 1 yr	> 500 hr	Regeneration of carbon-air electrode after 10–20 discharges
Remarks	Good overload capacity of 500 W, 27.5 V. Supply of air (not CO_2 free) up to 20 times stoichiometric, including air for heat and water removal: N_2H_4 A hr efficiency 65–75%.	N_2H_4 current efficiency, 40–55%. Much NH_3 formation. One fuel charge of 60 W unit = 0.52 kg ≡ 390 W hr			Reaction water evaporates $N_2H_4 \cdot H_2O$ hr^{-1}, i.e. 1 kg per kW hr. N_2H_4 current efficiency, 80%	No water removal. 2450 g $N_2H_4 \cdot H_2O$ hr^{-1}, 2.1 kg H_2O_2 per kW hr. Current efficiencies: N_2H_4 80%, H_2O_2, 50%	To date, ca. 50% CH_3OH current efficiency on a charge for 15 hr	10 mW cm^{-2} at 20°C; 2 mercury dry cells for start-up and stand-by conditions	1 charge provides 775 hr continuous operation

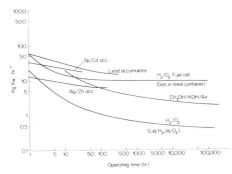

Figure X.1. Battery weight (including fuel) per
kW hr, as a function of operating period.

for operation times of more than 1000 hr. At 5000 hr, their $kg\,kW\,hr^{-1}$ are about one third of that for hydrogen–oxygen cells supplied with gases from steel cylinders. The weight of the cylinders imposes a lower limit of about 10 kg $kW\,hr^{-1}$* (cf. data on storage of liquefied gases).

Table X.3 contains the data for conventional batteries[8] required for comparison (first costs are, of course, included in costs per kW hr). The more important differences from fuel cells are as follows:

1. Current densities for continuous use are relatively low.
2. Weight and volume per kW hr are practically independent of operating time in the case of accumulators. For working periods of more than 5 to 10 hours, fuel cells are, in general, superior to accumulators in this respect (although the Ag/Zn accumulator requires only 10 kg per kW hr, its working life amounts only to 3 to 6 months, and its initial cost is high).
3. Conventional batteries (except the Ag/Zn accumulator) are superior in working life.
4. The kW hr costs for primary cells and the Ag/Zn accumulator are higher, but those for the lead and Ni/Cd accumulator are lower, than the costs of fuel cells.

There has been development recently of so-called 'metal–air cells'. These consist of combinations of the oxygen or air electrodes developed in the course of fuel cell research with negative electrodes of customary primary and secondary cells. Table X.4 shows performance data of such primary and secondary batteries, working with zinc, magnesium or iron as active materials for the negative electrodes (cf. also chapter V). These batteries are, however, mostly in the laboratory stage of development, and how far they may be suitable for technical application is yet to be seen.

---

* With special low carbon chrome molybdenum steel and welded construction 3.7 kg per kW hr (at 0.8 volt terminal voltage) and 2.1 kg per kW hr for hydrogen alone have been obtained[10].

Table X.3. Data for accumulators and primary cells, compiled by Schmier (primary cells), Thurm (Ag/Zn and Ag/Cd accumulators) and Witte (Pb and Ni/Cd accumulators) (cf. S. W. Wagner, Stromversorgung electronischer Schaltungen und Geräte, Hamburg, R. v. Decker's Verlag, 1964).

	Lead accumulator	Nickel–cadmium accumulator	Silver–zinc accumulator	Silver–cadmium accumulator	Air–zinc cell	MnO_2 dry cell
Construction and characteristic features	A number of parallel positive and negative plate electrodes in alternating sequence separated from each other by insulators, in closed vessels provided with filling vents				Compact electrode assemblies in closed type of cell	
	Cell with reinforced positive plates and negative grids in plastic case	Cell with positive and negative pocket-type plates (TN) in steel housing	Cells with positive and negative pressed-powder plates in plastic housing		Plastic housing	Outer electrode (zinc) serves as container
Nominal voltage	2.0 V	1.2 V	1.5 V	1.1 V	1.5 V	1.5 V
Electrolyte	H_2SO_4 in aq. solution	KOH in aq. solution	Alkaline solution	Alkaline solution	NaOH in thickened solution	NH_4Cl in thickened solution
Positive electrode	PbO_2	NiO(OH)	Ag_2O and Ag_2O_2	Ag_2O and Ag_2O_2	Air/carbon	MnO_2
Negative electrode	Pb	Cd	Zn	Cd	Zn	Zn
Wt (kg kW hr^{-1})	45 ($\pm 30\%$)	44 ($\pm 30\%$)	8.5–32	13.4–25	9.6	17.0
Vol (l kW hr^{-1})	18 ($\pm 25\%$)	19 ($\pm 20\%$)	4.6–20	6–12	6.9	8.3
For discharge of	10 hr	10 hr	10 hr	10 hr	8 hr daily	8 hr daily
No. of discharges	1500	2500	150	25 000 if only 25 % of the capacity is used per cycle	3 to 200, according to cell type	
DM per kW hr Withdrawn	0.24–0.31	0.24–0.31	10 to 20	0.2 to 2	88.0	150.0
	With addition of charging cost 0.10–0.15 DM kW hr^{-1} and interest					
Current density (mA cm^{-2}) max.	ca 9.5, 10 hr ca 350	ca 5.5, 10 hr ca 160	ca 8.0, 10 hr ca 1200	ca 8.0, 10 hr ca 1000	15 relative to Zn electrode	10 to 25 relative to Zn electrode

Table X.4. Data for metal–air cells.

	Magnesium–air	Zinc–air	Iron–air
Construction and characteristic features	Compact construction. Air electrodes (2 per anode) integral with cell housing. Anodes and electrolyte easily renewable	Zinc in fabric pocket. Air electrode on either side of zinc electrode, firmly pressed to it, 130 cm² effective cross-section. Outer nickel grids for current collection. 50 ml electrolyte per cell	Probably similar to zinc–air
Nominal voltage (V)	1.7	1.45	1.1
Electrolyte	7% NaCl	30% KOH	20–45% KOH
Positive electrode	Air electrode catalysed with Pt or C	Teflon-bonded Pt, 5.4 mg Pt cm^{-2}	Air electrode as for Zn
Negative electrode	Mg alloys with Al + Zn	Sintered Zn	Sintered Fe
kg per kW hr (theoretical) at 1 V, referred to anode material	0.45	1.2	1.04
Wt., (kg per kW hr)	6.5 for 7 charge, ca 1 for 60 charges	8.2 for 1 charge	ca. 10
No. of discharges	30		> 200 (at 65% depth of charge)
Max. current density (mA cm^{-2})	50 at 1.1 V	20–30 at 1.2 V	20 at 0.9 V

To complete relevant comparisons, typical data for two Diesel generators are:

Nominal rating		4 kW	15 kW
Weight		750 kg	1100 kg
Dimensions:	length	1.35 m	1.50 m
	height	0.97 m	1.30 m
	width	0.70 m	0.90 m
Nominal hp of Diesel		7.5	22.5
Fuel consumption		190 g per hp–hour at nominal load	
Spec gravity of fuel		0.856 kg l^{-1}	

Disadvantages of the Diesel generator as compared with batteries:

1. necessary attendant,
2. fallibility of moving parts,
3. higher risks; breakdown cannot be excluded,
4. continuous running for periods of about 12 hours on 5–10% load is detrimental.

X.3   APPLICATIONS AND PROSPECTS OF FUEL CELLS

Tables X.2(a) and (b) summarize several fuel-cell units that are now quite technically advanced and extensively tested. A general observation applicable to fuel-cell systems, rather more than to accumulators, is that their possible uses are not confined to existing conventional applications. This is exemplified by space vehicle application*, for which the working life of fuel cells is already adequate and economic factors are of minor significance.

The principal problems that arise in adaptation to more normal terrestrial purposes are associated with lack of adequate metal catalysts (quite general for cells with acid electrolytes, and also for alkaline methanol and formate cells) and with auxiliary equipment. There is a lack of low voltage dc motors of the smallest possible weight and size, and suitable inverters have yet to be developed, although under favourable conditions, 85% efficiency in dc–ac conversion has been attained. Similar problems arise in relation to gas and electrolyte circulation pumps and control systems.

X.3.1   Application in Space Vehicles

The hydrogen–oxygen cell is at present predominant for use in space vehicles because of its high current density, simplicity and its welcome production of drinking water.

* Because of stringent requirements in relation to volume and weight, the hydrogen–oxygen cell is superior to conventional batteries for operating periods longer than a few hours (cf. Fig. X.1).

Both hydrogen and oxygen are stored cryogenically, (i.e., at low temperatures in either liquid or gaseous states), to avoid the weight involved in containers for storage of gases at high pressures (high pressure tankage).

Representative figures[12] for reactant/tankage weight ratios for low temperature storage of gaseous hydrogen and oxygen are 0.3–0.4 and 3–3.5 respectively, for quantities stored in the range of 20–30 lb (9–13.5 kg) of hydrogen and eight times as much oxygen. These figures could be improved by the development of lighter structures for the outer shells of the Dewar vessels used for such storage. A projected single container for both reagents (as gases at low temperature) gives a ratio of 4.25, possibly raised to 5 by storage of the oxygen in the liquid state. Theoretical storage volumes for the liquid reagents are, for hydrogen, 4.4 lb ft^{-3} (0.07 g cm^{-3}) and, for oxygen, 71.2 lb ft^{-3} (1.14 g cm^{-3}), both at their respective boiling points under atmospheric pressure. Some 95% of the stored reactants can be made available to the fuel cells.

X.3.1.1   *The Fuel-cell System of the Gemini Spacecraft*

One of the most successful demonstrations of the possibilities of fuel-cell application is undoubtedly the 'Gemini' undertaking, equipped with General Electric batteries. In the cells used the electrolyte is confined in an ion-exchange membrane based on sulphonated polystyrene (see section IV.1). The electrodes (10–40 mg cm^{-2} of platinum and palladium) are directly in contact with the membrane, and the whole battery is assembled on the filter-press principle (Figure X.2 and X.3). Each membrane is separated from a titanium collector plate by a ring gasket; these plates serve not only for current collection, but also lend mechanical stability to the assembly and provide for heat abstraction by means of cooling coils in contact with them. The space behind each negative electrode is closed, and is provided with ducts for entry and exit of hydrogen. The positive electrodes, however, are open to oxygen flowing through the whole of the battery enclosure. The oxygen stream controls the water content; the reaction water (0.4 l H_2O per kW hr at 0.84 V) is collected by a wick system (Figure X.2) and is available for drinking as the battery operates during space travel.

A stack (9 × 9 × 7 in) of 32 such cells is held together by end-plates and tie-bolts, giving a very compact structure. Each stack is manifolded for parallel flow of hydrogen and coolant to each cell, and also has its own wick/porous-plate system for the removal of water. Oxygen is fed by a manifold to the bottom of the stack and fills all the free volume accessible to it.

Optimum data for a 32-cell module (stack) are 26 V at 13 A (36 mA cm^{-2}; ~ 300 W) and 20–30°C. Three of these modules form a battery system, which is enclosed in a titanium pressure vessel (diam. 30 cm, length 60 cm, weight 31 kg), lined with foamed elastomer for mechanical insulation, i.e., isolation from vibration (see Figure X.3). Only pipe-lines and systems for regulation and control are external to this vessel. Liquid oxygen and hydrogen are carried in separate containers (see Figures X.4 and X.5). Each of two *batteries*, operating in parallel, has its own pressure regulators for both gases, gas filters, feed shut-off valves and purge valves. All pressures were regulated to maintain a set differential with respect to that of water, the gas pressures being in the range 20 to 30 psi.

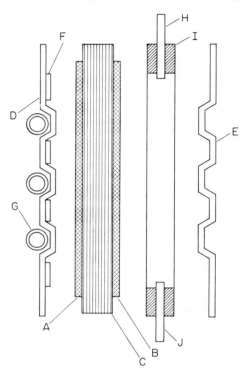

Figure X.2. Schematic diagram of the hydrogen–oxygen cell of the General Electric Company, Schenectady, N.Y., U.S.A. (after Cohen[11]). (A) Oxygen electrode, (B) hydrogen electrode, (C) membrane (electrolyte), (D) and (E) current collectors for hydrogen and oxygen electrodes respectively, (F) wick for water collection, (G) cooling coils, (H) H_2 feed tube, (I) frame, (J) H_2 purge tube.

Waste heat of about 0.4 kW is transferred by circulated coolant to the cooling system of the space capsule, but also serves for evaporation of the liquefied reactant gases. Additional electrical heating is however provided for regulation of evaporation and for starting up. There was initially some trouble with the Gemini-V mission because of damage to the evaporator system during blast-off, and the battery did not become fully serviceable until it was able itself to assume control of evaporation of the reactants.

Figures X.6 and X.7 show the actual performance of the batteries during the spaceflights of Gemini-V and Gemini-VII [11]. The batteries were brought into operation twice before the start (activation), once shortly before. The working voltage improved by 2–3 V between these activations, but the reason for this is

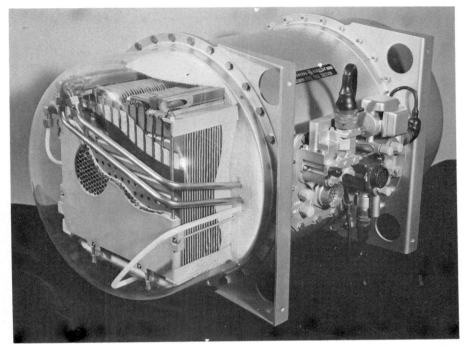

Figure X.3. Cutaway view of General Electric fuel battery for Gemini spacecraft (after Lieb-hafsky and Cairns[13]). A stack $= 32$ cells in series; the battery $= 3$ stacks in parallel.

not yet clear. The current-voltage curves plotted in Figure X.6 show that for battery 1 the activity slightly improved between the first and the eighth day of flight, but the data for battery 2 do not seem to be quite as favourable.

Finally, Figure X.7 shows the outputs of batteries 1 and 2 at a fixed terminal voltage of 27 V (0.84 V per cell), as functions of the flight time of Gemini-VII. The average output of about 10 A corresponds to a current density of 10 mA cm^{-2}, a total load of 0.54 kW and a production of 215 cm^3 of water per hour. It can be seen that the activity of the electrodes, and the performance of the battery declined continuously, to an extent corresponding with 3–5 mV hr^{-1}.

After 127 hours, the first of several rapid deteriorations in performance were experienced; they occurred at 127, 149, 201 and 220 hours after the start of the mission, and lasted for 15, 24, 9 and 13 hours respectively. All were followed by recoveries to the normal 'decayed performance' level. This consistent trend of returning to the same level indicated that there was no abnormal membrane degradation. This marked irregularity in battery current was compared by Cohen[11] with the rate of production of water. He found that the deviation of the latter below the theoretical rate was particularly large during the low output periods.

Since the fuel cell itself is the only place in the system where these quantities of water could have accumulated, it was concluded that the periods of poor performance were caused by partial flooding of the cells. The average deterioration in cell performance was attributed to unsatisfactory stability of the ion-exchange membranes.

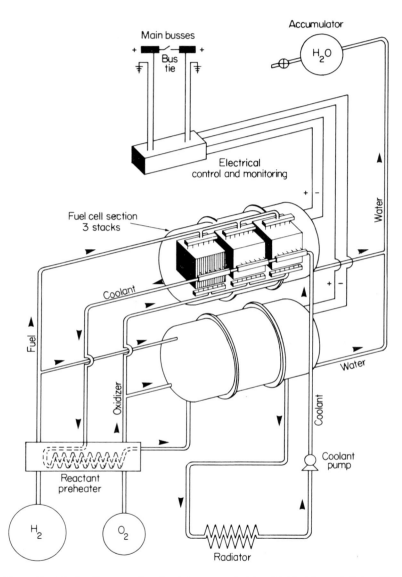

Figure X.4. Fuel cell system for Gemini spacecraft (after Liebhafsky and Cairns[13]). Two batteries, each of 96 cells, are connected in parallel, and provide a maximum power output of 1 kW each.

Figure X.5. View of interior of Gemini capsule, showing both batteries (A) and the pressure vessels for hydrogen and oxygen (B). Photo: McDonnell, U.S.A.

X.3.1.2   *Bacon-type Battery for the Apollo Project*

The robust Bacon cell developed by the Pratt and Whitney Company (see section IV.2) uses 85% KOH as electrolyte at temperatures between 200 and 250°C. Using gases at relatively low pressure (50–60 psi), the cell attains a terminal voltage of 0.9 V at 200mA cm^{-2}. The battery designed for the Apollo project is illustrated in Figure X.8. The power plant consists of monitoring, reactant- and operation-control instruments, together with water and heat removal sub-systems, located above the 31-cell series-connected battery. The battery operates at 27–31 V, supplying 500 to 1500 W for more than 400 hr. It has a maximum power output of more than 2 kW; further data are supplied in Tables X.5 and X.6. The reaction water is adjusted to pH 6–8, and is potable.

For such batteries of high rating/volume ratio, the critical adjustment of heat balance is facilitated by the relatively high working temperature. Since, however, 85% KOH is solid at room temperature, the electrolyte must be melted before operation of the battery and the heating-up time of 60 minutes

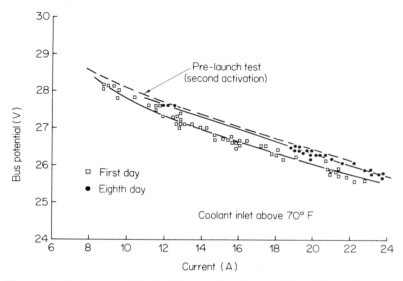

Figure X.6. Performance of fuel cell battery 1 during the Gemini-V mission (after Cohen[11]).

(cf. Table X.6) is relatively long. The reactant gases must also be preheated before entering the battery. The conception and particularly the technical evolution of the auxiliary equipment of this battery have been outstandingly successful.

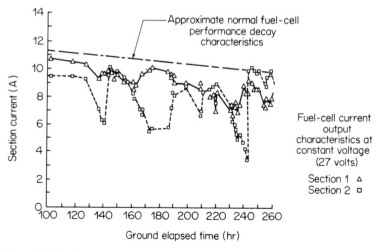

Figure X.7. Performance of fuel cell batteries 1 and 2 during the Gemini-VII mission (after Cohen[11]).

Figure X.8. 2 kW hydrogen–oxygen battery of the Pratt and Whitney Aircraft Company, East Hartford, Connecticut, U.S.A., for the Apollo project.

The battery (Figure X.8) has been successfully tested under space-simulating conditions for more than 1000 hr at loads exceeding requirements. This period of operation is about thrice that envisaged for the project (14 days = 336 hr).

X.3.1.3   *The Allis-Chalmers Hydrogen–oxygen Cells for Use in Space*

The Allis-Chalmers Company, commissioned by the Air Force and NASA, has also developed hydrogen–oxygen batteries for space travel[3], including equipment for moon vehicles. The Company is now engaged in the development of a battery, equal to not less than 2500 hours of continuous running, for use in the Apollo programme. The construction of a module can be seen in Figure X.9.

All these units use the capillary matrix type of assembly, with KOH electrolyte at 90–95°C, and the static moisture removal system (Figure IV.32) adapted to produce potable water.

Each cell of the new 2 kW unit[15] consists of two porous electrodes separated by an asbestos capillary membrane that holds the KOH electrolyte. The anode is constructed of porous nickel, activated with platinum–palladium catalyst; the cathode is of high surface area silver[16]. Slotted magnesium cell plates provide the cavities for distributing the reactant gases and removing the water vapour.

The standard battery (30 cells, 28 V) can be loaded at 2 kW in continuous operation. The terminal voltage of each cell is between 0.9 and 0.8 V at 250 mA cm^{-2} and 90°C. A large number of batteries have been subjected to prolonged

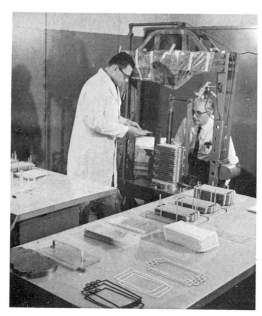

Figure X.9. Construction of a hydrogen–oxygen
module for use in space (Allis-Chalmers).

tests. In addition to leakage, loss of KOH may occur by attack on the asbestos membrane—silicate deposition in the pores of the electrodes is the most frequent cause of collapse of cell voltage. Because of this difficulty, recent experiments have been concerned with free electrolyte and strongly hydrophobic electrodes.

X.3.1.4   General Considerations and Technical Problems [3,12]

In general, the state of development of hydrogen–oxygen cells for application to space travel can be described as already satisfactory in relation to electrical data, reliability and working life. The following points, however, are of particular significance to further improvement.

(a) Improvement of electrode structure, to increase limiting current densities.
(b) Improvement of cathode catalysts, to minimize overpotential and hence inefficiency due to insufficient activation of oxygen.

At present, silver is used as the catalyst for the oxygen electrode. Platinum is possibly more suitable, and osmium has possibilities as an even better catalyst. For the anode, platinum seems always to be best. Use of these noble metals for civil purposes is of course an outstanding handicap.

Table X.5. Characteristics of individual hydrogen–oxygen cells for space application (after Cohn[12]).

	Gemini	Apollo	Allis-Chalmers
Current density (mA cm^{-2})	15	92	100
Initial voltage (V)	0.8	0.97	0.95
Power density (mW cm^{-2})	12	89	95
Active cell area (cm^2)	360	400	200
Cell power (W)	4.5	35.6	19
Reactant consumption [kg (kW hr)$^{-1}$]	0.41	0.36	0.36
Degradation rate (mV per 1000 hr)	3000–5000	60	40
Principal degradation mode	Membrane decomposes	Cathode corrodes	Asbestos reacts

Note: Data are for sustained, not peak power.

(c) Replacement of membrane or matrix (if used) by a more durable structure, to obviate deterioration in performance due to degradation of membrane materials.

Volumes and weights of the batteries themselves have already been brought close to a minimum, but there is much improvement yet to be made in auxiliary and additional equipment in this respect.

There is another important problem in the control of *water balance*. A wick system is used in the General Electric hydrogen–oxygen membrane cell. Pratt

Table X.6. Characteristics of fuel cell modules for space application (after Cohn[12]).

	Gemini	Apollo	Allis-Chalmers
Module size (kW) (sustained power)	0.5	1	1
Module weight (kg kW^{-1})	63.5	122	75
Module volume (l kW^{-1})	170	303	161
Stack volume (l kW^{-1})		134	56.6
Parasitic power (W per module)	0[a]	75[b]	50[c]
Life under load (hr)	400–800	400–1500	1000–2000
Storage at room temperature and below	Good[d]	Good[e]	Good
Heating time, room to operating temp., min	0	60	15–30
Auxiliary heater (kW)	0	3	0
Recovery from abuse	Poor	Poor	Good

Note: a. Integration of the fuel cell and environmental control systems complicate these estimates; a separate power system would have weighed more and required parasitic power.
b. This is alternating current for starting up from cold, and does not reflect losses in power conditioning.
c. This is the total power drawn from the dc supply.
d. This is before activation; once activated, the stored system degrades by 0.2–0.7 volt/1000 hrs.
e. Systems stores well even up to 200°F (94°C).

and Whitney use an electrolyte circulation system which is technically simple, but requires an additional pump. The Allis-Chalmers cells, with asbestos membranes as electrolyte carriers, utilize the difference of water vapour pressure between two electrolyte-saturated membranes for static water control (cf. section IV.1.2.2 and Figure IV.32).

Methods of *heat removal* vary considerably from one system to another, but usually a cooling medium is circulated along the cell assembly, or over suitable cooling ribs or fins. High temperature, as well as high battery voltage, eases the problem of heat transfer, but, on the other hand, low temperatures and pressures are advantageous when defective parts have to be replaced.

Reliability is of particular importance in space vehicle application, and presents an increasing problem when current densities are raised; high current densities are, of course, desirable on the grounds of economy of weight and volume. Opposed to this, reduction of current density, combined with increase of battery voltage, carries the advantages of higher efficiency of cell operation and facilitation of removal of heat and reaction water.

Module life is generally a fraction of cell life, apparently mainly because of lack of quality control. This means, not that the modules are being put together carelessly, but that the factors requiring close control and the severity in effect of slight variations, are often not recognized until a number of units have been built and operated.

An important characteristic of any device is its capability to withstand *abuse* and to recover from *overstress*.

A series of eight full-size fuel cell modules with asbestos membranes ('asbestos stacks') are being systematically mistreated[12] to find how they react to repeated start-stop operation, overload, pressure imbalance, and so on. No doubt the same kind of procedure will have to be applied to all the mechanical, electrical and electronic auxiliary equipment before it is fully mastered.

In the tests in question, it has already been discovered that an asbestos stack need not be filled with helium for storage, as had previously been supposed. This result effected an important simplification of procedure and equipment. It has also been found that power peaks of up to 5 kW can be tolerated without catastrophic damage, the duration of the peaks (ca. 15 sec) being limited only by the cooling capacity of the stack. Multiple starts and stops appear to have no effect. Other tests have involved starting at room temperature and allowing the stack to heat up by means of its own waste heat, operating either at constant voltage or at constant current. Times to reach normal working temperature varied from 6.0 to 6.5 minutes, with temperature differences within the stack varying from 1°C for the hour-long start to 15°C for the most rapid start. None of these tests have as yet been shown to have any deleterious effects.

For *longer residence in space*, such as will be necessary in space stations, it will be possible to use only the regenerative type of fuel cell, the primary power producer relying on nuclear or solar energy. In this connexion, a combination of water electrolysis with hydrogen–oxygen cells to form a 'gas accumulator' is under development (cf. section VIII.1). Whilst sunlight is available, photocells provide the station with electrical energy and simultaneously electrolyse the reaction water from the cells. Oxygen and hydrogen are stored and consumed in the fuel battery during periods of darkness.

For long space missions, fuel cells are likely to play a continued and more general role in the space programme[12]. They may be used as emergency and peak primary power sources in association with both solar and nuclear power plants. They may supplement or replace secondary batteries, with astronauts using the by-product water before it is electrolysed again. They may be the only practicable power sources for lunar surface vehicles, from which water would be returned to an electrolyser at the base for reprocessing.

X.3.2   Terrestrial Applications

As already emphasized in the Introduction, commercial success of the new electrical energy sources is to be expected only in a limited field at first. At present, application of fuel cells to the operation of signals, radio stations, emergency energy sources and, in particular cases, to traction seems to be technically and economically feasible.

Systems for which storage of fuel in compact and easily handled form is possible, and which require only air as the oxidant, will certainly be preferred. These include methanol and hydrazine cells, and hydrogen–air cells supplied with hydrogen from a reforming unit working with methanol, liquid or liquefied hydrocarbons, or perhaps ammonia, as the original fuel. Alkaline electrolytes are mainly used. Acid electrolytes, besides introducing corrosion difficulties, have the disadvantage of requiring the use of noble metal catalysts.

X.3.2.1   *Batteries of 50 W to 100 kW for Standby Power and Other Purposes*

Batteries within this range are adaptable to provision of reserve or emergency power supplies (in competition with Diesel generator sets—cf. comments at the end of section X.2) for hospitals, radar and weather stations, air and ocean travel signals, remote farms and the like. They also come into consideration for mobile as well as fixed power plants, mainly for military use, but also for driving industrial and delivery trucks, and other road or rail vehicles. The problem of application to transport is discussed in section X.4. The types of battery predominantly studied for these purposes incorporate hydrogen–air cells associated with hydrogen producers, or hydrazine cells.

Hydrogen cells with hydrogen generators

The Allis-Chalmers Company, in collaboration with Engelhard Industries Inc., Newark, N.J., U.S.A., has developed a complete 5 kW unit intended for military use[17]. It comprises

1. a reformer, generating hydrogen from liquid hydrocarbons,
2. the fuel-cell battery (Figure X.10), providing 7.3 kW at 28 V dc,

Figure X.10. Allis-Chalmers 7.3 kW hydrogen–air battery (after Engle[17]). (A) Fuel cell module, (B) Water recovery heat exchanger, (C) KOH cooler, (D) Start-up heater, (E) Start-up batteries.

3. an inverter, changing the dc battery output to ac at 120 V at, alternatively, frequencies of 50, 60 or 400 cycles per second.

In the reformer, the hydrocarbon and steam react together at about 800°C and 10 atm pressure in presence of a catalyst, to give a mixture of hydrogen, carbon dioxide, methane and excess steam. The hydrogen (yield $> 50\%$) is separated from the other products with a palladium–silver alloy membrane before being fed to the fuel cells. It takes 45 minutes to bring the system into full operation; hydrocarbon is burned for the initial heating and until the battery itself is capable of maintaining the required working temperature of the reformer.

The fuel cell assembly itself operates with hydrogen, air and an alkaline electrolyte at 70–80°C. The air is compressed to 2.5 atm and freed from carbon dioxide with a KOH wash before reaching the cells; 22.6 kg hr^{-1} of CO_2-free air required. A condenser removes water from exhaust air. 34 pairs of parallel-connected cells in series constitute a module; 0.83 V per cell provides the total of 28 V per module at an electrode current density of 135 mA cm^{-2}. Four such modules in parallel form the battery (Figure X.10), with a total power output of 7.3 kW. The alkaline electrolyte is circulated and serves to regulate the heat and water balance of the battery (Figure X.11). Reaction water is returned to the hydrogen generator to minimize the net water requirements.

Water control is effected with a second KOH solution-impregnated asbestos membrane situated in the hydrogen compartment. This is in contact with a porous nickel plate (Figure X.12) behind which is circulated a KOH solution of appropriately higher concentration. This serves to absorb reaction water (cf. Figures IV.32 and IV.33) and also to regulate heat content.

Finally, the inverter, 25 kg in weight, is non-mechanical and operates with 75–80% efficiency on full load. The whole installation, weighing about 500 kg and about a cubic metre in volume, attains an efficiency of 27%.

As yet, pilot-plant trials have encountered difficulties both with hydrogen generator and battery. Whereas single cells have given trouble-free service for more than a thousand hours, numerous complications appear when they are built together to form larger units. Apart from this, the noise of the air compressor would seem to be a disadvantage to military use, and the inverter is so costly that reversion to a conventional dc–ac converter may be necessary. It is hoped, however, that as soon as the technical difficulties have been solved, smaller, lighter and cheaper installations of this kind will become available.

In February 1967, 27 natural gas companies in the U.S.A. formed a research group TARGET—'Team to Advance Research for Gas Energy Transformation, Inc.'[24], to develop a civil version of this type of hydrogen–air cell with indirectly coupled hydrocarbon reformer. The Pratt and Whitney Aircraft

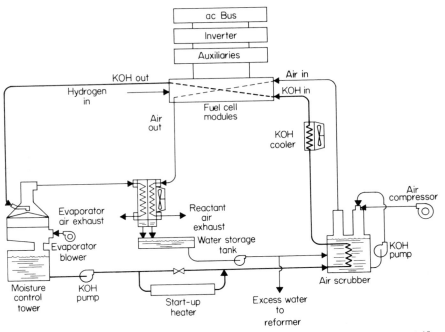

Figure X.11. Circulation system of the 7.3 kW battery illustrated in Figure X.10 (after Engle[17]).

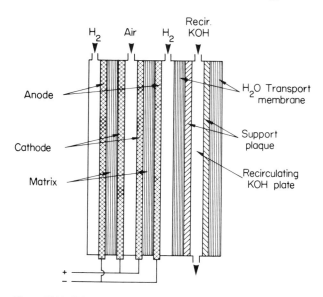

Figure X.12. Schematic representation of a cell construction with asbestos membrane as electrolyte carrier, with water and heat contents regulated by a KOH circulation system (after Engle[17]).

Company and the Institute of Gas Technology, Chicago, have undertaken the development of the fuel cell itself. The ultimate objective is said to be the so-called 'All-electric Home', i.e., the provision of all household energy requirements by means of fuel-cell generated electricity. Enquiries are said to have indicated that the average consumption of power per household is relatively small and that the periods of peak load are relatively short.

The Pratt and Whitney Company have built a 3.7 kW unit as the first battery of this kind, shortly to be tried out in domestic use. With a volume of 1220 l, it uses natural gas with 38% efficiency; the auxiliaries (pumps for air, oil and water and the electronic controls) require 300 W. Natural gas, freed from sulphur-containing impurities, and distilled water enter the reformer via a pre-heater. To maximize hydrogen yield, the gases from the reformer are somewhat cooled, and then pass to a second reactor in which carbon monoxide provides a supplement of hydrogen. Hydrogen is separated with a palladium–silver membrane and goes to the fuel cell; residual gas is burned to supply heat required by the reactor.

Water formed in the cell reaction is removed in the air stream. Two cell-packets of 36 cells each are connected in parallel and provide the total output of 3.75 kW. The electrodes are 18×18 cm^2 in area. Asbestos membranes serve as electrolyte carriers for the aqueous KOH solution used. Individual cells are separated from each other by cooling plates which also contain the gas distribution ducts. A complete cell is 4.5 mm thick and operates at about 150 to 200 mA cm^{-2}. The output of 28 V is converted to 120 V ac by an inverter.

Williams and co-workers[18] have studied a combination of *methanol reformer* and *alkaline hydrogen–oxygen cell* with porous plastic substrate electrodes (cf. section IV.1.1.3). Hydrogen is isolated from the products of the reaction

$$CH_3OH + H_2O \rightarrow 3H_2 + CO_2$$

by means of a palladium–silver membrane. The whole system (Figure X.13) works quite automatically. On starting up, lead accumulators provide operating power, and the reformer is heated by combustion of methanol. As soon as the reformer supplies hydrogen, a low pressure blower comes into operation to supply air to the cells; the cells begin to work and take over the auxiliary power requirements and begin to recharge the lead batteries. The fuel battery itself consists of two modules, each of 62 cells. The power required by the auxiliary components (hydrogen generator, electrolyte pump, cooler, air-blower, etc.) amounts to 900 W, and the net power output of the unit is 5 kW, with an overall efficiency of 20–22 %.

Energy Conversion Ltd., Sunbury-on-Thames (U.K.) have also developed a battery working with hydrogen from a methanol–steam reforming unit[19,20], but in contrast to the system just described, the *reformer is directly coupled with*

Figure X.13. 5 kW hydrogen–air battery (Shell Research Laboratory, Thornton, England). Hydrogen is generated, in an associated reformer, from methanol and purified by a Pd–Ag membrane. Over-all efficiency, 20–25 %.

the cell. A palladium–silver membrane is situated between the electrolyte and the reformer gas compartment, of which it forms one wall. It thus serves the double purpose of hydrogen-permeable barrier on one hand, and anode on the other. The advantage of this is that the heat of the cell reaction is transferred directly to the reformer, with an improvement in efficiency compared with arrangements in which reformer and battery are separate units. Temperature control is also simplified because there is only a single unit requiring control, and there is a further advantage in that the anode acts as a hydrogen reservoir— it can store appreciable quantities of hydrogen, accordingly available for the demands of short peak loads. Porous nickel electrodes of the Bacon type are used for cathodes, the electrolyte is ca. 75% KOH, and the working temperature is 200–250°C. A battery of this kind rated at 170 W has already been built; a 6 kW battery of 36 modules and terminal voltage 24.3 V is planned.

Other types of hydrogen–air cells in combination with reformers are under investigation[21-23]. One of these is a cell with *phosphoric acid as electrolyte*[22]. The associated reformer is a classical hydrocarbon–steam reactor that can be operated at 250°C; the cell works at 150°C. Supplied with a standard hydrocarbon fuel, the system has an overall efficiency of over 30%. The 500 W unit studied has a weight and volume comparable with present engine-generator power sources.

How far batteries of this kind, with integral or separate reformers, are able to establish themselves depends mainly on the possibilities of making adequate palladium–silver membranes. At present, those available—because they are necessarily thin—are suitable only for small units; larger membranes become leaky. Moreover, the as yet limited working life and high price of the membranes are generally unfavourable to their industrial use.

Hydrazine cells

Detailed reports on the design, construction and properties of modern hydrazine cells have been made in chapter IV (cf. also Table X.2b). Trials have been made of 60 W to 20 kW batteries. Air, oxygen or hydrogen peroxide are suitable as oxidants according to the type of application.

The hydrazine–air system is the *simplest* of the direct fuel-feed cell systems. Figure X.14 shows strikingly how all the auxiliary equipment (electrolyte tank and pump, heat exchanger, fan for cooling and 'chemical' air, etc.—cf. Figure IV.95b) for a 60 W unit can be assembled in compact and handy form. A circulating pump for anolyte is not essential because of *nitrogen evolution*. Figure X.15 shows a 300 W laboratory design on the 'bath tub principle': the whole cell is immersed in anolyte and nitrogen evolution alone provides fast enough circulation. Concentrated hydrazine hydrate is injected at the base of the cell under control of a sensor cell depending on the sensitivity of working electrode potential on hydrazine concentration.

Figure X.14. Portable 60 W hydrazine–air battery (after Salathe, Smith, Gallagher, Terry, Kozloff, Athearn and Dantowitz[25]); upper left and lower middle, openings for passage of 'chemical' and cooling air (photograph, Monsanto Research Corporation, Everett, Mass.).

A 100 W unit consumes up to 1 cm^3 of $N_2H_4.H_2O$ per minute; the Monsanto 60 W battery uses 0.5 kg of fuel in 6 hr, and the Union Carbide 300 W unit 4.5 kg in 12 hr. These figures correspond to 407 W hr per kg, including battery weight, for 24 hr operation, and are so favourable as to make numerous applications practicable, despite the high fuel cost of $1.90 per kW hr.

The *start-up time* of the 300 W hydrazine–air unit[27] depends on its initial temperature; it is 15 min for 75°F, 30 min for 32°F, and 60 min for 0°F. With hydrogen peroxide as the oxidant, however, the nominal rating can be attained within seconds at 20–30°C—but to avoid too rapid self-decomposition of H_2O_2, a working temperature of 40°C should not be exceeded. The N_2H_4–H_2O_2 system is therefore to be recommended for particular applications in which maximum load must be met after a short start-up period.

In addition to their simplicity, hydrazine cells have the advantage of obviating use of noble metal catalysts[28]. Raney-nickel and nickel boride furnish excellent current densities at favourable potentials (cf. Table X.2(b)), and ease the production of modules at competitive costs; silver or carbon as oxygen electrode catalysts are similarly advantageous.

Hybrid power supply for radio sets[29]

A combination of fuel cell and storage battery is especially suited to provide for intermittent, variable power levels, such as required by radio stations for receiving and transmitting. Nickel–cadmium and hydrazine–air batteries form

Figure X.15. 300 W laboratory version of a completely automatic hydrazine–air battery on the 'bath-tub' principle (after Dünger and Grüneberg[26]); upper left, hydrazine injection pump; upper middle, heat exchanger with ventilator; lower right, rectangular tank for cooling fluid (isopropyl alcohol).

an ideal combination, contributing respectively to good power and energy characteristics (W and W hr per kg). In this application, the fuel cell can handle electronic load, and acts as battery-charger in non-peak periods. Fuel cell operation is improved by use of the battery for an electrical start, and the power delay typical of fuel cells is eliminated. Such a hybrid unit is easily assembled; Figure X.16 illustrates a 30 W fuel cell in combination with a 0.45 A hr (12 and 24 V) Ni–Cd secondary battery—it can provide a peak load of 300 W.

Methanol as fuel

If, for a particular application, lower operation costs are essential, methanol is to be preferred to hydrazine as fuel. Running costs per kW hr are $1.90 for hydrazine cells, $0.12 for methanol–oxygen cells with acid (invariant) electrolytes, or $0.30 for the latter with consumable alkaline electrolytes; use of air instead of oxygen reduces costs by $0.06 per kW hr. It is, however, to be considered that platinum metals are necessary as anode catalysts for both acid and alkaline methanol cells. Use of acid electrolyte eliminates the difficulty of carbonate formation by the carbon dioxide content of air, but calls for more

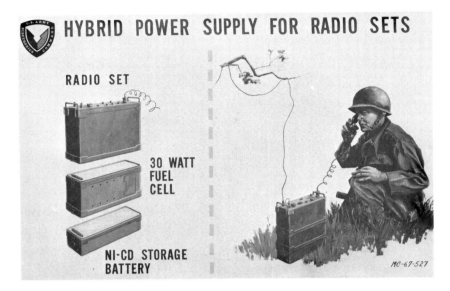

Figure X.16. Hybrid power supply for radio sets (after Wrublewski[29]).

careful selection of the materials used—tantalum gauze, for example, has hitherto been used as catalyst support (cf. Table X.2(b)).

The Esso Research and Engineering Company has demonstrated that completely automatic methanol–air batteries with *acid* electrolytes can already operate for over 100 hours (cf. Chapter IV), with an efficiency of 30 % at nominal rating and cell voltage of 0.4 V. The volume of the 80 W unit, 30 l, is yet rather large; a 120 W unit of 12 kg weight is planned. This battery is proposed as a military power source, being silent, highly efficient and operable at low temperatures. The high cost of the catalyst electrodes militates against commercial application. The use of *alkaline* electrolyte carries the necessity of periodical renewal of the circulated anolyte because of carbonate and water formation— this is the essential disadvantage compared with acid cells. On the other hand, according to the recent results of the Battelle Institut, Frankfurt[30], the relatively plentiful noble metal palladium can be used as the methanol catalyst.

Figure X.17 illustrates the 100 W methanol–oxygen battery conjointly developed by the Battelle Institut, Frankfurt and the Robert Bosch Company of Stuttgart. Platinum is still used as anode catalyst. Electrical and other data for the battery are included in Table X.2(b). It is notable that the oxygen electrodes have recently been replaced by alternative air electrodes working at normal atmospheric pressure[30], which have already given several months of trouble-free service at 50 mA cm^{-2} (terminal voltage 0.5–0.6 V). These electrodes consist of a silver–Teflon layer on the electrolyte side, and pure,

Figure X.17. View of a 100 W methanol/KOH/oxygen battery of the Robert Bosch Company, Stuttgart[31]. Below, two modules of 24 double cells; middle, 5 l oxygen pressure vessel; above, electronic controls.

porous Teflon on the gas side. Adequate air supply for the above performance is provided by the 'chimney action' of upper and lower vents of the air chambers.

Direct conversion of hydrocarbons and partially oxidized hydrocarbons in high temperature cells—the fuel cell power station

For large, stationary units of 100 kW or more, only cheap fuels and cells of long working life (more than 5 years) come into consideration. Since, furthermore, noble metals for electrode catalysis must be avoided, the only cells meriting discussion for such application are those with molten salt (500–700°C) or solid (> 1000°C) electrolytes. The development of batteries adequate to meet such requirements as yet remains very much in its infancy.

Hydrocarbons are also very desirable fuels for military applications because of their high energy content, ready availability and existing distribution facilities. For this reason, batteries of lower rating (100 W–15 kW) have been developed for these special purposes in the U.S.A.

Systems with molten electrolytes (cf. section IV.2) of, as yet, only small capacity have been tried out. Texas Instruments Inc.[32] is at present developing a 15 kW battery. The design is based on multi-fuel operation on both CITE and combat gasoline, with air as oxidant, and draws on earlier experience of single modules working with molten carbonate electrolyte at 700°C. The fuel preparation process, partial air-oxidation, is expected to supply a gas of the following composition:

Gas	Mole %
H_2	21.4
CO	23.4
H_2O	2.8
CO_2	4.8
N_2	47.1
CH_4	0.5

Nickel screen anodes and silver-plated stainless steel cathodes are used. The 15 kW unit should have a total weight of 1500 lb and a potential efficiency of 30%.

Of fuel cell batteries with *solid electrolytes*, only the 100 W unit of the Westinghouse Research Laboratories has reached the technical stage of development (cf. section IV.2.5 and Figure IV.69c). Subject to experience with this, a 100 kW unit is projected, in which it is hoped to eliminate platinum as catalyst. The main difficulty with these batteries at present is their severely limited working life, arising from poor electrode/electrolyte contact—the best so far attained for the 100 W unit is two months, whereas five years would be needed for practicable power station application.

X.3.2.2 *Batteries from 5 to 50 watt for Application in Buoys, Remote, Unattended Beacons and Weather Stations, Portable Transistor Equipment, etc.*

Considerably simplified types of fuel cell can be used to meet demands for relatively low power electrical supply, to be maintained for long periods—weeks or months—without attention. Hydrophobic air electrodes as used in primary cells (e.g., zinc–air) can be combined with fuel electrodes operating with electrolyte-dissolved fuel, working at ambient temperature (cf. section IV.3.1.2). Since only quite low current densities are normally involved, methanol or formate are suitable fuels, and have the advantage of high energy capacity (cf. Table X.1; 250–500 W hr per litre of fuel–electrolyte mixture). Nevertheless, hydrogen–oxygen batteries are also being studied for applications of this kind.

Alkaline methanol–air cells have for several years undergone successful trials in *energizing flashing light buoys*. Electrical systems are clearly superior in ease of control to mechanical systems, using propane or acetylene as illuminants, for such a purpose, and, with the new current sources to be described, they need less frequent attention.

The battery illustrated in Figure X.18 (the construction of the component cells has been described in section IV.3.1.2.2) provides for signal operation (2 seconds on, 4 seconds off), 5 A at a stabilized voltage of 6 V. Preliminary trials over 13 months showed voltage stabilization, at cost of enlarging battery dimensions, to be necessary because of the high temperature-sensitivity of the methanol electrode between 20 and 0°C[33]. 180 kW hr is provided by 400 litres of fuel-electrolyte mixture. In 'field tests' this battery has worked a flashing light continuously for 15 months, and operation for two to three years, on load for four hours daily, should be possible. The weight/energy ratio of the battery (with plastic container) amounts to 3–4 kg per kW hr, the operating costs being about $0.25 per kW hr. Further development has led to an improved battery with lower production costs (see the formate–air cell of Figure X.24).

A completely closed, self-contained, 5 watt hydrogen–oxygen power source with operating life up to a year, has been made by the General Electric Company[34], for application in *buoys, gasline monitoring units, oceanographic laboratories,* etc. (Figure X.19). The heart of the system (diameter 48 cm, height 46 cm, weight 113 kg; 32 V, 43.8 kW hr, 386 W hr kg^{-1}) is a hydrogen–oxygen module of series-connected ion-exchange membrane cells, for operation at ambient sea

Figure X.18. 30 watt (6 V) methanol–air battery with alkaline electrolyte for servicing the light buoy illustrated opposite (after Guth, Haase, Plust and Vielstich[33]). Capacity per filling, 180 kW/hr.

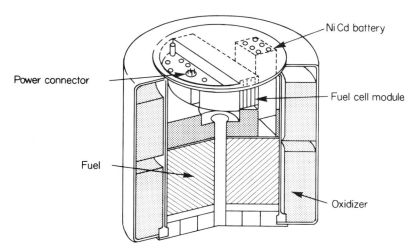

Figure X.19. General Electric 5 watt fuel cell power source, with hydrogen and oxygen generated by dry chemicals (after Liebhafsky and Cairns[34]).

temperatures. Hydrogen and oxygen are generated by the reaction of dry chemicals. The water produced by the cell is used within the system in the generation of hydrogen. With addition of a conventional nickel–cadmium secondary battery, short duration (2 sec per hr) peak power up to 500 W is possible.

Coastal beacons, with power requirements ranging from a few to several hundred watts, also invite the application of batteries operating for several months without attention as attractive alternatives to conventional batteries needing frequent recharging. Gillibrand and Gray[35] have investigated a hydrogen–oxygen battery (Figure X.20) in this connection, assuming a beacon power requirement of 60 W at 12 V for 14 hours a day. This is low enough to obviate the necessity for circulation systems (for water and heat balance), so that the only items of ancillary equipment needed are pressure reducing valves. The water formed in the cell reaction can be accommodated by increasing the size of the cell container. Seventeen 5 A cells (5 mA cm^{-2}) are connected in series. Two litres of electrolyte are required to cover the electrodes of each cell, with 5.9 l of additional free volume per cell left to take the water produced during 6 months of unattended operation. A tray, fitted at a height of 13 cm above the initial electrolyte level, contains 450 g of solid KOH, which is taken into solution when the electrolyte level rises to it. For 6 months operation the theoretical gas consumption would be 90 000 l of hydrogen and 45 000 l of oxygen. An ample reserve of gas could be stored in two hydrogen cylinders and one oxygen cylinder, each being 380 cm long and 46 cm in diameter, weighing 1000 kg. The battery

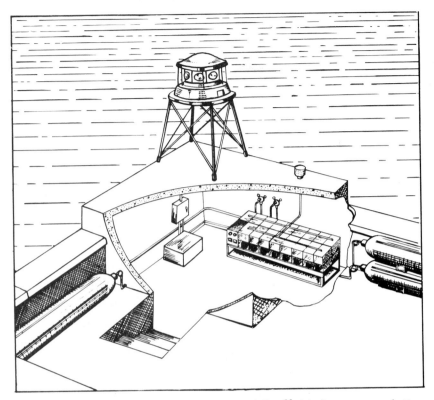

Figure X.20. Fuel battery beacon, after Gillibrand and Gray[35]. A hydrogen–oxygen battery, with gases supplied from cylinders, for operation of a 60 W signal light for 6 months without attention.

would weigh 100 kg, so that the overall weight to power ratio would be almost 10 kg per kW hr (see Figure X.1 for comparisons).

High altitude television relay stations have similar minimum energy requirements (20–40 W) to those of buoys and beacons, but need, of course, steady and not periodically fluctuating current, to be maintained in winter under conditions of extremely low ambient temperatures (Figure X.21). Swiss Television (PTT) has used fuel cells for this purpose[36]. In the winter of 1965/66, the main electric supply cable to the principal television station for the upper Valais had not yet been laid, but television programmes were nevertheless radiated by means of power from a methanol/formate battery installed by the Brown Boveri Company (Baden). This battery consisted of 320 cells (of construction described in section IV.3.1.2.2; cf. Figure IV.84), and, thanks to the addition of formate, was capable of supplying 20 W at very low temperatures for around

Figure X.21. Left: Television station over Visp (Oberwallis) in winter[36]. Energy supply by a 20 (40) watt methanol–formate battery (Brown, Boveri u. Cie.).
Right: Battery of 16 individual fuel cells, of box form, suitable for combination to form indefinitely large units. 20 such batteries (320 cells) were used at high altitude; each cell provided 0.2 W at 0.6 V for about 7000 hr.

7000 hours. A dc converter supplied the transmitter with the stabilized voltages (16 and 28 V) required. The battery provided uninterrupted service for 6 months, but the formate component of the fuel mixture was exhausted shortly before the end of the cold period towards the end of February.

Lemaignen and Demange[37] have shown how very special problems can be solved with aid of hydrogen–oxygen cells. The problem was power supply to *transmitter–receiver radio equipment in sounding balloons.* To be reckoned with were pressures of 350 millibars and temperatures between 214 and 243°K, and an average battery output of 100 mW (at 6 or 8 V) with short peaks (1 % of the total energy) of 10 W (at 28 V).

Figure X.22 shows a diagram of the well thermally-insulated installation used. The hydrogen–oxygen battery consisted of 13 cells (electrode area 10 cm², 5 mA cm⁻²) and weighed 240 g. The reactants, 15.1 g of hydrogen and 121 g of oxygen, were contained in fibre-glass reinforced plastic containers. Reaction water (136 g) collected at the bottom of the cell container. The 10 W impulse came from a nickel–cadmium battery (80 mA hr) charged from the fuel cells via an 8–28 V dc converter. The entire weight of the equipment, including the gas containers, remained less than 1.5 kg.

To conclude this section, reference may be made yet again to the special attributes of fuel cells in *powering transistorized equipment.* The simple construction and ease of operation of primary type cells with formate or methanol as fuel and air as oxidant, suggest their use in fields where, hitherto, dry cells or

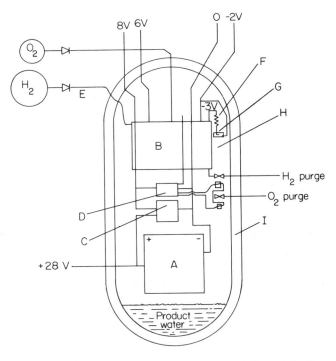

Figure X.22. Diagram of a hydrogen–oxygen unit for use in balloons (after Lemaignen and Demange[37]). (A) Ni–Cd battery, (B) Fuel cell battery, (C) Dd–dc converter, (D) Purge control, (E) Pressure reduction valve, (F) Heating element, (G) Thermostat, (H) Oxygen atmosphere, (I) Thermal insulation.

zinc–air cells have been employed. The advantages of these fuel batteries are excellent storage qualities, constant discharge voltage, high capacity in A hr and W hr, and repeated usage by renewal of fuel-electrolyte mixture. Disadvantages are relatively low terminal voltage and the use of palladium as anode catalyst.

A D-size air cell (see Figures IV.85 and 85a) with formate as fuel, offers 5.6 W hr, and has a power to weight ratio of 70 W hr per kg[38]. High loads of 100–300 mA can be sustained, as demonstrated by the operation of the transistor radio set of Figure X.23 by one cell only. The terminal voltage of the cell is transformed to the necessary 6 V by use of a small dc converter (efficiency 65%). A hindrance to the immediate general use of the cell described is the difficulty involved with a liquid electrolyte; a pressure valve is needed for gas- and liquid-tight operation. Using small, porous plastic cylinders for this purpose allows the cells to be used upside down only for a few minutes.

On the other hand, there is a genuine demand for the larger type of cell, illustrated in Figure X.24, with electrodes mounted on the lid of the container. The cell is intended for prolonged operation at 300–600 mA and 0.6–0.7 V. The 1100 cm^3 of fuel-electrolyte mixture contain 300 W hr for formate, and 700 W hr for methanol, as fuel. The total weight of the cell (with plastic container) is 2.35 kg (see also Table X.2b). For production in larger numbers it is

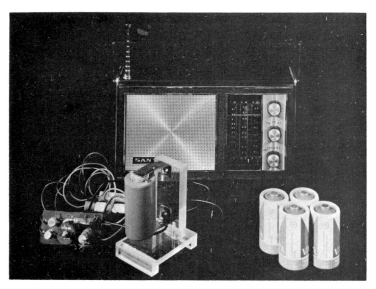

Figure X.23. Operation of a transistor radio by one formate–air cell and a dc converter (0.6 to 6 V); the conventionally used four dry cells are also shown (after Vielstich and Vogel[38]).

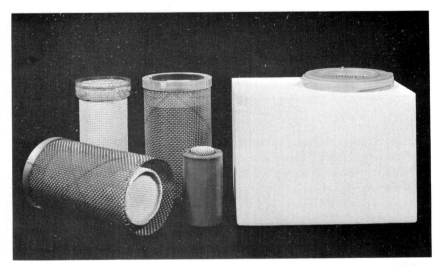

Figure X.24. 320 W hr formate–air cell for continuous operation at 300 mA and 0.6 V (after Vielstich, Vogel, Sendhoff and Stichnote[39]). From left to right: electrodes with protective screen, oxygen electrode with silver-plated nickel grid, protective screen with fuel electrode; D-size zinc–air cell for comparison; fuel cell with electrodes inserted, cap removed.

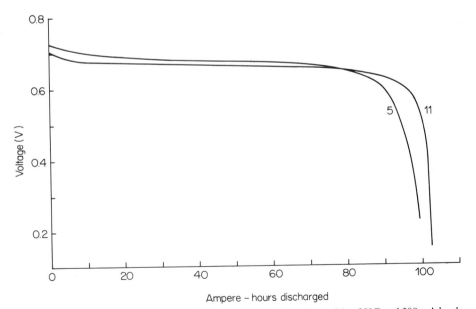

Figure X.25. Discharge voltage of the formate–air cell shown in Figure 24 at 20°C and 300 mA load; fifth and eleventh discharges. Electrolyte, 400 cm³ of 7.5M KOH + 5M HCO_2K (after Vielstich and Vogel[38]).

estimated that the selling price, despite the palladium content of the anode, could be only \$7.5; with dc converter, 0.6 to 6 V, fitted inside the cell lid, \$12.5. Economic use of this battery undoubtedly depends on its recharging facility by renewal of the fuel-electrolyte mixture. Renewal of the carbon–air electrode may become necessary after 10–20 discharges.

A disadvantage of batteries of this type is that not every consumer can be expected to handle caustic alkali safely. Nevertheless, they should not be limited in application to military purposes, and the following civil uses may be suggested:

(a) Long-period power supply for transistorized equipment in remote regions or for camping, e.g., radio, tape recorders, shavers, lighting.
(b) Electric gadgets for which a constant discharge voltage is important, e.g., electric clocks.
(c) Electrical devices which have to operate at low temperatures (below $-10°C$).
(d) Ready supply of electrical energy for long periods of time, e.g. in case of catastrophe or emergency.

X.3.2.3   Technical Problems

If fuel cells are to find wider terrestrial application, progress must be made in the following directions:

1. Reduction of the cost of electrodes, particularly in decreasing their noble-metal content*; reduction of costs of constructional materials.
2. Use of air as oxidant.
3. Reduction of fuel costs and improvement of storage facility.
4. Optimal development of ancillary equipment, dc motors and converters.

Only if *noble metals* can be completely excluded can costs of \$50 to \$150 per kW for units of more than 1 kW be expected. This can be foreseen, for cells working at temperatures below 100°C, only if alkaline electrolytes are used.†

Nickel boride has already proved to be a suitable catalyst for hydrogen and hydrazine in alkaline media, and its use reduces costs to \$10 per kW; in comparison, the cost of 1 mg Pt cm^{-2} alone comes to \$22 per kW. For oxygen, silver and carbon have long been known as suitable catalysts.

In the light of costs and mechanical properties of presently available *constructional materials*, the low temperature, alkaline cells seem to be the most suitable for general exploitation[40]. For low cost and long life (at least 5 years and 10 000 hr operation), the use of plastics in construction is obligatory because of the corrosion and parasitic electrolyses inseparable from use of metals.

* The use of *platinum* is limited not only by cost (at present ca. \$7.5 per gramme), but also by availability. On the plausible assumption that about 25 tons of platinum a year could be directed to use as fuel cell catalyst, then, on the basis of 1 mg cm^{-2} and 5000 cm^2 per kW, this would provide a total power production of 5×10^6 kW per annum. This is to be compared with an automobile production in all countries corresponding (at an average of 40 hp) to 1.7×10^9 kW a year. *Palladium* is, however, much more plentiful and also cheaper (currently ca. \$1.5 per gramme).
† Recently also an acid H_2–O_2 cell with non-noble metal electrocatalysts has become feasible: H_2–tungsten carbide/H_2SO_4/phthalocyanine–O_2 (cf. Appendix).

Polypropylene (similar to polyethylene) is very promising in low cost and resistance to acid or alkaline attack but its use is limited to temperatures below 200°F, and it is not easily bonded to itself or to metals. The combined attack of alkaline electrolyte and oxygen at 200°F causes severe deterioration.

Polysulphone has desirable temperature stability and resistance to acid or alkaline media. The cost is of the order of $1 per pound, but it has been estimated[40] that this could be reduced to about 15 cents per pound for commercial fuel cell exploitation.

Teflon is frequently used in fuel cells, especially low and medium temperature acid cells. It is, however, subject to cold flow, and cannot meet the dimensional tolerances of gas ports that must be maintained to ensure that pressure drop will not vary by more than $\pm 5\%$ from cell to cell. This is the prime reason for believing[40] that the medium temperature acid cell will not succeed because Teflon seems to be the only material usable over the range of conditions concerned.

Some *epoxy resins* have been evaluated, but cannot withstand temperatures of 230 to 250°F, and their resistance to strongly alkaline solutions under oxidizing conditions does not seem good enough for the 10 000 hr operational target.

The *metals* that can be used in medium temperature alkaline cells are nickel, silver, gold, or anything that can be plated with these metals. Since weight and cost prohibit use of the pure metals, magnesium or aluminium are plated with nickel and gold, but the plating costs alone exclude metals as constructional materials for commercial application. Corrosion imposes an upper temperature limit of about 200°F.

Materials used for the medium temperature acid fuel cell are usually *tantalum*, gold, platinum and Teflon. Although attractive as a fuel cell that can use hydrogen mixed with CO_2, CO, CH_4, etc., available from hydrocarbon reforming, the cost of materials and the poor dimensional control of Teflon are severe limitations to commercial application.

The *removal of CO_2* and other impurities from air is a major problem, especially for batteries with alkaline electrolytes, not because of technological difficulties, but because of the size and weight of the ancillary plant required. Apart from CO_2 removal, use of air carries other difficulties due to changes in atmospheric conditions. Air temperature may vary between -30 and $+50°C$, and humidity between 0 and 100%. The very large gas volumes associated with use of air can lead to extensive loss of water. Laboratory experiments have shown that not more than 70% of the oxygen content of air can be utilized in a cell[3].

Fuel costs and *methods of fuel storage* play a very prominent part in terrestrial applications, since, in general, provision must be made for storing large quantities of fuel. The ideal solution would lie in the direct use of liquid hydrocarbons, but at present there is no means by which this can be done. Under these circumstances, the reformer–hydrogen–air system appears to be particularly attractive. It also brings ammonia and methanol into consideration as possible fuels, since they are cheap and easily stored. In relation to hydrogen itself, good progress has been made in the problem of its storage in liquid form[12]. Finally, hydrazine might become significant for commercial application when an economic method of preparing it has been developed.

The *adaptation of equipment* ancillary to fuel cells, such as inverters, converters, pumps for electrolyte and gas circulation, heat exchangers and dc motors, to the special conditions and requirements of these cells is yet in its earliest stages. This applies particularly to dc motors, which are very suitable for traction because of their good working characteristics. However, in comparison with ac motors of equal output, they are as yet much too large and much too heavy.

X.4   FUEL CELLS AND ELECTRICAL TRANSPORT

There is nothing new about the propulsion of vehicles by electrochemically powered electric motors. Cars driven by lead accumulators were built at the end of the last century—the electric brougham plied the streets of London and the electric car 'Jamais Contente' established a speed record of 65 mph in 1899.

The advantages of an electrical prime mover over the piston engine (with its transmission) are silence, absence of exhaust, favourable torque-speed relation and relatively high efficiency. Demand on batteries at zero load is negligible. Relatively high torque at low speeds (Figure X.26) is perhaps the most outstanding

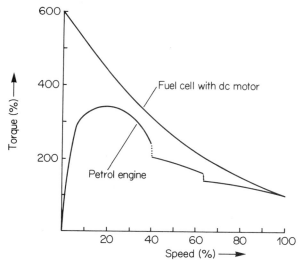

Figure X.26. Comparison of torques of a petrol engine and a fuel cell driving a dc motor (after Plust[45]). Ordinates: torque as % of that at maximum speed. Abscissae: speed as % of maximum speed.

advantage, favourable to good acceleration in moving from rest and in over-taking, and to good performance on hills. A further advantage would now be the low cost per kW hr or hp/hr permitted by battery charging from the national grid. In practice, the economics of electrical traction for road vehicles therefore depend on the initial cost and working life (number of charge and discharge cycles) of accumulators. A great disadvantage in 1900 was the considerable weight of batteries and of the still very heavy dc motors. It was because of these disabilities that the internal combustion engine took the decisive lead in technical development for road transport, despite its noise, fumes and low efficiency (about 20% compared with 50–75% for electrochemical power sources; cf. Figure X.27). Only in very special cases has the electric vehicle maintained its position—as for milk floats, street cleaning carts and electric trucks for luggage and posts at rail stations and elsewhere.

In the U.K. 40 000 electric vehicles are used for 85% of milk distribution to consumers. Typical data are; 30 cells up to 346 A hr, 60 V, 10 mph, 6.25 hp and 35–40 miles a day. Operating costs for such vehicles are less than half those for their petrol-driven analogues, but total costs are only 20% less because of the higher initial outlay.

The objectionable attributes of the Otto engine, the congestion of cities, and progress in development of electrical current sources, have contributed to the return in recent years of the question of electric traction to a centre of interest. Especially in the United States intensive efforts have been directed to study-ing the possibilities of electrochemical propulsion of vehicles—the basis of a

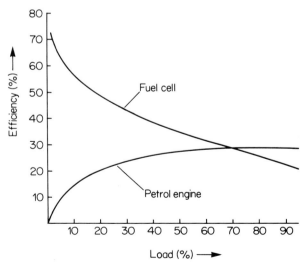

Figure X.27. Comparison of efficiencies (%) of petrol engine and fuel cell with dc motor as functions of load (% of maximum load) (after Plust[45]).

Senate Committee report (September 1967) on progress in the field of battery research*.

Studies of the traction problem have been supported both by government and private industry. The experiments of General Motors and the Union Carbide Corporation have become well known. The performance of a conventional minibus equipped with a 32 kW hydrogen–oxygen battery has been studied (see below). The U.S. Army has supported a project for testing a jeep driven by a 20 kW hydrazine–air battery. Such experiments have shown that the fuel battery—with its essential economy of weight—cannot alone provide the optimum system. For most applications there is no doubt that it is expedient to use a *hybrid system*, i.e., fuel and secondary batteries coupled together (or two secondary batteries of different type so coupled; see below in relation to the General Electric 'DELTA' vehicle). Short period heavy loads (acceleration, overtaking, hills) are met by the secondary battery; the fuel battery recharges it, and can of itself provide for continuous moderate loads.

Silicon diodes (thyristors) have provided an elegant solution to the problem of load variation. With no mechanical device, they are used to build impulse generators providing voltage pulses of any form or sequence. Connexion of such a generator between battery and motor permits continuous regulation of average operating voltage, leading to loss-free control of drive.

Regenerative braking schemes discussed at various times, in the interests of energy economy, are not as yet to be recommended because of the large, complex and costly control systems required.

The manner in which the examples that follow illustrate the present state of investigation of electrochemically powered transport, can perhaps be clarified by the following summary:

1. In some special cases where the weight of the fuel battery is not of main significance, a fuel system as the sole source of energy is quite practicable. This is so, for example, for submarines or rail-cars, and hydrogen–air or hydrogen–oxygen systems are proposed, with hydrogen from ammonia cracking, and oxygen carried (in submarines) as liquid oxygen.
2. For driving cars, trucks and buses, the hybrid system is suitable on purely technical grounds. The nickel–cadmium accumulator is advantageous for the secondary battery; although the lead–acid battery is cheaper, it is very much less suitable to meet the requirement of 'deep discharge'. At present, the hydrazine–air and hydrogen–air systems (*without* noble-metal catalyst) provide the most suitable fuel batteries.

At present, however, commercial applications of such hybrid systems are hardly practicable, mainly because of the high initial costs of the fuel cells (see section X.3.2.3), but also on account of high costs of operation (cf. Tables X.1 and X.2b). This restricts them as yet to military and other special applications.

* R. Jasinski, *Research Trends in Batteries and Fuel Cells*, Report to the US Senate, September 1967.

In this situation the most recent experiments of General Motors with the so-called Stirling-electric drive are significant. This consists of a combination of a combustion engine with, for example, helium as working gas with an induction motor and also a secondary battery (see below).

3. For less exacting requirements in performance, as for trucks, for rail station, post and factory use, two other means of improvement come into consideration. The first is to so modify the lead accumulator as to reduce its weight/energy ratio (25% reduction is not impracticable) and, in addition, to decrease its sensitivity to deep discharge. The second possibility is to use a hybrid system of two kinds of accumulator, one being suitable, for instance, for prolonged load and for the recharging of a nickel–cadmium battery of high load rating. This should furnish a battery system more elastic in meeting the necessary wide fluctuations in load, without incurring impracticably high initial costs (see below—the General Electric DELTA).

Studies already carried out are briefly described in the following sections.

200 kW hydrogen–oxygen battery of the ASEA Company for a Swedish Navy submarine

Fuel cells are of particular advantage where a store of energy in kW hr, large in comparison with battery rating, is needed. This is particularly important for prolonged space travel undertakings, but is also valid for submarine propulsion. Fuel cells appeared to be advantageous for this purpose in the Baltic Sea, where atomic-powered submarines would be unnecessarily expensive. On grounds of safety and minimization of costs, ammonia and liquid oxygen were chosen as primary reactants.

Figure X.28 shows in outline the drive system adopted for the Swedish submarine. The ammonia is cracked and the hydrogen formed is supplied to the hydrogen–oxygen battery; the nitrogen is condensed in a refrigeration system. The cells (described in chapter IV; see also Table X.2a) operate without noble metal, but with nickel boride and silver catalysts. The modules (Figure X.29) are of 20 kW, 20 kg kW^{-1} and each cell has an internal resistance of about 1 ohm cm^2. The complete battery has not yet been in operation, and it is not

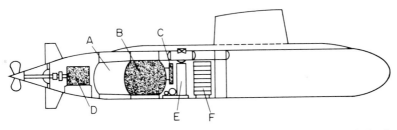

Figure X.28. The ASEA submarine for the Swedish navy, powered by an Amox fuel cell system (after Lindström[41]). A, Ammonia store; B, liquid air; C, heat exchanger; D, screw propulsion motor; E, ammonia cracking unit; F, fuel cell battery.

Figure X.29. A 20 kW module of the 200 kW hydrogen–oxygen battery for propulsion of the submarine of Figure X.28 (photograph by ASEA, Västeras, Sweden).

yet known what technical difficulties remain to be overcome. Nevertheless, it appears that this relatively economical AMOX fuel-cell system has potentialities for non-military applications.

During the 1939–45 war, the ASEA Company built an electric vehicle equipped with a lead battery of 850 kg, corresponding to an energy store of ca. 20 kW hr; it had a speed of 14–15 mph and an action radius of 28–34 miles. With 14 kg per kW hr for hydrogen–oxygen and 1.4 kg per kW hr for ammonia–air batteries, the stored energy for batteries of the same weight (including propellants) would be 60 kW hr for the Hydrox system and 600 kW hr for the Amox system. The nominal ratings would be 21 kW and 10.6 kW, with respective speeds of 28 and 12.5 mph and action radii of 93 and 850 miles (on one charge of fuel).

The Amox system has recently been studied by the Allis-Chalmers Company[40]. With a cracking unit and an 8 kW hydrogen–air battery, this should be adequate for town use of a car of Volkswagen size; 12 kW output can be sustained for 5 minutes and even 24 kW, for 15 seconds. An additional secondary battery would certainly be desirable, since at least 10 minutes warming-up time is required. Even the most optimistic assessment, however, indicates that the financial hurdle is yet too high—even for quantity production, the cost of the 8 kW unit would exceed $1200.

The Union Carbide motor cycle with hybrid hydrazine–nickel–cadmium drive
Dr Karl Kordesch of the Union Carbide Corporation has converted an Austrian motor cycle (type Puch) to electric drive (Figure X.30). Two Union Carbide hydrazine–air batteries (see chapter IV and Table X.2b) producing about 800 W at 50 mA cm^{-2} are connected in parallel with nickel–cadmium batteries, so that peak output of 2 kW can be obtained. The original engine of 2.3 hp weighed 125 lb; after the conversion the new power unit weighed 175 lb. The two fuel cells (without electrolyte) were 9 lb each, and the nickel–cadmium battery was 22 lb in weight. A steady speed of 25 mph could be maintained with a fuel consumption of 150 miles per gal of $N_2H_4.H_2O$. Hydrazine feed was controlled

Figure X.30. Light motor cycle (80 kg) with electric drive by the hybrid system hydrazine–air (800 W) and nickel–cadmium (after Kordesch[42]; photograph by Union Carbide Corpn., Cleveland, Ohio).

by the voltage of a sensor cell (N_2H_4 concentration in the battery ca. 0.5%), air was purified by soda lime, and an operating temperature of 60°C maintained by air cooling.

Speed regulation was effected without use of resistances, by alternative connections of the windings of the compound motor, with additional steps by series-parallel switching of the batteries. Although this 'electrocycle' is yet very far from being a practicable or economic means of transport, it has proved valuable as a way of studying vehicle propulsion by means of an electrochemical hybrid system.

The Monsanto 20 kW hydrazine–air battery for driving $\frac{3}{4}$ ton army vehicles

The Monsanto Company, in cooperation with the U.S. Army, has studied the drive of a $\frac{3}{4}$ ton vehicle exclusively by fuel batteries—hydrazine–air aggregates. It can be seen in Figure X.31 that 4 modules, each of 5 kW, are assembled in the front of the vehicle. Twelve electric motors mounted above the modules are used to supply the 'chemical air' required. Construction and data for these batteries are dealt with in section IV.3.2.2 and Table X.2b. The fuel is hydrazine monohydrate; 64% hydrazine, 36% water.

The regular 94 hp engine of this vehicle is replaced by the fuel battery, a static dc voltage controller (General Electric), and a standard industrial 3900 rpm, dc series traction motor. At least 40 kW (53 hp) would be needed to equal the original performance, and a hybrid system would still be advantageous for economy in space requirement and weight.

Figure X.31. View of power plant of U.S. Army jeep, with 20 kW hydrazine–air battery by Monsanto (photograph by Monsanto Research, Everett, Mass).

The General Motors 'Electrovan'

A minibus built by the General Motors Corporation, Indianapolis, has been experimentally equipped with a 32 kW Union Carbide hydrogen–oxygen battery, liquid reactants being carried in vacuum tanks (Figure X.32). In spite of a doubling of weight over the normal vehicle, the same maximum speed was attainable without greatly lessened acceleration (30 sec instead of 23 sec to 60 mph from rest). It is remarkable that 32 kW (formally, 43.5 hp) is adequate to this extent, since the original GMC vehicle was driven by a 225 hp engine. It is, however, to be remembered that the hp rating of the motor should be related to its short-term maximum output, and the hydrogen–oxygen battery can sustain a large overload—as much as five-fold for 10 sec. Thus 160 kW peak loads, or 90 kW overload for two hours, could be tolerated, giving the vehicle much the same performance as its petrol-driven analogue. Containers for hydrogen and oxygen (6 kg liquid hydrogen, 48 kg liquid oxygen) were designed to give a range of 150 miles.

Experience with this vehicle has shown[42] that the ac motor originally fitted is unsuitable. A high speed motor of lower voltage (zero load voltage is 525 V) would be preferable. Furthermore, the 32 kW battery is larger than required and was installed only to secure acceleration equal to that of the normal,

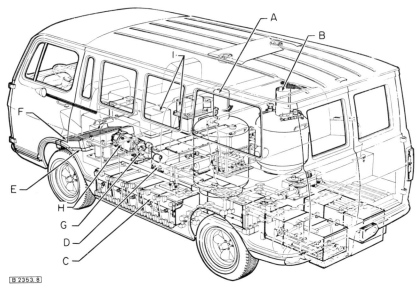

B 2353.8

Figure X.32. The General Motors minibus, 'Electrovan', with 32 kW hydrogen–oxygen battery; weight, 3.2 ton, maximum speed, 70 mph; range, 100–150 miles; weight of power plant, 1650 kg (after Irwin[43]). A, Liquid hydrogen tank; B, liquid oxygen tank; C, 32 fuel cell modules; D, electrolyte reservoir; E, electrolyte radiator; F, water condenser; G, gearbox; H, ac induction motor; I, motor controls.

piston-engined version. There is no doubt that in the light of experience a hybrid system would now be used, and also operation with air in place of oxygen. According to Kordesch[42] these changes could lead to a reduction in weight of 500 kg.

The General Electric electric vehicle DELTA (Developmental Electric Town Auto[44])

An interesting experiment aimed at improving the characteristics of electric drive for vehicles, has recently been conducted by the General Electric Company. In this, the experimental vehicle DELTA (Figure X.33) was equipped with a hybrid electrochemical power source consisting of lead and nickel–cadmium batteries.

The limited drain-rate capabilities of lead–acid, zinc–air, and lithium metal-halide cells require a hybrid battery system to provide adequate acceleration and sustained power at high speeds. Figure X.34 contains simplified plots of power density (W lb^{-1}) against energy density (W hr lb^{-1}) for leading batteries; it can be seen that energy density deteriorates rapidly as power density requirements increase.

Figure X.33. General Electric 'DELTA' vehicle (after Laumeister[44]).

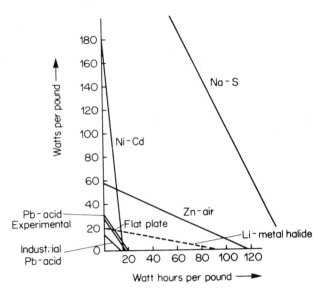

Figure X.34. Comparison for traction purposes of the batteries discussed by Laumeister[44].

Details of the power plant of this vehicle are as follows:

Motor—10.9 hp, dc series type.

Control—SCR 'chopper' with push-button reversing and 'passing gear' (field weakening).

Main battery—72 V experimental lead acid battery, 225 A hr, 18 W hr lb^{-1}, 5 hr discharge rate, 964 lb.

Acceleration booster battery—72 V nickel–cadmium vented type battery, 11 A hr, with automatic discharge on high current demand, and automatic recharge from main battery.

Charger—30 A, solid-state, 'on board' unit.

Drive—front wheel drive, trans axle, with 4 ratio selection for motor speed variation.

Brakes—4-wheel hydraulic with standard shoe-drum action.

Body—fibre glass-epoxy, with steel reinforcement.

The design was tailored to the power system needed for limited range and speed needed for urban–suburban use. Performance data for the 2300 lb vehicle are as follows:

Top speed—55 mph
Acceleration—0–30 mph in 6 sec

Range: city cycle—40 to 50 miles in 3 to 4 hr; operation, 4 stops per mile, 0–30 mph in 8 sec, 30 mph cruise.

Range: steady cruise at 30 mph, 100 to 120 miles; at 40 mph, 50 to 60 miles.

Recharge—8 to 10 hr using on-board charger, 2 to 3 hr using stationary charger (80% recharge).

Figure X.35 illustrates the operation of the two-battery system in normal driving. The upper plot shows the vehicle speed during acceleration, cruise, deceleration and wait for two start-stop cycles of town operation. The lower plot is an approximation to the power demand on each battery during these cycles, the booster unit delivering power above a certain power level, and accepting power below it.

The operating costs for the DELTA vehicle, including battery amortization and power at 1.55 cents per kW hr, are estimated, in cents per mile, as follows:

Battery	Steady speed		30 mph top speed
	30 mph	50 mph	with 4 stops mile^{-1}
Standard lead–acid	1.0	3.2	2.4
High output lead–acid (projected)	0.5	1.5	1.1

The estimated maintenance cost, based on experience of electric lift and delivery trucks, would be $\frac{1}{3}$ to $\frac{1}{4}$ that of petrol-driven vehicles, or about 0.3 cent per mile.

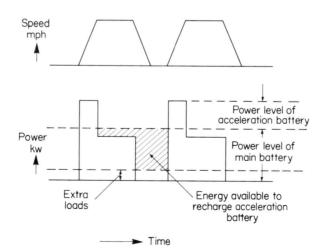

Figure X.35. Operation of the General Electric hybrid system of 'DELTA' (after Laumeister[44]).

The General Motors experiment with Stirling-electric drive[46]

A further study of a possible electrically driven vehicle of the future is represented by a combination of a combustion engine of the Stirling type with an electric motor as well as a battery.

In the Stirling engine the combustion takes place in a separate burner, in which a variety of fuels can be used—petrol (without additives), diesel oil, mineral oil, etc. The fuel does not enter the working cylinder in which the pressure-exerting gas (which can, for example, be helium) is enclosed, and, because it can remain long enough in the active combustion zone, complete oxidation is favoured. The exhaust gas of the engine therefore contains insignificant quantities of carbon monoxide or smoke-forming hydrocarbons, and the engine has the considerable advantage of odourless operation. In the experimental vehicle, the Stirling engine drives an alternator, which in turn supplies power to a 3-phase induction motor, and also feeds a rectifier which charges a battery in which additional energy is stored to be used for acceleration. It can therefore be argued that the assembly as a whole constitutes a hybrid system.

General Motors have used an Opel Kadett car as a test vehicle for tests of this Stirling-electric drive. The total weight with driver and one passenger was about 1400 kg, some 500 kg more than that for the regular Kadett. The modified car accelerated from rest to 50 mph in 10 seconds and had a maximum speed of about 56 mph. The primary power unit was an 8 hp Stirling engine which itself would give the car a towing speed of 31 mph and 36–47 miles per gallon of fuel. The considerably improved acceleration and maximum speed available from the hybrid system was due to the energy store of 5 kW hr in the lead battery incorporated in the system. If, however, the battery is drained faster than the Stirling engine can recharge it, the range of the car at full performance becomes considerably restricted. This would occur if the maximum speed of 56 mph were maintained for about 35 miles, but a steady cruising speed of 30 mph would allow the car a range limited only by the capacity of its fuel tank (in this case, 20 l).

REFERENCES

1. B. R. Stein, Status Report on Fuel Cells, U.S. Department of Commerce, O.T.S., PB 151 804 (1959).
2. E. Justi and A. Winsel, *Fuel Cells—Kalte Verbrennung*, Steiner, Wiesbaden, 1962.
3. R. Jasinski and T. G. Kirkland, Fuel Cells—State of the Art, Research Division Allis-Chalmers, Milwaukee, Sept. 1963, Dec. 1963.
4. W. Mitchell, *Fuel Cells*, Academic Press, New York, 1963.
5. K. R. Williams, *An Introduction to Fuel Cells*, Elsevier Publishing, 1966.
6. *Les Piles à Combustibles*, Publication de l'Institut Français de Pétrole, Editions Technique Paris, 1965.

7. D. R. Adams, P. Y. Cathou, R. E. Gaynor, R. D. Jackson Jr., J. H. Kirsch, L. L. Leonhard, G. S. Lockwood Jr., W. P. Warnock and R. E. Wilcox Jr., *Fuel Cells—Power for the Future,* Fuel Cell Research Associates, Cambridge, Massachusetts, Oct. 1960.

8. H. Bode, *Chem. Ing. Techn.* **35**, 367–371 (1963).

9. G. Sandstede, 'Elektrochemische Brennstoffzellen,' in *Fortschritte der chemischen Forschung,* Volume **8**, Springer, Berlin, 1967. p. 171–221.

10. F. T. Bacon in B. S. Baker, *Hydrocarbon Fuel Cell Technology,* Academic Press, New York, London, 1965, p. 4.

11. R. Cohen, *Proc. 20th Power Sources Conf.,* Atlantic City, 1966, p. 21.

12. E. Cohn, 'Primary Hydrogen–Oxygen Fuel Cells for Space,' paper presented at 29th AGARD-meeting, Liège, 1967.

13. J. H. Russell, *Proc. 19th Ann. Power Sources Conf.,* Atlantic City, 1965, p. 35; H. A. Liebhafsky and E. J. Cairns, *Fuel Cells and Fuel Batteries,* John Wiley, London, New York, Sydney, 1968.

14. C. C. Morrill, *Proc. 19th Ann. Power Sources Conf.,* Atlantic City, 1965, p. 38.

15. J. Platner, D. Ghere and P. Hess, *Proc. 19th Ann. Power Sources Conf.,* Atlantic City, 1965, p. 32.

16. J. E. Schroeder, D. Pouli and H. J. Seim, in B. S. Baker, *Fuel Cell Systems—II,* American Chemical Society, Washington, 1969, p. 93.

17. M. L. Engle, 58th Session of the Amer. Inst. Engrs., Philadelphia 1965.

18. K. R. Williams, *An Introduction to Fuel Cells,* Elsevier, London, 1966, p. 307.

19. C. G. Clow, J. G. Bannochie and G. J. W. Pettinger, *5th Intern. Power Source Symp.,* Brighton, 1966, Pergamon Press, 1967; M. A. Vertes and A. J. Hartner, *Proc. Journ. Intern. Étud. Piles Combust.* Brussels, Ed. Serai, Brussels 1966, Vol. I, p. 63.

20. M. Barak, address to AGARD (Advisory Group of Air Space Research Development, Nato), 29th Propulsion and Energetic Panel Meeting, Liège, 1967.

21. St. J. Bartosh, *Proc. 20th Ann. Power Sources Conf.,* Atlantic City, 1966, N.J., p. 31.

22. G. R. Frysinger, *Proc. Journ. Intern. Étud. Piles Combust.* Brussels, 1966, Ed. Serai, Brussels, 1967, p. 312.

23. G. R. Frysinger, *Simplified Fuel Cell Systems for Army Application,* lecture to 154th meeting Amer. Chem. Soc., Chicago, 1967.

24. M. V. Burlingame in B. S. Baker, *Fuel Cell Systems—II,* American Chemical Society, Washington, 1969, p. 377.

25. R. E. Salathe, J. O. Smith, J. P. Gallagher, P. L. Terry, J. Kozloff, L. V. Athearn and P. Dantowitz, address to the American Chemical Society, Chicago, 1967.

26. J. Dünger and G. Grüneberg, unpublished.

27. E. A. Gillis, *Proc. 20th Ann. Power Sources Conf.,* Atlantic City, 1966, p. 41.

28. M. Gutjahr and W. Vielstich, *Chem. Ing. Techn.* **40**, 180 (1968).

29. F. J. Wrublewski, *Proc. 21st Ann. Power Sources Conf.,* Atlantic City, 1967, p. 80.

30. H. Binder, A. Köhling, W. Kuhn, W. Lindner and G. Sandstede, *Chem. Ing. Techn.* **40**, 171 (1968).

31. H. Jahnke, private communication.

32. F. L. Gray, *Proc. 2me. Journ. Intern. Étud. Piles Combust.,* Brussels, 1967, Ed. Serai, Brussels, 1967, p. 385; *Proc. 21st Ann. Power Sources Conf.,* Atlantic City, 1967, p. 29.

33. E. Guth, G. J. Haase, H. G. Plust and W. Vielstich, *Proc. VIIth Int. Conf. on Lighthouses and other Aids to Navigation,* Rome, 1965; W. Vielstich in B. S. Baker, *Hydrocarbon Fuel Cell Technology,* Academic Press, New York, 1965, p. 79.

34. H. A. Liebhafsky and E. J. Cairns, General Electric Report Nr. 67-C-235, July, 1967.

35. M. I. Gillibrand and J. Gray, *Proc. Journ. Intern. Étud. Piles Combust.,* Brussels, 1967, Ed. Serai, Brussels, p. 118.

36. *Funkschau* **5**, 1966 'Die provisorische Fernsehstation Gebidem.'

37. J. Lemaignen and Ph. Demange, *Proc. Journ. Intern. Étud. Piles Combust.*, Brussels, 1967, Ed. Serai, Brussels, p. 94.
38. W. Vielstich and U. Vogel, in B. S. Baker, *Fuel Cell Systems—II*, American Chemical Society, Washington, 1969, p. 341.
39. W. Vielstich, U. Vogel, H. W. Sendhoff and H. Stichnote, unpublished.
40. N. A. Cook, 'Analysis of Fuel Cells for Vehicular Applications, Automotive Engn. Congr. Detroit, 1968.
41. O. Lindström, *ASEA's Tidnung* **58**, 171 (1966).
42. K. Kordesch, private communication.
43. R. W. Irwin, *The Detroit News,* 28.10.1966.
44. B. R. Laumeister, General Electric Report 68-C-128, April, 1968.
45. H. G. Plust, 'Aussichten der Brennstoffzelle als Automobilantrieb,' *Automobil Revue* Nr. 25, 1966.
46. *Industriekurier,* 13.7.1968, p. 10.

APPENDIX

Phthalocyanine as an Oxygen Catalyst

In the human and animal organism, it is the haemoglobin in the blood corpuscles that transports oxygen from one place to another. This process probably involves an activation of the oxygen molecule. Since haemoglobin is not stable in acid media, Jahnke and Schönborn* have investigated a class of compounds that also contains a metal atom centred in a porphyrin ring: the phthalocyanines (Figure 1). These compounds are perfectly stable in acids. The corresponding polymers have a conductivity 10^3 to 10^4 times higher than the monomers.

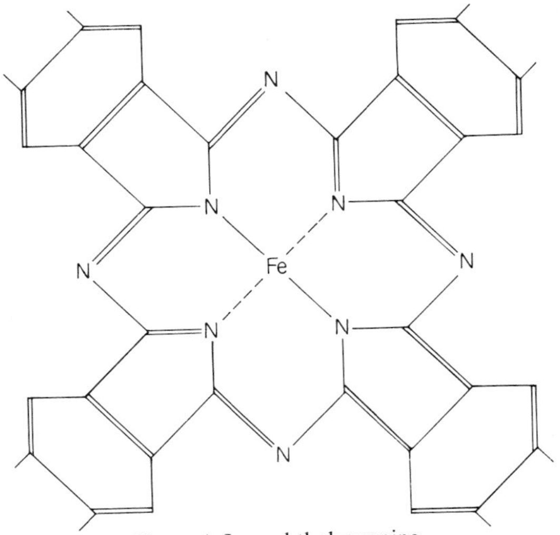

Figure 1. Iron phthalocyanine.

* H. Jahnke and M. Schönborn, *Proc. Journ. Intern. Étud. Piles Combust.*, Brussels 1969, Presses Académiques Européennes, Brussels 1969, p. 60.

For electrode preparation, phthalocyanine powder was dissolved and carbon powder suspended in concentrated sulphuric acid. Stirring this suspension into water resulted in a finely divided phthalocyanine deposit on the carbon. With polyethylene as a binder, mechanically stable electrodes were fabricated. They were tested in a half-cell set-up.

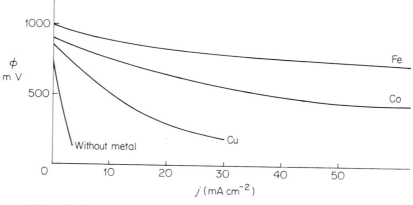

Figure 2. Catalytic activities of phthalocyanines with different central atoms.

Figure 2 shows test results. Clearly, iron phthalocyanine is the most suitable choice. The current densities obtained with these first electrodes promise future success for this type of electrode not containing any noble metal. It is furthermore important that this new oxygen catalyst is inactive toward fuel (e.g. methanol) dissolved in the electrolyte.

Sources of Figures

Allis-Chalmers Manufacturing Company, Milwaukee, Wisconsin, U.S.A.
IV.30, IV.94, X.9, X.10
Arbeitsgruppe Dr. Grüneberg, Aachen, Germany
IV.85c, IV.94b, IV.96b, IV.96c, IV.96d, X.15
ASEA, Allmänna Svenska Elektriska Aktiebolaget, Västeras, Sweden
IV.9a, X.29
F. T. Bacon, Westfield, Little Shelford, Cambridgeshire, U.K.
IV.50
Robert Bosch GmbH, Stuttgart, Germany
X.17
Brown, Boveri u. Cie., Baden, Switzerland
IV.5, X.18, X.21
Chloride Technical Services Ltd., Clifton, Manchester, U.K.
IV.87
Compagnie Francaise Thomson-Housten, France
IV.25a
Electric Storage Battery Company, Yardley, Pennsylvania, U.S.A.
IV.12
General Electric Company, Schenectady, New York, U.S.A.
IV.15a, IV.25, VI.8, X.3, X.33
M. W. Kellog Company, New Market, N.J., U.S.A.
V.5
Magna Corporation, Anaheim, California, U.S.A.
V.16a, V.18
McDonald Aircraft Corporation, U.S.A.
X.5
Monsanto Research Corporation, Everett, Mass., U.S.A.
X.14, X.31
Pratt and Whitney Aircraft, East Hartford, Conn., U.S.A.
IV.54, IV.55, X.8
Ruhrchemie A.G., Oberhausen-Holten, Germany
IV.83

Shell, Thornton Research Centre, Chester, U.K.
IV.82a, X.13
Société Alsthom, France
IV.96f
Texas Instruments, Dallas, Texas, U.S.A.
IV.39, IV.49
Union Carbide Corporation, Cleveland, Ohio, U.S.A.
IV.14, IV.14a, IV.14d, IV.35, IV.95c, X.30
Institut für Physikalische Chemie der Universität Bonn, Germany
IV.85a, X.23, X.24
U.S. Army Electronics Command, Fort Monmouth, N.J., U.S.A.
X.16
Varta AG, Forschungs- und Entwicklungszentrum, Kelkheim, Taunus, Germany
IV.4a, IV.4b

Author Index

Subject Index

DE PAUL UNIVERSITY LIBRARY

30511000146944

621.359V661F C001
LPX FUEL CELLS CHICHESTER